Toward a Grammar of
Biblical Poetics

Toward a Grammar of Biblical Poetics

Tales of the Prophets

HERBERT CHANAN BRICHTO

New York   Oxford
OXFORD UNIVERSITY PRESS
1992

Oxford University Press

Oxford   New York   Toronto
Delhi   Bombay   Calcutta   Madras   Karachi
Petaling Jaya   Singapore   Hong Kong   Tokyo
Nairobi   Dar es Salaam   Cape Town
Melbourne   Auckland

and associated companies in
Berlin   Ibadan

Copyright © 1992 by Herbert Chanan Brichto

Published by Oxford University Press, Inc.
200 Madison Avenue, New York, NY 10016

Oxford is a registered trademark of Oxford University Press.

Library of Congress Cataloging-in-Publication Data
Brichto, Herbert Chanan.
Toward a grammar of biblical poetics:
tales of the prophets / Herbert Chanan Brichto.
p. cm.   Includes bibliographical references and index.
ISBN 0-19-506911-0
1. Bible. O.T.—Criticism, interpretation, etc.
2. Prophets. I. Title.
BS1171.2.B75
1992   221.6'6—dc20   91-3644

1 3 5 7 9 8 6 4 2

Printed in the United States of America
on acid-free paper

To M. P. B.

whose book,
in so many ways,
this is.

Preface

Criticism consists of analysis, synthesis, and evaluation or appreciation. In respect to literature, *poetics*—as I understand the term—is part of, or an aspect of, *literary criticism.* It is the isolation of the elements, features, and techniques employed by a creative author to bring a story to life. *Rhetorical criticism,* properly speaking, is a preoccupation with the elements and features that make for a persuasive presentation of an argument in speech or in writing; it, too, may therefore be viewed as a part or aspect of literary criticism. When biblicists use *rhetorical criticism* or *poetics* as the label for the procedure they are practicing, it is usually because in the area of Bible study *literary criticism* has taken on a connotation that it has nowhere else: *source analysis.* Whether in the classical Graf–Wellhausen hypothesis, or with the various emphases of form criticism, tradition history, or the Scandinavian school, literary criticism in Bible study has been either virtually synonymous with source analysis or based on an acceptance of its *genetic* premises.

The interpretive essays that constitute the bulk of this volume (and a companion one to follow) were originally written or sketched in outline as *explications du texte,* employing the tools and techniques exercised in the enterprise that is normally called literary criticism. Upon presentation of these essays to fellow biblicists, I soon learned that a number of factors made it difficult for these colleagues and friends to accord them a sympathetic address. For one, fine scholars who are essentially philologists may only rarely have recourse to fiction, for recreation; and the recourse to the conventions of the composition of fiction (or the exposition of these conventions, literary criticism) may involve for them a language or mode of discourse foreign and incomprehensible. This factor is intensified in the case of biblicists for whom *literary criticism* means source analysis, which has become the keystone for methodologies that have become synonymous with "scientific Bible study."

Scholars of such orientation (and they yet constitute a large majority of the world's biblicists) may greet with equanimity or even with mild approval suggestions that principles of literary criticism developed in connection with extrabiblical literature may be deployed to enhance our appreciation of biblical texts. But when the deployment of these principles results in a challenge to the methodologies that have become synonymous with "scientific Bible study," the legitimacy of their application to a unique literature, the Bible, is impugned; and the entire approach shunted aside as unworthy of further serious consideration.

The foregoing explains in part why *poetics* rather than *literary criticism* appears in this work's title, for all that the latter term, being more comprehensive, may be a more accurate description of my enterprise. It also will help me explain the nature of the first two chapters (part I) and the selection and order of appearance of biblical texts for the interpretive essays in part II (and the companion volume).

Perhaps I might best start with a declaration as to what these volumes are (or are intended to be) and what they are not. They do not constitute a guide, manual, or handbook of either literary criticism in general or biblical poetics in particular. Some of the elements discussed in chapter 1 are of a rather elementary nature as far as general literary criticism is practiced. They are presented here so that the antiquarian philologist may appreciate how ubiquitous these elements are in modern literature and, by contrast, how absent—largely or in part—in the exposition of ancient literary texts, particularly Scripture. These and the rest of the material in these discursive discussions are presented for their particular and crucial relevance to issues of interpretation of biblical texts as primarily creative, imaginative, and fictive, for all that their essential purpose is didactic (in keeping with an overall view of the nature of the human and the divine) and that they are presented as narratives of an essentially historical nature or as ideology in the form of narrative or precept. The *poetical grammar* posited in the title of this work is thus a hypothesis to be demonstrated, namely, that there is a set of rules that will, when uncovered, show that the Hebrew Scriptures, as a whole and in its constitutive units, constitute a unitary design and a single "authorial voice," even though the several or many authors who contribute to that voice and design may have lived centuries apart.

The elements and arguments thus discussed in chapters 1 and 2 and illustrated or featured in the exegetical essays are arguments for the existence of such a grammar and not the whole or even the larger part of this hypothecated grammar. The elements, categories, and syntax of part of this grammar presented in chapters 1 and 2 are those which I have found particularly relevant for the investigation of biblical texts—a discussion of the problems such as a grammar of rhetoric raises when it is applied to an ancient literature (particularly one preserved as sacred) that more resembles a library than a book. Contemporary scholars engaged in literary criticism of Scripture are cited because they address the same features or problems that are critical to my discussion. If such as Alonso-Schökel, Auerbach, Bar Ephrat, Fishbane, Fokkelman, Gros Louis, Kugel, and Polzin (to name but a few) fail of citation, it is not for lack of appreciation of their contributions to the burgeoning enterprise of biblical poetics.

In respect to the selection of biblical texts and the order of their presentation, the composition of most of the material in the companion volume antedates the exegetical essays in this first one. The "Stories and Structures in Genesis" essays in the companion volume were originally composed as presenting a literary-critical alternative to interpretations deriving from source-critical presumptions. They were not intended as a methodological challenge or rebuttal of the source-critical schools. The reaction of colleagues to whom these manuscripts were submitted convinced me of two things: (1) that despite recent and incisive assaults on the genetic approaches, source analysis continues to enjoy a regnant position in "scientific Bible study"; (2) that its occupation of the high ground in literary analysis almost requires a rebuttal of it of any Bible scholar who would proceed to a literary treatment based on altogether different asssumptions.

Inasmuch as poetical analysis was my chief concern and rebuttal of documentary hypothesis a less-than-vital personal priority, I determined to publish first a series of shorter narrative pericopes—narratives that did not figure significantly in the cruxes of source-analytic theories and that in their shorter compasses and (relative) independence of one another might serve to focus the poetical lens. The prophetic theme lying at the heart of Scripture—the thought of stories featuring prophets—suggested itself naturally, and the question became how many stories might be handled feasibly in a single volume and whence the selection should be made.

The block of chapters in 1 and 2 Kings featuring Elijah and Elisha (along with a few other prophets, some anonymous) turned out to require more pages than I was prepared to allot to it. I chose rather to include the freestanding Book of Jonah and the three chapters of Exodus featuring a Moses who in many ways (as the reader will have concluded for himself) represents Jonah's polar opposite. These two stories, too, feature an almost identical formulation of God as merciful and rigorous; and each narrative presents parade examples of the poetic techniques and strategies exploited by the scriptural author to drive his lesson home. The reader will not fail to make the connection between Moses and Elijah, then between Elijah and Elisha as embodiments of the Divine Presence. The last story, extracted from the Book of Jeremiah, shows, I believe, that even in the context of the oracles of a writing prophet, the inclusion of an incident in his career is due not to its biographical weight (that is, its critical import for the career of that prophet) but for the heuristic value of its lesson, which is critical for every prophet or would-be prophet.

One story that I have slipped in (for reasons I shall not explain now) is one that features no prophet at all; its inclusion in the overall design will become apparent in a later volume. I do regret the limitations of space that precluded treatment of the remaining Elisha narratives. In particular, readers may be intrigued by the failure to treat Elisha's restoring to life a child who has died, a tale in remarkable contrast to the lesser feat of Elijah's restoring to health a child who is barely breathing.

In this connection, I should be neither surprised nor dismayed if I were suspected of ignoring many a story that I cannot adequately explain. Many of the biblical stories, vignettes, and seemingly otiose or out-of-place glosses are a challenge to our problem-solving talents. While the awareness of certain poetic techniques and strategies may provide keys to the solution of many problems, we may not know which of many keys are needed for a given lock; and many a key may remain to be discovered. The operational term in this work's title is *toward*. This is not a grammar, but a beginning *toward* one. As in the case of all beginning enterprises, it may turn out that our treatments of the elements in the framework discussions will seem uneven, overweighty, or scanted in one part or another. In like manner, the illustrative essays may in many instances fail to be convincing to some or many readers.

I shall, as I have said above, be neither surprised nor dismayed; for none of these eventualities will altogether vitiate either the poetic effort or the enterprise to construct a grammar for it in regard to Scripture. Language and literature may be studied with disciplined rigor, but neither they nor their grammars lend themselves to the experimental criteria of the empirical sciences. Biblicists, especially, would do well to remind themselves that the claim to ''scientific study of the Bible'' is either a disclaimer of dogmatic prejudice or a wishful attempt to ride the prestigious coattails of the empirical sciences.

But the deductive method is far from contemptible. The arguments in favor of the grammar toward which I shall move are presented in the hope of persuading and convincing, without any claim or illusion as to scientific objectivity. In the empirical sciences a chain of reasoning is, indeed, no stronger than its weakest link. In the case of a deductive investigation such as ours the chain of evidence may lead to conviction by the sheer mass of the links, by the logic of their connections. A weak link here or there, a less-than-cogent connection, a less-than-convincing conclusion or inference—all these may be set aside by the reader as vagaries of an overzealous interpreter. The purpose of this work will be served best by the reader who, correcting and refining the grammar, extends its applicability, expands the possibilities of interpretation, and thus vindicates the methodology of Scriptural poetics.

Cincinnati   H. C. B.
August 1990

Acknowledgments

It has been my rare good fortune to have had a succession of great teachers, at home and at school, subscribing to a variety of religious or ideological confessions, each one steeped in varying degrees in the literatures of classical antiquity and/or the Judaic and Near Eastern civilizations. Whatever of merit inheres in this work is largely owing to the gifts of mind and heart they shared with me: my father Shlomo Brichta, my grandfather, M. M. Frankel-Theumim; Henry Slonimsky, John J. Tepfer, Gilbert Highet, H. L. Ginsberg, and E. A. Speiser.

Debts of gratitude are freely confessed to the colleagues on the faculty of Hebrew Union College–Jewish Institute of Religion (HUC–JIR) who read and made invaluable suggestions for the improvement of large sections of these chapters: Samuel Greengus, Stephen A. Kaufman, Mattitiahu Tsevat and David Weisberg. To the president and deans of HUC–JIR I extend deep appreciation for their generous and enlightened support in encouragement of research and publication.

To Cynthia Read, Paul Schlotthauer, and Michael Lane—my editorial team at Oxford University Press—my deep thanks for a level of editorial competence, assistance, and support that I had been led to believe no longer existed in the enterprise of academic publishing.

Finally, every teacher knows that there are colleagues and students who have by reason of their work, character, or generosity contributed significantly to his or her own growth yet fears to begin an enumeration that will inevitably overlook a name to whose bearer he owes much. I must take that risk, lest in fearing injustice to a few I fail to acknowledge the many: Uri Herscher, Jay Holstein, Jacob Milgrom, Shlomo Morag, Aaron Shaffer, Robert Ratner, and—*aharon aharon haviv*—Yochanan Muffs.

Contents

I

On Poetics

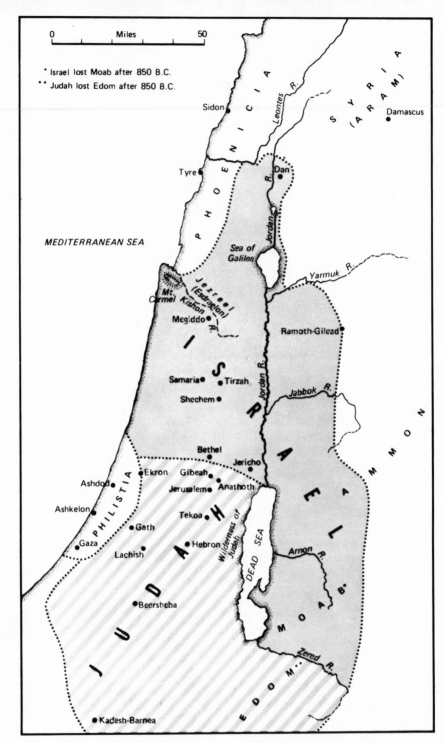

The kingdoms of Judah and Israel. (Henry Jackson Flanders et al., *People of the Convenant: An Introduction to the Old Testament*, 3rd ed. [New York: Oxford University Press, 1988], p. 267).

1

Introduction

The grammar of a language consists of the elements of speech and the proper relationships among them. The elements are, on the most basic level, the range of sounds we divide into vowels and consonants and the vowel–consonant combinations we call syllables. More precisely, however, the basic elements on the level of sound are not simply consonants and vowels but the sounds that are irreducible units, namely, phonemes. (Thus, in biblical Hebrew the sounds represented by [b] and [v] constitute a single phoneme whereas in English they constitute two separate phonemes; the sounds represented by [k] and [q] are separate phonemes in Hebrew but a single phoneme in English.) Phonemes alone or in combination may constitute units of meaning we call words; and words, in turn, may be classified in terms of meaning function under the rubric *parts of speech:* nouns, pronouns, adjectives, verbs, adverbs, prepositions, articles, and conjunctions. These parts of speech may then be grouped together to produce meaningful statements, the meaning being dependent on the words themselves and upon rules of ordering, declension, pitch, or tone that we call syntax.

To construct a grammar of poetics (that is, literary composition) on the model of a grammar of language, we shall have to isolate the constitutive elements and the varieties of interrelationships among them that constitute the syntax. At this point it is of vital concern to keep in mind that *language* (literally, "tongue") is first and foremost an oral-vocal-audial phenomenon—in other words, speech. Literature, by contrast, is a visual phenomenon. It depends on a representation of speech in writing, a reduction of sounds to a set of symbols that at best represents phonemes adequately and that, as a rule, is woefully inadequate in conveying such meaningful elements of speech as volume, stress, pitch, tone, and rhythm—not to speak of eloquent vocal accompaniments such as ejaculations, sighs, groans, gasps, hisses, and grunts. Not all writings are literature; for instance, inventories. But literature as we experience it is composed speech, which is not only preserved in accurate detail by but even composed in large measure dependent on the possibilities opened up by, the writing process.[1] For example, the longer the composition and the more varied its subunits or the more the composition partakes of the nature of *riddle* (with scattered clues vital for the conclusionary resolution), the less likely that the composition could have been achieved without writing.

The concern of this book is primarily with the poetics of narrative, that is, the telling of a story. And here again we need to remind ourselves of some essential differences between storytelling (oral narration) and story writing (written narrative). Take, for example, the identities of the author (story creator) and of the narrator (the storyteller). In the case of oral presentation the storyteller is present before our eyes, and that presence may preclude such questions as whether author and narrator are one person or two and whether the narrator is trustworthy. If, for example, the storyteller is Mother, and she begins, "One lazy summer afternoon I was half-asleep and half-awake, playing with my cat Dinah, when she leaped from my hands and dashed across the lawn just as a bunny rabbit popped into his hole," our relationship to Mother will implicitly answer many questions which might be asked and keep others from even coming to mind. If the same sentence opens the story in a book it may be some time before we learn whether the speaker is the author, whether the speaker is the narrator, whether we ought to like and therefore trust the speaker, dislike and distrust her, or perhaps trust her despite our dislike. Numerous additional examples could be cited of the necessary differences between storytelling and story writing, such as those which follow from the storyteller's being a performing artist (like an actor on the stage), who will employ change of voice and tone to indicate different speakers or sarcasm as opposed to straightforward statement. The effects thus conveyed by the spoken word are effects that the story writer must somehow present by other techniques, techniques that would figure prominently in a poetics.

In attempting our grammar of biblical poetics it would be an exercise in pedantry to proceed from the simplest to the more complex. For all that, it is not amiss to note that such matters as sentences and paragraphs, superscriptions and chapter headings, quotation indicators, and punctuation conventions in general, are all elements of a poetics. Nor may we ignore such elements as figurative language just because they are common both to oral and literary art. We shall, however, be required on more than one occasion to ask whether the simplest of figures, common in ordinary speech, are necessarily an earlier development than "sophisticated" literary metaphors.

The introductory discussion that follows features the elements that I have found of particular or pervasive importance for biblical poetics.[2] These elements are grouped under four headings: Foci of Literary Analysis, Categories of Interpretation: Narrative Genre, Factors in Interpretation: Metaliterary Conventions, and Figures of Speech. These headings do not, however, represent hard-and-fast categories. So complex and fluid is the syntax of poetics that many elements will require discussion under all four headings. Thus, for example, metaphor is a figure of speech and falls under the fourth heading. But its deployment in one place rather than another would render it a *focus;* it may feature a cultural convention and require consideration as an *interpretive category;* and such consideration and the conclusions it yields may render it a crucial *factor in interpretation.* The introductory treatment in this chapter is not intended to be definitive, since, of course, it cannot be exhaustive. Its purpose is to map out the topology of the poetical terrain and to scout out the kinds of problematics that terrain will pose and the varieties of strategies it opens up.

The reader should be warned, however, that these four headings do not constitute categories of a single order. They do not, in my mind, represent categories of poetic analysis that are equally legitimate. Only the first category is altogether legitimate.

In regard to the second head, *genre* is a misleading tool for interpretation. To say that a piece of literature is a satire, for example, is to offer an interpretation (which may or may not be correct); but the genre label is in itself no tool at all. This methodological error leads to even more absurd results when there is no general agreement as to the definitions of the genre. A generally perceptive critic of this discussion in the manuscript altogether missed the point of my discussion of this second category by suggesting that I get my "definitions for myth, legend, etc. from authoritative and serious studies about these subjects and not from a dictionary!!" He thus provides an example of the almost universal habit of reifying literary abstractions—*abstractions* in that they are concepts, not facts and *less-than-useful* abstractions due to the absence of "authoritative" definitions of what they do and do not represent.

The third head deals with presuppositions about various aspects of an author's intent or the level of competence of author or audience in antiquity. Such presuppositions, largely shared by critics and their audiences, often provide a treacherous basis for judging literary phenomena.

Finally, the fourth head comprises a miscellany of figures of speech to which too little attention is paid in addressing such questions as historiography versus fiction, literal versus metaphoric intent, and so on. They are particularly important for meaningful translations of Scripture's Hebrew original, translations that replace (and often traduce) the original text that we are purporting to criticize. Here, too, the subheadings are not necessarily mutually exclusive. This entire heading was included to provide part of the skeletal framework to which reference might be made in the poetical reviews or to obviate the need for detours in the arguments on the individual exegetical essays. A worthwhile project for literary investigation would be to explore the extent to which the relative presence and absence of such figures in the biblical texts might be indications for judgment as to the questions raised in the third category.

Foci of Literary Analysis

While all stories are narratives, not every narrative is a story. A story has three features: characters, plot, and setting. Characters, situated in settings of time and place, act, or are acted upon, in a connected series of events. (By contrast, some narratives may contain the stuff of story without achieving—or even aspiring to—the status of story, for example, the diary of a shipwrecked mariner, the log of a ship, the chronicle of a military campaign, the minutes of a meeting, the victory stele of a conqueror.) Story is an art form, unlike other sequences of events that are adequately described by terms originally bespeaking "number," such as *account, recounting, tale* (see also *tally, teller, toll*). Such tales and accounts, whose purpose is to communicate information, will more often than not eschew such artistic frills as word-pictures of scenery and weather, probings of character in monologue or dialogue, and revelations of relationships between events and the characters of the protagonists and antagonists. The author of a story may, in keeping with his artistic purpose, give prominence to all three features or may scant on one or two of them.

No perceptive literary critic has failed to note the artistic economy of biblical narrative. One is tempted to replace *economy* with *thrift*. Compared to the literature of almost any other people, time, place, or clime, biblical narrative is remarkable for the sparseness of descriptive detail in the case of all three features: setting, characters, and, yes, even plot.[3]

Setting and Description

A writer who is in love with language may out of sheer ebullience wax poetically lyrical or grim in describing landscape and season, shapes of buildings and sounds of stir, shades of light and darkness, features of body and face, dress of characters and their idiosyncratic gestures. Critics are generally agreed that all such details should ideally be avoided in fiction unless they serve to heighten a mood, sharpen a character, or advance the action. If descriptive detail serves none of these purposes, the description will be judged as superfluous or excrescence no matter how beautifully the word-picture is drawn. A random sampling of today's popular fiction will reveal, I believe, that this standard is rarely honored. It is not enough for the narrator to say, "Framed in the doorway stood the most beautiful woman in the world." It is almost a canon of today's fiction to specify color of hair, eyes and complexion, tilt of nose, curve of cheek, droop of eyelid, jut of chin, and thrust of breast or hip. All this when what is crucial for the plot is the beauty of the woman in the eyes of the narrator and not at all whether the narrator prefers lithe blondes or plump brunettes.

The contrast with biblical practice could not be more stark. Such is the economy of this art that no descriptive detail is devoid of significant purpose; and the careful attention of the reader is called for when a descriptive detail seems merely ornamental or a detail is absent where one would most expect it. In regard to setting, then, why the specific detail of where Noah's ark grounded and no mention of the vessel's home port? Why is the peak in Moriah land two days distant from Beersheba (north or south?) while Horeb's peak is (for Elijah at least) a walk of forty days and forty nights from that settlement at the northern fringe of the Negev? Why do we learn that YHWH–God makes his appearance in Eden's garden "at the breezy time of day" but not how old Adam was at the time?

The presumption in the foregoing paragraph is bold both in regard to the high level of Scripture's purposive art and in regard to the sweeping implication of the consistency of this feature throughout the many and varied biblical narratives. If convincing (or at least plausible) reasons for inclusion or omission of such details are forthcoming in the varied texts that will be examined, it will constitute rather conclusive evidence for this rule of biblical poetics and one link in the chain of argument for a grammar of poetics as a legitimate method—if not the method of choice—of Scriptural hermeneutics.

Character and Characterization

The range of a term's metonymic extension and the naturalness (as contrasted with the artistic or literary nature) of the metonymic process will be readily apparent to the eye that scans the score of entries in an unabridged dictionary under *character*. Deriving from a Greek root meaning "mark" and usages such as "engrave" and "describe," the term in English can stand for a trait or an aggregate of traits, an individual distinguished for such trait or traits, or (in its most neutral sense) a person represented in a drama or story, that is, a member of the cast, one of the dramatis personae.

Robert Alter and Adele Berlin each devote a chapter to this focus of narrative art.[4] Both tackle the problem of the degree of characterization in biblical story as compared with what is customary in other literary traditions and examine the means by which the Bible achieves these effects. Both treatments are highly instructive and often original and on occasion represent invitations to productive debate. Here, too, I want to stress the significance of the uniquely thrifty features of biblical characterization. One question that deserves fuller investigation is the extent to which individuals in biblical story are merely members of the cast, mere functionaries denied all characterization (Berlin calls them *agents,* or *round characters,* that is, full-fledged individuals, or flat characters, that is, types). Berlin cites a view that in primitive heroic narrative there is no aspiration to the kinds of complexities of characterization found in later narratives. To the extent that this applies to biblical narrative Berlin takes issue with it, arguing that all three of the foregoing categories of character "can be found in biblical narrative and the same person may appear as a full-fledged character in one story and as a type or agent in another."[5]

Berlin is certainly correct in this last observation. Yet the second part of her statement points to a dimension of truth in the source with which she quarrels. Without granting that the term *primitive* applies to Homer or the Bible except in a chronological sense, I would stress that for good (and different) reasons neither Greek epic not biblical prose narrative evinces the kind of interest in the kind of depth characterization that we would generally regard as sine qua non in a fine novel, be it a comedy of manners, a psychological fiction, or a verisimilitudinous historical romance. If a Moses, say, can be a mere agent in one story, a type in a second, and a round character in a third, what can be said of the character of Moses in the literature as a whole? The answer, I believe, points to the difference between a central character in the Bible and one in modern fiction. In today's serious fiction the central characters are in a sense the very raison d'être for the fiction; the fiction is what it is only because the characters are what they are. Without denying the profound lessons in psychology and philosophy present in modern fiction, it is yet safe to say that entertainment is most basic to its motivation, or (to put it less crassly) it gives a high priority to delectation over edification. The Bible, however, is in fine and large the development of a religious ideology—a theology that is paradoxically anthropocentric, a literature of preachment. That the expressions of this vision in generalities and particularities are achieved in a highly artistic manner (to the extent they can be shown to be so) makes the biblical achievement even more wondrous. Robert Alter is neither religious ideologue nor professional biblicist. It is therefore with gratitude for his insight (and perhaps envy of his diction) that I cite the following paragraph in full:

> Since art does not develop in a vacuum, these literary techniques must be associated
> with the conception of human nature implicit in biblical monotheism: every person is
> created by an all-seeing God but abandoned to his own unfathomable freedom, made
> in God's likeness as a matter of cosmogonic principle but almost never as a matter of
> accomplished ethical fact; and each individual instance of this bundle of paradoxes,
> encompassing the zenith and the nadir of the created world, requires a special

cunning attentiveness in literary representation. The purposeful selectivity of means, the repeatedly contrastive or comparative technical strategies used in the rendering of biblical characters, are in a sense dictated by the biblical view of man.[6]

Moses exists for the Bible, not the other way around. What makes him so supreme a literary accomplishment is that in his greatest glory he is (in Berlin's term) merely an agent (God's); in his full-fledged roundedness as a human sorely taxed by God's call and overtaxed by human perversity he is yet a flat character, that is to say, a type of the prophet, who embodies God—voice and person—all the while he shows how great the abyss between human and Deity. The biblical patriarchs, as has often been appreciated, exist on two levels: they are, and must be shown to be, credible individuals of flesh and blood, working out their own destinies, experiencing fear and faith, exultation and despair. They are at the same time, as the rabbis recognized, foreshadowings in person and event of their descendant clans and tribes and nations. The more central and developed (rounded) a character such as David, the more typical as model and warning for dynast and peasant, philosopher and boor.

Plot

Basically, plot is a matter of locating a series of events (or two series of events and their convergence) in time and place in a way that suggests other meaningful relationships (such as causality) between the events and the characters who figure in them. Some critics prefer the term *action* to *plot*, but there is reason to doubt that one term creates less problems than the other. Such elements as viewpoint and dialogue, which can scarcely be sundered from considerations of plot, I shall soon treat as independent foci. But consider the question whether a dialogue is better seen as an action or an event. It is clearly, in a literary sense, both. What shall we say, however, of a genealogical list or a chronological table? They hardly constitute actions, but as data supplied by the author in the context of plot and character they can be construed as literary events; that is to say, they must function in some meaningful way to promote the general theme, thesis, or thrust of the literary composition.[7]

As far as action proper is concerned, poetical analysis cannot admit of superfluous action any more than it can such superfluous description as discussed above—not, that is, without raising questions as to the competence of the literary artist (or, in the case of Scripture, editor). And the greater the number of seemingly unrelated side actions or subplots, the greater the doubt that the text is amenable to poetical treatment.[8] I would stress, however, that the proposition that even Homer nods occasionally is more often than not the self-serving recourse of an unperceptive critic: failing to account for something present (or missing) in the design, he covers his own failure by charging the bard with inattention.

After his treatment of Genesis 38 (see note 8), Alter expresses some astonishment "that at this late date literary analysis of the Bible of the sort I have tried to illustrate here in this preliminary fashion is only in its infancy." (My only quarrel with this, as I hope to show in a future volume, is that the statement is equally applicable to such epics as the *Iliad* and the *Niebelungenlied;* dramas such as Marlowe's *Dr. Faustus;*

histories as varied as those of Herodotus, Suetonius, Bishop Gomara, and Corporal Diaz and even Cortez's letters to the king of Spain; political tracts like Machiavelli's *Prince;* and philosophical treatises such as Plato's *Dialogues.*) Alter goes on, "By literary analysis I mean the manifold varieties of minutely discriminating attention to the artful use of language, to the shifting play of ideas, conventions, tone, sound, imagery, syntax, narrative viewpoint, compositional units, and much else." Aside from providing a fair definition of what I mean by *poetics,* this sentence includes the foci *point of view* and *compositional units,* which I shall discuss later. Nor, I think, would Alter disagree that his "much else" might include the foci *telling, showing and dialogue, treatment of time,* and *dramatic effects.*

Point of View

Voice. Perhaps no focus of literary analysis has received and continues to receive the attention devoted to this one. In part this is due to logical and chronological priority of this element: Here, as I suggested earlier, is the point or line that marks off oral from written narration, storyteller from story writer. Who is talking? Or rather, whose voice am I to understand is addressing me in these written lines? In part this attention is due to a continuing experimentation with voice as authors realize how many narrative strategies are opened to them by the many choices as to point of view.[9]

Narrator. The author is the creative artist who stands outside the creation for which he is responsible. Thus, the author is not to be confused with the "person" whom he chooses as narrator. The narrator is the person (or voice) who tells the story. The narrator will ususlly speak in the mode of third person or first person. In the third person mode the narrator speaks as one who stands outside the story. He may be—as the author will artfully convey to the reader—altogether omniscient (like God, so to speak) or partially omniscient, that is, omniscient in regard to some of the characters or a part of the action. When the narrator is made to speak in the first person, omniscience is necessarily denied him and he becomes to some extent a character in the story. As first-person narrator he may feature in the action as the central character or as a major one, or he may be an unidentified and hazy observer and commentator on the characters and actions he reports. The third-person narrator refers to all characters as *he, she,* and *they* while the first-person narrator also speaks of himself, as *I.*

In similar fashion we may address the question of the audience to whom the narrator's voice is addressed. This audience may be explicit and characterized to one or another degree, or it may be the narrator's self, or it may be as hazy as the hazy narrator himself. The first-person narrator is characterized in criticism as intrusive when he addresses his reader directly, as *you.* When this *you* is not permitted to remain an anonymous reader outside the story but becomes one of the characters, the narrative mode is sometimes called the second person. While this form of narration is still rather rare, experimental, and generally regarded as a modern device, it is basically a variation on the first-person mode; it is of particular interest because it is the mode for some 95 percent or more of the Book of Deuteronomy. The very rarity of this mode makes Deuteronomy a fascinating subject for poetical study. Source-analytic approaches to this book focus, as we know, on the repetitions of and discrepancies between material common to this book and other pentateuchal books on the assumption that these stem from substantively different and often conflicting traditions. Poetical analysis would begin with an assumption of editorial, if not authorial, harmony in the five books and ask how this point-of-view stratagem in Deuteronomy serves to advance the thrust expressed outside this book in a different way, a way made possible or inevitable by reason of a different point of view.

Narrator's Omniscience and Reliability. To the degree that the narrator is omniscient, he can know what is happening in many places at one time or what is going on in the minds of the characters (perhaps even better than they do); he may know what had led up to the present state of affairs and what the future holds in store; he may interpret actions, supply motivations and make objective judgments of all kinds: moral, esthetic, pragmatic, and so on. The consistency with which the author endows the narrator in these respects will at least in part determine for the reader the extent of the narrator's reliability. The author, however, just as he may limit the narrator's omniscience, may choose to impart information to the reader through the device of impugning the narrator's reliability, altogether or in some respect. The author may, for example, let the first-person narrator impeach himself as a consistent and bare-faced liar or establish himself as a faithful reporter of the events to which he is witness even as he betrays the implausibility of his interpretation of the meanings of the events and the motivation of the characters.

The reliability or unreliability of the narrator is one of the trickiest stratagems available to an author and poses equally tricky questions for the literary critic. Let us expand the example presented earlier, of Mother's recollection of the time she witnessed a rabbit popping into his hole. Mother goes on to describe the rabbit as wearing a waistcoat, consulting a watch and fretting over being late for an appointment. Neither Mother as storyteller nor the author of *Alice in Wonderland* are likely to be understood as unreliable narrators. In either case we will recognize the story as belonging to a genre where such events are to be assumed as credible. We will willingly suspend our disbelief, will grant reliability to the narrator in the interest of learning what delectation or edification the author has in store for us. But how shall we go about deciding a similar question in regard to some Scriptural narratives—to mention but a few, the events in Genesis 1, which could not have been witnessed by any human; Genesis 22, where God is reported as bidding my great-grandfather slay my grandfather (an act that if completed would have precluded my own existence); the prophet Jonah's composing hymns in the dank ambience of a fish's belly.

Showing and Telling: Dialogue As Showing

In a sense everything between the first and last pages of a piece of literature represents a telling on the part of the author or narrator. The art of narration, however, lies largely in the narrator's suppressing the ubiquity of his own voice, the consciousness of his presence behind the scenes he is evoking, by focusing the reader's attention on the specifics, the concreteness, the spontaneity of characters and events. This aspect of narrative art is characterized as showing, as over against (artless) telling. The omniscient narrator, presuming our faith in his judgments, will tell us explicitly about his protagonist's eyesight and habits of thrift: "Like all peasants M. Corot was near with a penny, and with all his myopia rarely missed an opportunity to add to his rat's nest of gleanings." Both the physical nearsightedness and the miserly trait could be conveyed by an action in which the character show these traits in himself: "Clad in his Sunday morning best, M. Corot yet dropped a knee to the dirt, brought his nose close to the ground, lifted the object to a squinting eye, then stuffed the two-foot piece of string into a pocket." The character here is *shown* by his *actions* as he can be further revealed in his direct discourse. "Stuffing the string into a pocket, he muttered, 'Never can tell when you'll need to tie somethin' up.'" The artistic control of the biblical narrator is, I believe, revealed by the observation that his deployment of direct discourse is never accidental or capricious.

Dialogue or Direct Discourse

From the above example one can get a preliminary idea of why a character will be made to show himself by the specificity of his direct discourse, as against our having to take the word of the narrator in the less concrete formulation of indirect discourse, for example, the innocent-sounding question of God to Cain about his sibling's whereabouts, "Where is Brother Abel?" The response of Cain shows that to his ears—after all, why would Abel be so quickly missed—the question's import was, "Whatever has become of your brother Abel?" And the sense of guilt is shown in the response which, denying knowledge, expressing indignation that he has been troubled for information, and gratuitously denying of a guardian's role, also reveals the psychology of every murderer, a rejection of responsibility for the welfare of his fellow: "Am I my brother's keeper?"

Interior monologue, like spoken dialogue, will similarly be featured when the narrator wants to make a character's motivation more specific, vivid, and immediate. But even such interior thought (that is, unexpressed in speech) can be reported in indirect or direct discourse. At every point, then, where the biblical narrator reports a character's thought (even God's, or perhaps especially God's), we should ask what subtle purport is conveyed in the direct discourse, as contrasted with a possible formulation in indirect discourse.

Berlin discusses speech and action under the rubric of *character and character-ization* and then again under *point of view*.[10] I find myself in agreement with her general observations even as I would interpret the author's intent differently in specific examples of direct discourse she presents. Alter devotes an entire chapter to this subject.[11] One must be grateful to him, here as elsewhere, for the theoretical poetical questions he raises—here, specifically, about the proportion of narration to dialogue in the Bible. Nor is the debt of gratitude diminished when one is inclined, as I am, to wonder whether his answers may not be less than definitive or applicable in only some instances. By the same token, I would not be surprised or dismayed if neither of these critics were convinced by many of my interpretations of specific instances of direct discourse. I want to emphasize that disagreements on particulars of interpretation on the part of literary analysts in no way brings into question the conviction that the meaning of a literature, biblical as well as nonbiblical, is most fruitfully investigated by recourse to poetical analysis.

I shall now proceed to discuss as an independent focus (due to its frequent appearance in biblical narrative) the phenomenon (I do not know who first recognized and so properly labeled it) of *free direct discourse.*

Free Direct Discourse

Direct discourse is, supposedly, exactly what a person expresses in words. Not quite. What a person expresses in words is speech. Direct discourse is the speech not of a person as such but of a character, that is, the representation of a person in a dialogic narrative. It is, so to speak, speech at a remove. Direct discourse is thus an artifice, a category of art rather than of life itself. It represents *in comparatively few words* a much larger complex of thought and temperament. As Mark Twain once demonstrated, there is no art in conveying the image of a bore by letting him run on for page after page of dreary dialogue; the art is in showing him (that is, allowing the bore to show himself as such) in as few words as possible.

Free direct discourse is the speech of a character that in some way must be understood as being *either more or less* than what the person portrayed as a character would have said in that particular circumstance in real life. When free direct discourse in the Bible is not recognized as such, it will operate to confuse or perplex the alert reader; it will impel the Bible scholar to posit an error in the transmission of the text or to find fault with the artistry of the biblical writer. Even when recognized for what it is, a purposive literary device, it often seems an idiosyncracy (and a clumsy one, at that) of biblical art. In trying to justify the artistry of this device it will be helpful to contrast the dialogue or direct discourse of a character in narrative with the speech of a character in a drama.

Direct and indirect discourse are options for a narrator in a narrative. In a drama, there is no narrator (although intrusive elements more proper to narrative may be present in the forms of prologue, epilogue, and chorus) and so no option as between direct and indirect discourse (except, of course, in the mouth of a character). While it is tempting to refer to all speech on a stage as direct discourse, it is noteworthy that we simply do not do so. The speech of a character in a drama are his "lines." There is no "direct discourse" because there is no option of indirect discourse.

It was noted above that free direct discourse is what it is by virtue of its inexactness as speech, the inexactitude being by nature either superfluous or elliptical—too much or too little. Let us consider a few examples, first, the dialogue between God and Cain. Even for those who report, or credit reports of, God's appearing to mortals and speaking to them, the likelihood is scant that God would, like a bolt from the blue, speak up to ask a mortal for a bit of information which even a less-than-omniscient deity would be expected to have already. There is no setting for the scene, and God's words on the printed page present an interpretive challenge for translator and critic comparable to the challenge to player or director were they lines in a drama. Is the unnecessary identification of Abel an offhand epithet as in my earlier translation, "Brother Abel"? If so, then the scene and the speech are clearly elliptical. God's question, for all its being an opening ploy, a leading device, and essentially a rhetorical question is that of a human companion who in the course of a sociable chat would have meant by the three-word Hebrew sentence, "By the way, haven't seen B'rer Abel in a spell, wherever is he?'' And if so, Cain's five-word reply is an equally casual, cool, and sneering repudiation of the sneaky sleuth posing as backyard gossip: "Don't know, ain't about to set a watch on him." Against the above interpretation it will be argued that the three Hebrew words rendered "Am I my brother's keeper?" cannot be reduced to the casual rendering I have suggested and neither in gist or tone provoke God's indignant response, "What *did* you do? Listen: your brother's blood, shrieking from earth!" These arguments are correct. But the alternative would have God's initial question as a request for information (which makes little sense) or as containing an implied accusation, in which case, Cain's response would have to be taken as indication that he thinks he can fob God off with a lie and then anticipate a question that God, along this line of logic, would not have the sense to ask, namely, *"Why* don't you know?"

The above lines are not "playable" on a stage, nor were they meant to be. They are part of a mythopoeic fable on the nature of God as creator of a world of men whom he grants freedom of choice while requiring of them a morality they reject. And the use of direct discourse (and in a sense, unrealistic free direct discourse) by the altogether-forgotten narrator makes for a theological statement so fraught with meaning, so simple in its directness, so pointed in its address to the Cain in each of us, so beguiling in its permitting us to identify with the slain victim rather than the perpetrator who survives to hear the story—in short, such a triumph of rhetorical art in service of theology—that it will forever mock all attempts to exhaust its meaning in propositional formulations.

For all its seeming rarity in Western literary tradition, free direct discourse, at least in its elliptical mode, is a natural and probably an inevitable feature of narrative. It is, for example, implicitly present whenever a night-long conversation is reduced to a few pages of dialogue. And the line between indirect and direct discourse is not as clear as we would imagine. Just as free direct discourse, bespeaking in it the presence of the narrator reporting and commenting, is almost direct discourse, the kind of dialogue identical to the *lines* spoken on the stage, the poetical corollary to this is that the narrator may employ indirect discourse yet suppress himself as reporter as he gives greater specificity and more color to the character. For example, instead of "He swore a mighty oath never to knuckle under to the king," he might write, "He said that he would be damned to boil for aeons in caldrons of pitch before he bent a craven knee to the whoreson usurper on the throne."[12]

The Synoptic/Resumptive Technique

Compositional units and treatment of time would normally receive separate treatment in a poetics of narrative. I am compelled to treat them together by my discovery in biblical narrative of an artful device, traceable in part, I believe, to Hebrew tense usage. This device I have labeled the synoptic/resumptive or, compounding the clumsiness, the synoptic-conclusive/resumptive-expansive. Essentially it is the treatment of one event two times. The first narration of the event (and an event may be simple or compounded of a number of actions) is usually *briefer* (hence *synoptic*) than the second, is an independent, freestanding literary unit. The second treatment or episode, usually longer than the first, may or may not be able to stand by itself. If the first treatment is blanked out, the second may reveal a gap between it and the preceding narrative or an esthetically less-than-satisfactory transition from the latter to the former. The second treatment seems to go back to the opening point of the first episode and, resuming the theme of that treatment, provide a more detailed account (hence *resumptive-expansive*) of how the bottom line of the first episode (hence *conclusive*) was arrived at. Awkward and full-blown as this nomenclature is, it yet fails to describe the technique. Were the two episodes altogether consistent with one another (as they sometimes are), the term would at least be accurate. Often, however, episode 2 is not merely a filling in of details but will so differ in some respects from the first as to constitute another version. When the inconsistencies or contradictions are blatant enough, they pose the question of how an author who has proven himself so fine an artist can suddenly lose his capacity to choose between two possible versions and in sudden ineptitude decide to keep them both.[13] The variety and richness of effects made possible by this technique are such that a full appreciation can only be achieved by examining each instance in situ. Before speculating on a possible explanation for the rise of this literary phenomenon it will be helpful to consider the questions of compositional units and treatment of time.

Compositional Units. The basic compositional unit in speech is a sentence—a subject accompanied by an intransitive verb, a predicate nominative or predicate adjective, or a transitive verb and object—although half-sentences may in context be highly effective. The basic unit of a literary composition is a number of sentences that come to a temporary rest, a rest that functionally marks them off as a unit. How both the smallest units (clauses and sentences) and the larger compositional subunits and units are indicated in writing or in print can vary in importance from helpful to crucial, from procedural to substantive. Commas, colons, periods, marks of quotation, question, and exclamation are small but not mean ingredients in a work of literature. Antiquarians, particularly, have reason to be grateful to their predecessors who labored to supply these aids to decipherment or interpretation for writings whose authors knew not even of word-separators, unfilled spaces, or lower- and uppercase letters. And for all this gratitude it behooves them ever to reexamine the text with an eye to its pristine jumble and to reconsider the spellings and the divisions of sentence, verse, or clause, all of which represent interpretations of what the author intended.[14] In the case of poetry the very question as to whether the piece before us was or is intended as prose or poetry to begin with may depend for answer on how the words are arranged on consecutive lines. In longer works of fiction such as the novel, the absence of indicators of compositional units, page after page of unparagraphed lines devoid of punctuation marks will be the giveaway for such experimental techniques as Joyce's stream of consciousness. Elsewhere, chapters, each representing the interior monologue of a different character, may offer no clue to the voice behind the words beyond the name of the character as a chapter heading.

In biblical scholarship critics use the term *pericope* for a passage of text isolated by one criterion or another from what precedes and what follows it; these pericopes may defy editorial demarcations of verse and chapter. Just as verse and chapter are late editorial demarcations and themselves open to question, so is the matter of where a pericope begins and ends. Source analysts at their most modest claim to be able to mark off such authors or schools as J, P, H, E, and D in the Pentateuch. And the biblicist who confesses that he knows no J or P will find that most of his colleagues will still look askance at him. Yet we all know that the biblical texts as first composed and recorded were interspersed with few chapter headings or paragraph markers and that even in regard to the Pentateuch, that block of material transmitted for centuries as a unit, the only "book" that is meaningfully separated from the others is Deuteronomy. There is perhaps a perceptible break in theme discernible in the first sentence of the book called Exodus but certainly none at all in the case of the two books named after their opening chapters' first significant word: Wayyiqra (Leviticus) and Bemidbar (Numbers).

One of the chief tasks of poetics in regard to an ancient literature is to determine the smallest subunits of a composition; the larger subunits, the "chapter" and "book" divisions, with a view to a better appreciation of the focus of the subunits; the frame of the larger units; and the interconnection of all the parts into one architectonic whole. I shall not in these volumes explore the limits of this architectonic whole. But the smaller units and subunits are a must for poetical demarcation. And we shall have to pay close attention to the rubrics or other sundry techniques that the biblical writers employed, in the absence of punctuation, as demarcative indicators. I have found it useful to use the term *episode* for a literary subunit that deals with one event; and the two treatments of one event that I call the synoptic and resumptive are, in keeping with this usage, episodes.

Treatment of Time. Events in a series are ordered sequentially. Events in two different series may be simultaneous. Events occurring simultaneously in two series in widely separated locales may be organically related, for example, in physical terms the relationship between the tidal movements on earth and the movement of the moon at a remove of thousands of miles and in human terms the physical events in a star going supernova at a remove from earth of hundreds of light-years and the impact on human history of the visual observation of that event millions of years later. Our visual experience of a brilliant new star in the heavens and our cognitive experience that the event we are viewing took place aeons before our race evolved are themselves two series of experiential events of different orders yet organically related. Physical time—time measured by events external to us such as the fall of sand in an hourglass or the movement of a celestial body in space—we regard as objective. Psychological time, time crowded with or empty of events, we regard as subjective. Yet both are real, both exist as aspects of human experience. If time flow is a phenomenon we do not experience consistently, measuring it at one time by sand running in a glass and at another by the pace of events, how problematic the representation of time flow in words! The omniscient author, capable though he is of witnessing events occurring in two different places at one and the same time cannot tell of both events at the same time. One episode must precede the other, at least in the telling. A complicating factor is that one objective (perhaps a chief objective) of literary art is that of not boring the reader. And nothing is so wearisome to an audience as the conformation to chronological order: "And then, and then, and then. . . ." It should not, therefore, be surprising that even in the earliest "most primitive" literatures we find examples of the explicit technique in Western romances ("Meanwhile, back at the ranch. . . .'') or more subtle ones such as the flashback technique in the voice of the narrator or the voice of a character recollecting events. (Many a modern writer, experimenting with time flow as a narrative strategy, will seem to have rediscovered the synoptic/resumptive technique, since he has a first-person narrator shuttle back and forth in a past time, mentioning a critical event briefly, skipping to a later one and then returning to the former for a more detailed exposition of the event and clarification of its significance for the latter event.) In biblical narrative, the treatment of time in flashbacks and in the synoptic/resumptive techinque are strategies related to certain peculiarities in the function of the Hebrew tenses as these tenses relate to past, present, and future time.

Biblical Hebrew Tenses, the Waw-Conversive, and Parataxis versus Hypotaxis. The more familiar we are with a phenomenon, the less likely we are to be stimulated to imaginative thinking about it. The more the biblicist is at home in the Hebrew of his texts, the more he must distance himself from it to appreciate how peculiar are its idiosyncracies, the kinds of potentialities owing to these peculiarities, and the achievements realized by exploitation of some of them. Consider the vagaries of the biblical Hebrew tenses in their basic functions and in their strange narrative transpositions. In virtually all Western languages perfect and imperfect are used in description of tenses that relate to the completion or incompletion of the verb's action and may be further classified in relation to that time in terms of present, past, and future for both. Among the *Semitic* languages only biblical Hebrew has two finite tenses, perfect and imperfect as designations for tenses that function not in terms of the time of the verb's action but in terms of the action's state of completion or incompletion in present, past, or future time.[15] And this anomaly of Hebrew tense function is complemented by another anomaly. In normal narrative style the verb precedes the subject noun or pronoun (or contains the subject), and the imperfect tense with a proclitic waw attached expresses the function of the perfect, while the perfect with a proclitic waw expresses the function of the imperfect. The same waw serves as a copula with these verbs in their pristine unreversed functions, namely, when the verb is governed by a negative or when a parenthesis is intended. The proclitic waws are characterized by grammarians as *waw-conversive* or *waw-consecutive,* while the copulative waw is called *waw-conjunctive.* This classification is essentially functional, not morphological; for only the waw-conversive, with the imperfect tense, affects the *form* of the verb.

We know nothing about the everyday speech of ancient Israel, inasmuch as all our records of it are literary. Guesswork about it is doomed to remain just that. Yet I offer the following speculation for its heuristic value. Inasmuch as the waw-conversive verb construction is, as far as we can tell, a feature only of narrative prose, might it have arisen as a peculiar literary convention, one that never became a feature of everyday Hebrew speech?[16]

What stimulates this speculation is the observation that the resumptive-expansive episode frequently functions as a flashback recapitulation of the preceding synoptic episode. And biblical Hebrew regularly employs the nominal sentence, subject followed by verb in the perfect tense, to interrupt the narrative flow with a flashback. Such flashbacks appear in Genesis 1:2 and 2:5–6. A particularly interesting series of flashbacks appears in the first chapter of 1 Samuel 1. Verse 3 is a flashback. Verse 4 begins with four words in the narrative past (two nouns and two verbs in the imperfect with waw-conversive), which are in turn succeeded by a long and involved flashback continuing through verse 7a. If the reader will try reading this passage aloud, he will realize how unlikely it is that this passage was meant for the ear rather than the eye. [17]

The availability of the nominal sentence with waw-conjunctive for flashback scenes (actually for parenthetic asides, of which one type is the flashback) points to the resumptive episode as a highly nuanced and flexible device, not limited to parenthetical or flashback functions. The modalities of the synoptic/resumptive technique must be investigated in its various occurrences with a view to such questions as, Does the resumptive merely expand on the synoptic without displaying any discrepant features? Or does it represent an alternative version of the synoptic episode? Or does it inform on the synoptic episode even as it seems to contradict it in one or another detail? I think it important for us to remember all the while that it is the waw-conversive narrative convention that makes this literary technique possible. [18] It will be useful in this connection to cite in part Alter's discussion of parataxis and hypotaxis in biblical texts. I begin with his note 4 in chapter 2 in full:

> Parataxis, we should recall, means placing the main elements of a statement in a sequence of simple parallels, connected by "and," while hypotaxis arranges statements in subordinate and main clauses, specifying the relations between them with subordinate conjunctions like "when," "because," "although." Thus the sentence "Joseph was brought down to Egypt and Potiphar bought him" is paratactic. The same facts would be conveyed typotactically as follows: "When Joseph was brought down to Egypt, Potiphar bought him." (My example is actually an abbreviated version of Gen. 38:1. The first version is the way the original reads, the second version, the way some modern translations, avoiding parataxis, render it.) (ABN, 26)*

This admirably clear exposition and illustration of two formidably named syntactic alternatives elicits our gratitude. Its citation here, however, is owing primarily to the possibly misleading formulation in the last sentence, in parenthesis. "The first version is the way the original reads" is true only if we add "literally" (and assume further that the literal is what the original intends) and if we assume that the waw-conversive (or any waw, for that matter) means "and" first and foremost. Now it is true, as Alter goes on to say on the next page that "biblical narrative prose exhibits a good deal of variation from parataxis to hypotaxis, according to the aims of the writer and the requirements of the particular narrative juncture." But this observation poses a challenge to the literary analyst: Why or how are the writer's aims or the requirements of narrative at certain juncture better served by one rather than the other? Why the relative specificity of hypotactic formulation in one place and the richly ambiguous paratactic formulation in another? In other words, hypotaxis and parataxis are syntactic variations that in biblical Hebrew present the author with options (not available to authors in other languages); but a paratactic formulation is not one whose meaning is accurately or adequately rendered by rendering the waws as so many *and*s. [19] As I shall have reason to stress under other headings, poetics is a study of texts

*See Abbreviations, p. 259.

for meaning(s); and one of the first problems for the literary analyst studying literature in a language not his own is that of producing a faithful translation. And the art of translation requires a grasp of the full range of the possible functions of syntactic options.[20]

Achieving Dramatic Effect in Literary Art

The means by which writers avoid boring their audiences are so varied that one may be tempted to classify all of these means under one heading. But that would render the heading almost congruent with literary art itself and thus defeat the purpose of the categorization. Fine writing, even artistic writing, can be achieved in areas where one would least expect it, such as legal briefs, judicial decisions, scientific essays, and even literary analysis or biblical scholarship. Stylistic elegance can by itself sometimes hold a reader even when his interest in the subject is less than vital (Dare one confess, in this day, to reading Macaulay with pleasure?), although others may find the very sonorities a soporific. Even the pall of too-long-sustained elegance can be pierced by the alternation of sentence length, lapse into whimsy, resort to wit or anecdote, graphic illustration, and so forth. But outside of narrative, even the most graphic of formulations do not properly fall under this heading. Justice Holmes is deservedly famed for excluding from freedom of speech the right to yell "Fire!" without due cause in a crowded theater and for limiting one citizen's freedom of action to the point where another's nose begins. Yet for all the vividness of imagery and trenchancy of expression these formulations are not *dramatic*.

Dramatic techniques are a category peculiar to narrative. It is not a category of playwriting. In the theater dramatic techniques are a matter for actor and director, not playwright. For the playwright drama is the sum total of the characters' lines; and, whether monologue or dialogue, no line can be more dramatic than another. In this respect, dramatic technique is to drama as **direct discourse*** is to dialogue. This last observation points, indeed, to the parade example of dramatic technique—direct discourse or dialogue, the one technique in narrative that is the quintessential mode of drama. By extension in one direction, the various "voice" strategies employed by an author may be seen as dramatic techniques; so also, by extension in another direction we may view the kind of indirect discourse that in its vividness borders on dialogue (see p. 13). Indeed, we may discern dramatic technique in the ordering of information provided by the narrator vis-à-vis what he reports in a dialogue exchange. (An example of this in Jonah 1:10 will be discussed below.) The synoptic/resumptive device, bringing an episode to a swift conclusion—sometimes with what might best be called a "punch line"—and then offering the same episode with an alternate punch line is another (biblical) dramatic technique. The variety of dramatic techniques possible in narrative is in proportion to the ingenuity of writers and the skill of critics in recognizing and classifying them. As *static, the same,* and *changeless* evoke associations with lifelessness and dullness, so *dynamic* and *contrasting* suggest the dramatic. But contrast is made possible only by the existence of the "foil" element. In the discussion that follows we shall see that even repetition can be deployed as a highly effective dramatic technique.

*Boldface expressions in text are keywords referring to sections of Part I (see the Contents).

Repetition

Both Alter and Berlin treat the problems posed by various types of repetition in biblical narrative; both offer explanations of some types of repetition and their rhetorical uses.[21]

Like the *dramatic effect,* this heading, too, subsumes so large a range of devices with such different functions as to put into question the usefulness of lumping them all together because they share one feature in common. Even in so narrow a category of repetition as the repetition of a single word the purpose of an author may be so different from one instance to another that the discernment of one use of verbatim repetition may blind us to the existence of radically different uses. Take, for example, Alter's enthusiastic adoption of the Buber–Rosenzweig focus on the *keyword,* or *Leitwort,* as one of the distinctive conventions of biblical prose, a convention whereby the recurrence of word or root enables the decipherment of a text or a more striking revelation of its meaning (ABN, 93).

As against this function of verbatim repetition I shall anticipate my discussion of the Book of Jonah, where, in a total of forty verses of narrative, the term *gdl* (big) appears fifteen times and the term *rʿ* (bad) appears ten times. I very much doubt that the characterization of either of these two terms as keywords, or *Leitwörter* in the sense discerned and defined by Buber will lead to any meaningful insight. If, however, the nature of this particular narrative and its meaning are discerned by the use of other poetical categories, one can then (as I shall argue) go back over the story with an eye to discovering why the author chose to use two simple terms, each in a striking range of meanings, when his lexicon certainly offered him quite a few alternatives.

Similarly in the case of other categories of repetition (of information by narrator or by a character in direct discourse or in several instances of direct discourse; doublets, of theme or story; repetitions of motifs; and what Alter calls sequences of actions and type scenes), it is my belief that their poetical functions are to be sought in each instance according to the nature and problems of the particular narrative event rather than in a defined pattern that can then be applied to the instances. In the **synoptic/resumptive technique,** the pattern of repetition is there; the function of the pattern must be separately determined in and for each occurrence. Repetition is not sameness. Along the lines of the example for Heracleitus' principle of the constancy of change (You cannot step into the same river twice, for the double reason that neither are you, nor is the river, "the same"), so in the case of repetition. The exact words in two different places are not the same; for their ambience or context is part of their meaning, or their meaning is what it is only in context. *Le plus c'est la même chose, le plus ça change.*

Categories of Interpretation:
Narrative Genre

Literature, as we have seen, is composed speech reduced to, or recorded in, writing. Literature is divided into two categories, poetry and prose. Without attempting a definition of poetry, which lies for the most part outside of our present poetical concern, I shall define prose as any literature that is not poetry. Prose literature can be classified as imaginative or fictional on one side and nonfictional, practical, or utilitarian on the other. From the Near East we have, as early as a-millennium-and-a-half before classical Greece, nonfictional literature that can be classified as legal (laws and contracts), epistolary (on matters of state and commerce), scientific (such as medical, mathematical, astronomical), religious (cultic and divinatory), annalistic (reports of statecraft in war and peace), and historical. While the contents of these genres of nonfiction are debatable (for example, one might argue that the divinatory properly belongs under scientific, that medical belongs under religious, or that both divinatory and medical belong under a separate genre, magical), the genres are generally easily recognized and not likely to occasion debate. (None of the above appear in the form of *essay* or *treatise*, that is, shorter or longer expositions (respectively) describing, analysing, or interpreting a particular theme or subject; essay and treatise are themselves genres that do not appear before the classical period in Greece.)[22] The debate as to genre is hot and continuing in regard to the genres of fiction—in particular, in regard to the genres of fiction that have reached us from antiquity.

The Problem of Genre in Fiction. Neither Alter nor Berlin addresses the question of genre as a key problem in biblical poetics. In avoiding this question they also avoid confrontation with assumptions and conclusions that are staples of biblical scholarship, especially in regard to narrative. The words *assumptions* and *conclusions* in the previous sentence point to one of the most treacherous aspects, if not to the most basic aspect, of the problem of genre assignation: Is the perception of a narrative's genre an assumption that will necessarily shape or determine the interpretation of that narrative; or is the reverse the case, namely, that a given interpretation of a narrative will then determine the genre to which it is assigned? The kind of debate that this question provokes among critics and historians frequently makes the protagonists on either side appear to the spectator like dogs chasing their own tails or debaters arguing over whether an egg is the means by which a chicken produces a chicken or a chicken is the means by which an egg produces an egg.

Ludicrous as these perceptions make us appear, they are close to the mark. A certain circularity of argument is perhaps intrinsic to the kinds of investigation that are featured in the humanities. Recognition of this should serve as a cautionary flag for researcher and student alike; but it need not signal an end to our enterprise. All science begins with classification, that is to say, with the imposition of pattern on the material that has been collected. In the case of our concern (that of classifying biblical narrative according to genre or questioning the applicability to biblical narrative of genres imported from other literatures), our chief task is to discern the limits of these genres, the extent of their overlap, and the various purposes to which any genre may be put in one or another setting.

Genres of Early Fiction. It will be helpful to review the genres of fictional narrative that have been discerned in various ages and cultures and applied to biblical narrative: *fairy tale, myth, legend, fable,* and *folktale.* The definitions in quotation marks are those of the *Random House Unabridged Dictionary.*

A fairy tale is defined as ''a story, usually for children, about elves, hobgoblins, dragons, fairies or other magical creatures.'' Thus, this category stands at one end of the fictional spectrum in terms of credibility. Designed for the entertainment of young minds, its plots are the stuff of fantasy, and it is peopled by creatures as far removed from everyday reality as the imagination may devise.

Myth is ''one of a class of stories, usually concerning gods or semi-divine heroes, current since primitive times, the purpose of which is the attempt to explain some belief or natural phenomenon.'' Myth, then, in terms of the intent behind it, is more serious than fairy tale. Its aim is more to edify than to entertain; its characters are accordingly less whimsical, and its plots are more expressive of the concerns of adults in the real world.

Legend, in our dictionary's definition, is at one remove from the fantasy element in myth: ''any fictional story, sometimes involving the supernatural, and usually concerned with a real person, place or other subject.'' The supernatural need not be involved in legend; and the focus is on person, place, or subject that is in some undefined sense ''real.'' Legend, therefore, is essentially credible in terms of what may actually have occurred to real people. Its intent would seem to be to inform rather than to entertain or explain. Its claim on our belief is greater than that of myth.

Fable is ''a fictional story designed to teach a moral.'' Credibility, therefore, is a factor that is of very little relevance for this category of fiction. Moral conviction is what this genre of fiction bespeaks; its characters and plots are confessedly constructs to help the audience to learn about itself by viewing itself from a distance.

In regard to legend, fable, and myth our dictionary points out that all three ''refer to fictitious stories, usually handed down by tradition (although some fables are modern).'' The qualification in parenthesis might equally apply to myth, especially if we agree with D. M. Thomas in his foreword to *The White Hotel:* ''By myth, I mean a poetic, dramatic expression of a hidden truth.'' It would seem, then, that fables and myths, though ancient genres, may be invented today, while the same could be said of legend only with poetic license, the first dictionary meaning of legend being, ''a nonhistorical or unverifiable story handed down by tradition from earlier times and popularly accepted as historical.''

There is left the genre of early fiction that I am hard put to define: folktale. Apparently differing from myth and fable in not having a didactic purpose, from fairy tale in its remove from fantasy, and from legend in not featuring heroic personages or historically significant feats, a folktale's essential elements would appear to be its mundane plot and characters and its reaching us through tradition.

The Problem of Genre in Biblical Narrative. For those fundamentalists of any branch of Judaism or Christianity who insist that the Bible is either literally ''the Word of God,'' or ''inerrant'' whether in prescriptive demand or narrative discourse, there can be no question as to biblical narrative genre. It is all history, nonfiction, a report faithful and true in its depiction of the past—place, people, time, and events. For most others, religionists and secularists alike, the Bible is a mélange of nonfiction and fiction, a curious hodgepodge of theology and law mixed into a batter with various kinds of poetry in the frame of a narrative whose main thrust is to trace the origins and vicissitudes of the people that produced it. If *hodgepodge, batter,* and *frame* represent a particularly incongruous mix of metaphors, the incongruence is in keeping with even the best attempts to hold the mirror of art up to the nature of the Bible. What in the remarkable range, sweep, and scope of biblical narrative is fairy tale, what myth, what fable, what folklore, what legend, what history? And is it possible that some parts of the narrative may share some or all of the attributes of these genres? Consider the story of Jacob: fairy-tale-like in his encounters with super-natural agents of the divine or the demonic; mythic in its encapsulation of romance and war in the family, nuclear and extended; prophetic in adumbration of characters and events in the lives of generations of his posterity; most, if not all, of it handed down by tradition (true legend, hence history); real in the particular personhoods of Jacob, Judah, Joseph yet preserving them also as three-dimensional prototypes or avatars, in the racial memory of the people Israel—itself emblematic of the best and worst, the transiency and continuity, the fate and destiny of the human race.

Literary critics, as they stand before the monolithic mass of biblical narrative, attempting to grasp the essential nature (or genre) now of this part, now of that, may indeed be like the blind men running hands under, over, and round the elephant's parts and proclaiming their not-to-be-denied apprehensions of the phenomenon: it's an eel, a python, a horn, a tree, a leaf, a baseball, a mountain. But the literary critic is one, not many; he is not blind (hopefully), so he need not mistake a part for the whole; and he has time to study each part, to examine its construction, to make inference from design to designer's intent, and (when the anatomy of the parts has been adequately explored) to recognize the whole as greater than the sum of its parts and the parts as more than textbook categories. A leg is a leg, to be sure. But the articulation of the grasshopper's leg is as different from that of the elephant's as the leg of a table is from the leg of a journey.

Once again, in this connection (as in other junctures of our discussion) I would emphasize that in poetics, as in any other investigation, we must go from the particular and particulars to the formulation of definition or rule. A review of our earlier catalogue of ancient fictional genres will reveal, I believe, that genre is determined more by the intent with which it is deployed than by formal outlines or surface characteristics. Fairy tale can be transfigured into myth and myth reduced to fairy tale. As concerns the study of biblical narrative and the application to it of categorizations borrowed from other ancient literatures or—for that matter—of categories invented for it (like "saga" and etiology), interpretation must precede pigeonholing.

One vital problem of genre remains to be discussed, the problem of a genre of literature that in its modern exemplars may (although not always) properly be classified as nonfiction. This genre is *history* (more properly, *historiography*). Whether this genre exists at all in the period of antiquity with which we are concerned is not often recognized as a question for poetics. Yet for biblical and cognate literature, history—or, rather, literature construed as history—is a crucial category of interpretation.

History in Antiquity: Fact or Fiction? To begin with, let me state that *history* as used herein refers neither to the events of the past themselves nor to the record of these events but to the recordings in writing of these events, that is, to literary compositions ordering and relating events to one another. History, then, is a certain kind of narrative, a narrative about the past (despite our objectifying it, personifying it, and speaking of it in the future as in "leaving the verdict to history"). But all stories are narratives of past events; even when the narrator speaks in a continuing present or, as in science fiction, projects events into a far-distant future. (In *Star Wars,* for example, the events occur in a far-distant future; but narrator and audience exist in an even more distant future, which renders the time frame of the events into their past.)

There is also a kind of natural redundancy in language whose existence must be remarked, even if it require no apology. The master of magical sonnet can muse, "When to the sessions of sweet silence thought / I summon up remembrance of things past . . ."; and even the most captious of critics will not object that memory must by definition be "of things past." Similarly, in my use above of the words *record* and *recording* in regard to past events, there can be no "record" of future events. Future events are the stuff of prophecy, not of narrative; and any record we have is not of the events to come but of the words the prophet spoke. Yet the very use of the term *record* bespeaks a narrative that is not fictional. The storyteller does not record. He spins a tale of the past, an artist who invents his material; the matter of his story derives from his imagination. Not so the historian. The historian does not create fact or artifact; these exist outside him, and he records—even as he reads—the record.

The records of the distant past that the historian reads are either nonliterate (such as kitchen middens or such artifacts as tools and weapons, pottery, buildings and fortifications) or written. Written records (such as diaries, logs, annals, stelae) are historical "documents," the stuff that the historian evaluates for veracity and weaves together to compose a particular kind of story, a history, that is to say, a story featuring people who actually lived and who participated in real, rather than imagined, events. The difference, then, between fictional story (realistic and veri-similitudinous though it may be) and history lies in the mind of the author, in the intent that guided his pen. There is not, however (much to the frustration of the humorless historian), a clear line of demarcation by which a reader may distinguish between story as fiction and story as history. The historian, if he is not to be mere archivist or chronicler, must, like the author of fiction, draw upon his imaginative powers to create (or, as he would claim, to recreate) the personalities of the people and the interconnectedness of the events constituting the *who,* the *which,* the *what,* and the *why* of his history. Even if the historian is contemporary to the events he is recounting—even if he is an eyewitness to some of them and acquainted with their personae—he must draw on memory to present his characters' lineaments or on intuition to read their traits; engage in guesswork (more or less informed) to divine and discriminate between their motivations and compulsions; and, finally, trust his philosophic disposition when he judges the degree to which personae were swept along by "history's" currents or were free to effect their own headway—to determine their own course and even "the course of history" itself.

These observations are not presented here for their novelty; the high proportion of subjective ingredients—the imaginative, the artistic, the speculative—is (at least) in ancient historical writings too obvious to occasion debate. Rather, they are made here to sharpen the question *which if any* of the writings from antiquity were intended by their author as the kind of historiography we have in mind today when we classify history with the social sciences rather than with the humanities or the arts.

Let us take, for example, Herodotus, the pioneering Greek master of the storytelling art, hailed as the Father of History and, on the grounds of that epithet, impeachable as the Father of Lies. He deserves neither title. He never claims to be more than a storyteller; his word for stories, *historiai,* means only (as far as we can gather) the answers he has got to his inquiries. In his opening sentence in book 1, he avers that his particular hope in his narrative is to show how the Greek and Asiatic peoples came into conflict. This hope is commonly interpreted as referring to the origins of the Greek–Persian wars; yet we must read two-thirds of the work through before we get to Xerxes' crossing of the Hellespont, and of the remaining third—if fanciful dialogue and irrelevant digressions are excluded—it is doubtful whether one-tenth informs on that war. As for Herodotus himself, the conflict is not as between Greeks and Persians but between Greeks and Asiatics, traceable to woman stealing on both sides, culminating in the sack of Troy—for which act, the Persians claim, they are now exacting revenge.

A typical example of the seriousness with which Herodotus expects his tales to be taken can be found in his account of how the Phrygians kill a bull by strangling it by means of a tourniquet, and "because there is no wood in Scythia to make a fire with, . . . an ox or any other sacrificial beast is ingeniously made to boil itself." The method described is the skinning of the animal, the stripping of flesh from the bones, placing flesh in a paunch made of the skin together with some water, and setting this improvised cauldron over a fire, the kindling of which consists of the animal's own bones; Herodotus takes pains to inform us that "the bones burn very well." It may be possible to believe that Herodotus has no inkling as to the kindling temperature of bones; it passes belief that he thought that water can be boiled in an animal skin suspended over a fire. Lest we miss the point of this story, our author immediately goes on to tell us that ceremonies in honor of Ares (to whom alone of all the gods the Scythians "build altars and temples") are conducted differently:

> In every district, at the seat of government, Ares has his temple; it is of a peculiar kind, and consists of an immense heap of brushwood, three furlongs each way and somewhat less in height. On top of the heap is a levelled off square, like a platform, accessible on one side but rising sheer on the other three. Every year a hundred and fifty waggon-loads of sticks are added to the pile, to make up for the constant settling caused by rains. (4.62)

Imagine a temple of twigs in every district of a land that has no wood, each pile almost a perfect cube, a third of a mile in every dimension, sheer on three sides, sloped or stepped on the fourth. Surely a construction that outdoes the pyramids! Replenished every year to compensate for settling caused by rains! Rains will erode a rock, they will not settle a brush pile. And imagine climbing this rickety tower sixty stories high, with bull in tow, to offer to "an ancient iron sword, which serves for the image of Ares, . . . a greater number of victims than (is offered up to) any other of their gods."

I have chosen this rather lengthy example from a nonbiblical piece of literature in order to help us gain perspective on the problem of the supposedly historical narrative as a legitimate concern for biblical poetics. For all that no claim of sanctity has ever been made for Herodotus' writings, his *Histories* are constantly cited today by historians of antiquity in support of one or another generalization; yet even the most naive of these historians is not likely to argue that a literary or poetical treatment is inapplicable to Herodotus on the ground that his text constitutes, or was intended to constitute, an objective history. Similarly, in the case of biblical narratives that are generally regarded as being in some sense historical, the possibility or the likelihood that characters and events were not made up out of whole cloth, so to speak, does not exclude such texts from poetical treatment.[23]

In general then, any and all of the **foci of literary analysis** and **factors in interpretation** may play a significant role in helping us to determine whether a given narrative work (or part of it) from antiquity is to be assigned to the fictional or the historical end of the spectrum of narrative genres. In particular, the **showing** technique, featuring a high proportion of **dialogue,** would be one indication of imaginative and fictional intent; while the telling technique (featuring little or no dialogue) would be an indication of historiographic intent on the part of the narrator.

Genre As Literary Convention: Introduction. The question just discussed, the problem of determining whether a piece or a body of narrative prose belongs in the category of fact or the category of fiction, should serve as a red flag to the practitioners both of historiography and of poetical analysis. That the problem is far more critical for historians than for literary analysts should not be for the latter an occasion for self-congratulation. While historians must be embarrassed by the revelation that their craft has often treated fiction (humorous, imaginative, or propagandistic) as reliable sources for reconstruction of the past, literary critics have long been engaged in enterprise equally questionable if not so obviously risible. This problem, that of genre, is particularly acute for literary critics of the Bible; but in the hope that the sharing of misery with other literary company will let us take a more searching look at ourselves, I shall have recourse first to a parade example from nonbliblical poetics: the deployment in drama criticism of the genre tragedy.

The earliest drama we have is that of ancient Greece, where it is already divided into the genres of tragedy and comedy. Nobody has ever provided a very satisfactory definition of either genre, even as limited to these ancient exemplars, although many attempts have been made, championed and challenged with a fervor matched only, perhaps, in the clash of conflicting views on the interpretation of Scripture. The most prestigious analysis of tragedy (also the first) is that which has come down to us in Aristotle's *Poetics.* Yet for all the value of his insightful constructs (such as the mechanisms of peripeteia and discovery) and the functional definition of tragedy as a form that arouses in the spectator the emotions of pity and fear, very few of the Greek tragedies that have come down to us are conformable to Aristotle's analysis. Yet over the centuries various elements, many of them derived (often incorrectly) from Aristotle have been built into definitions of dramatic tragedy; and the champions of these definitions have tortured classic examples of tragedy to fit them into these Procrustean molds or have criticized other examples for features they either possess but should not or do not possess but should. And in our own century, tragedy has become more an accolade than a descriptive category, with admirers and detractors of a given drama debating whether it merits this label-palm or not.

   In view of this sad history of literary-critical gaucherie; in view of the varied
examples of dramatic tragedy as represented in ancient Greece, in Elizabethan
England, in the neoclassical exemplars of seventeenth- and eighteenth-century
France, in the romantic and postromantic German exemplars from the eighteenth and
nineteenth centuries and in the "liberal" tragedies of Ibsen, what purpose is served
by raising the question as to whether the Book of Job constitutes a tragedy or not? I
shall not pursue this question any further here. Nor have I raised it with a view to
denying any relevance at all to the comparison of a bliblical book's genre with genres
as defined elsewhere. My objective is rather to keep before us the awareness that the
various literary genres are themselves conventions, sometimes simple conventions,
sometimes complex conventions; in other words, a genre may be a convention that is
constituted of a number of conventions. Unless we can come to an agreement as to
what are the sine qua non features of a genre and what are the features that a genre may
not show, we are perhaps better off abandoning genre altogether as a tool in literary
criticism.

Factors in Interpretation:
Metaliterary Conventions

Genre As Literary Convention

I have been focusing on genre as a category of interpretation and last discussed the
dangers that this convention holds for literary criticism. These dangers are traceable
to three questionable assumptions. First, it is assumed that a given genre is a
descriptive category whose parameters have been well defined, rather than (as is often
the case) a label bespeaking an impressionistic characterization of a composition that
may also and more usefully be placed in another genre. For example, in the
deployment of the dramatic genre, tragedy, one critic will assume that tragedy
requires an unhappy ending; another that it requires a protagonist of stature, station,
or renown; another that regardless of the play's ending or of the standing of the
protagonist, the audience should be stirred to pity or terror. Second, it is assumed that
the literary convention of genres is basically an analytical tool whose usefulness
guarantees the legitimacy of its deployment whereas in actual practice literary critics
more or less consciously assume that the varied genres may be ranked in a hierarchy
of aesthetic valuation and on the basis of such assumption will debate not the nature or
effectiveness of a given play but whether it deserves the accolade *tragedy* or not (an
analogous example: the characterization of the epilogue of the Book of Job as a *fairy
tale* ending will require its sundering from the preceding chapters if the composition is
to constitute a tragedy). Third, it is frequently assumed that the historicity of the
characters and events in the plot of a given drama have a bearing on the question as to
whether one genre label is more fitting than another. For example, is Shakespeare's
Julius Caesar a tragedy or a history? Is *Anthony and Cleopatra,* a tragedy or a
history? In respect to all three questionable assumptions, it should be clear that the
dangers they pose for critics is in the measure of the assumption as to the existence of a
consensus on what a given genre is and what it cannot be when, indeed, there exists no
such consensus.

The foregoing, however, in no way dictates the abandonment of genre as a tool of poetics. Classification is an indispensable tool of science, be the science deductive, inductive, or empirical. It does, however, dictate a review of the categories of interpretation in the interest of more precise definitions, of awareness of their contiguity and overlap, and of the legitimacy of the uses to which we are putting them. Thus, the validity and utility of genre as a *category* of interpretation is likely to be in inverse proportion to its deployment as a *factor* in interpretation. Especially in regard to biblical narrative, it is my contention that categorization of genre must follow exegesis and not serve as a tool for exegesis. Only when a consistent and satisfying exegesis of a given narrative has been achieved should we ask which, if any, genre label is most apposite to our narrative. Perhaps a biblical narrative will contain significant elements not found in any genre of extrabiblical literature and so require the coining of a new genre term.

It would be tempting, at this juncture, to consider a narrative genre that, discovered in the Bible, has remained virtually restricted to the biblical literature: the etiological narrative. Before we take up this matter, however, it behooves us to consider a variety of conventions that, nonliterary in themselves, do yet operate in the critic's mind as factors in interpretation. These conventions are concepts or ideas in the realms of psychology, anthropology, political science, and the like that— accepted virtually as axiomatic in these disciplines—underlie the critic's address to literature and are particularly operative when that literature derives from antiquity.[24] Although these concepts or ideas exist independently of the poetical enterprise, their significance for poetical judgments alongside other conventions that are intrinsically literary ones will explain our labeling them as *metaliterary conventions*.

Primitivity and Naïveté Versus Modern Sophistication

I have chosen to treat this metaliterary convention first because it exemplifies the functions of, and the borders between, the literary and the metaliterary convention. Just as the questions of narrator versus author and the reliability or nonreliability of the narrator are essentially categories and foci of literary criticism, so, too, is the question of audience, or readership: Is a literary work (or part of one) addressed to children or adults, boors or philosophers, groundlings or gentlefolk? In such questions raised by the critic, the intellectual and artistic competence of the author is assumed and the choice of seemingly peculiar expressions or narrative strategies is a clue to the limitations of the audience the author is addressing. But there are other questions as to audience or readership that bespeak prior judgments on the part of the critic as to the competence and capacities of the audience (and often equally, perhaps by a recoil of the audience judgment, as to the intellectual capacities of the author). A parade example of the operation of such a metaliterary convention is the assessment of an audience in antiquity as "primitive," childish (rather than childlike), and naive; this in implicit contrast to ourselves as a "modern" audience, intellectually adult, and sophisticated.

I have argued elsewhere that the presumption of so great and general a gap between "them then" and "us today" would logically dictate that we altogether abandon the study of the literature of the ancients, for the assumption of such a gap is equivalent to the denial that we share a universe of discourse with the ancients:

If we do share a universe of discourse, then it is part of our task to ascertain what the elements are that we hold in common, and what are the different ways in which these elements may be expressed by them, by us, by both. Elements of advanced technology and scientific theory whose introduction into the world we can confidently trace to a time and place subsequent to and remote from that of the ancients we may presume not to have been known to them unless we find evidence to the contrary. But the canons of logic, a sense of humor, the subtleties of rhetoric—these we may not deny them *ab initio* as beyond their level of sophistication. Nor may we summarily rule out the possibility, nor even the likelihood, that an ancient author may have formulated a message in such way as to be addressed simultaneously to the most naive and the most sophisticated of his generation, to be comprehended by each according to his level. That *Gulliver's Travels* is read by children as naive fantasy leads no one to exclude a deeper, more serious intent on the author's part. It is not the reputation of the biblical authors but our own understanding of them that suffers when we arbitrarily accord them less of a hearing than we grant Jonathan Swift.[25]

The convention of the naïveté of the audience—if not of the authors—of antiquity will be considered again in connection with the reading of ancient texts on the literal versus the figurative level. For the present, however, we may now proceed to consider that biblical literary convention, the etiological narrative.

The Etiological Narrative: A Poetical No-no for the Appreciation of Scripture

Any number of narratives, particularly in the Book of Genesis, present the reader with problems in regard to the point of the story as a whole, or in regard to the inclusion in it of informational details that most readers today would not consider especially edifying. The labeling of these narratives as etiological implies that the primary purpose of the narrative in question is to explain something to the reader—the rise of a phenomenon or the origin of a name. The term *etiology* basically means the study of or the attribution of origins, causes, or reasons; and it is especially and most commonly domiciled in medicine, in connection with the causes or origins of disease. I must confess that to me the use of this term as a category of interpretation by students of folklore and of literature constitutes a disease in itself. Essentially the malady is to be seen in a chain of illogic, in which a metaliterary convention of dubious merit leads to the creation of a literary convention (genre) of questionable meaning or sense, a literary convention comparable to an iron maiden mold into which a narrative is pressed and pinned. It is, indeed, a fine example of that methodologic distortion: the metaliterary convention, an unfounded supposition borrowed from a different discipline, disguises itself as a legitimate literary convention, a *category or focus* of interpretation when it is in reality a *factor* in interpretation.

To get another perspective on this methodological distortion, let us consider the *petitio principii* (fallacy of circular reasoning, begging the question), one of the most insidious snares for the investigator of any phenomenon. Part of the reason for its insidiousness lies in that the argument, while it is not advanced, is not, however, formally contradicted; thus, the a priori assumption retains its weight in the mind of the reasoner or audience who shares that assumption. What often fails of notice is that one of the most seductive forms of circular reason is the label. A question of *what, why,* or *how* is asked; and the answer given is a label, a name, which explains nothing yet is astonishingly effective in cutting off further inquiry.*

By labeling a myth or a legend an etiology the scholar is not, in his own mind, merely pointing out the obvious; he thinks, and we more often than not agree, that he is explaining something. But is he? Is he explaining anything at all? Or is he achieving nothing but the foreclosure of further and possibly productive inquiry? Let us consider one of the biblical narratives classified as an etiology.

Nine verses in Genesis 11 tell the story of a united mankind's enterprise to rear turreted Babel and YHWH's frustration of that design by engineering their disunity and dispersion, this achieved by splintering their one tongue into many; in the course of the story the name Babel is explained by reference to YHWH's action. If the story makes any clear statement, it is that YHWH was displeased either by mankind's unity or by the expression of this unity in the building project and chose to disrupt this unity by scrambling the one original language of mankind into many different ones. The underlying question of this story for the critic is, What is its meaning? Why is it told? The scholar makes an elating discovery. It is an etiology! It is told to explain why there are so many languages. What has he discovered? Nothing. What has he explained? Nothing. He has merely made a judgment on the central meaning of the story, for the author and for his intended audience. And his judgment amounts to the same thing as to classify it with Kipling's *Just So Stories,* tales for the amusement of children. But in so doing he assumes that the storyteller, if not himself a child, is addressing an audience of children or an audience so childish that it will be altogether as satisfied with this answer here as it was with the answer to *How Did the Camel Get His Hump?* But are even young children so naive (much less an audience of childish "primitives")? Consider: Had we not all the built-in capacity for "the willing suspension of disbelief," there would be no theater. Theater is no more reserved for children than are childlike stories; and even young children or childlike adults might well go on to ask, "Why did God want to mix up the languages?" And if the answer to this question is, "Just because" (= "It's an etiology"), even the grade school child of today might well ask, "Well, why is this story in the Bible?"

A child of seven may appreciate the humor of a story that attributes the sea's saltiness to a salt-making machine that, dropped overboard by a sailing vessel's careless crew, has been grinding away in the long centuries since. On the other hand, a child of seven may just miss the humor of the sea's salinity being attributed to its population of salt herring. In like manner religious naifs in the millions can still accept the label God as answer to such metaphysical questions as, Why the world? Why life? Why death? But it is condescension to our grandparents and a childish generational conceit to attribute to the best of yesteryear's minds the kinds of childishness that we attribute only to the youngest children or to the least sophisticated adults of our own generation.

The mythopoeic tale—myth, legend, or myth–legend—operates in another way altogether. It does not ask why the sea is salty, why gods live and die, why humans

---

*Perhaps we are conditioned to this in childhood. My grandchild, not yet three years old, pointed to an ungainly monster outlined against the sky. "What's that?" "A steam shovel." "Ah, a steam shovel," he repeated sagely and turned his attention to other matters.

speak in different tongues. It starts with the known or the assumed. The sea *is* salty, gods *are* immortal (by definition), all men die, and humanity is divided by differences in speech or—in the medium of the social sciences with their gift for transmuting the gold of metaphor into the lead of jargon—by failure in communication. If the sea's salinity has no great significance for philosophy, it will be ignored. But the nature of the powers that brought the world into being, that endowed man with a longing for immortality and destined him for a life so short, the divisions of mankind into warring peoples and speech-separated families—such are the stuff of tales told by philosophers and theologians. Yesterday as today.

The Literal/Metaphoric Dichotomy

Stephen Jay Gould is rightfully celebrated for his contributions to our understanding of the history of science, the philosophy of science, and the quirky investigatory misadventures in each of these two disciplines as well as in the interpenetration of the two. He is also highly endowed with imagination and literacy, gifts that enable him to recognize and explore the gray areas beteen science and literature. In a recent study (or meditation) on conceptions of time as a physical dimension, *Time's Arrow, Time's Cycle,* he writes, ''Any scholar immersed in the details of an intricate problem will tell you that its richness cannot be abstracted as a dichotomy, a conflict between two opposing interpretations.'' For all this, he goes on to acknowledge how generally true it is that ''the human mind loves to dichotomize.''[26] I suspect that this love to dichotomize may be an aspect of a broader propensity of the human mind, namely, an inclination, perhaps out of laziness or impatience (which may come to the same thing), to oversimplify. And, aside from this natural impulse to make things easy for ourselves in the matter of cogitation, oversimplification offers an irresistible tool for polemics. (The appositeness of this observation to the dichotomizing oversimplification *literal/metaphoric* in doctrinal debates over interpretation of Scripture will be evident to all.)

As we explore the vagaries of usage in regard to the general term *dichotomy* and the specific dichotomy *literal/metaphoric,* let us remember that the enterprise of a dictionary is not to determine what a term should mean but rather to reflect the separate and different usages of a term in a given linguistic population. The first definition of *dichotomy* in the *Random House Unabridged Dictionary* is, ''division into two parts, kinds, etc.; subdivision into halves or pairs.'' According to the first part of this definition a chicken with its severed head next to it would be a dichotomy; according to the second part the chicken would be a dichotomy only if split down the middle from head to tail. If we should come to a conclusion that neither of these divided chickens is a dichotomy, we might have to conclude further that this definition simply does not reflect usage and must be abandoned.

The second definition of *dichotomy* is ''classification by division into two mutually exclusive and exhaustive groups.'' By this definition a biped's left foot and right foot would not constitute a dichotomy (and rightly so) if for no other reason than that a foot is not a group. What then would serve as an example of a dichotomy? Any set of opposites such as North and South, left and right, masculine and feminine, plus and minus? The problem will become clearer if we consider how close this second definition comes to the definition in logic of a *contradiction:* "a set of two terms or propositions that are mutually exclusive and divide the universe between them." Logic, however, deals with form and abstraction and not with specifics of reality. The only contradiction of A is not-A and the only contradiction of not-A is A.

(Merriam-) *Webster's Collegiate Dictionary* offers the following entry for *dichotomy:* "*Logic.* Division of a class into two subclasses, esp. two opposed by contradiction, as *white* and *non-white.*" I cite this entry for its illustrative example; as far as definition is concerned, the two dictionaries seem to be essentially in agreement: *classification by division* (Random House) corresponds to *division of a class* (Webster's) and *two mutually exclusive and exhaustive groups* (Random House) corresponds to *two [subclasses] opposed by contradiction* (Webster's). If we focus on the illustration, however, we shall note that *white* and *non-white* are logical contradictories in that they divide (''exhaust'') the color universe between them (various shades of white, for example, would be classed as either *white* or *non-white*) while *white* and *black* are not contradictories inasmuch as they do not exhaust the universe of color between them, omitting red, yellow, blue, and so on. It would appear, then, that a dichotomy is a set of two opposites that may (''especially''— Webster's) be contradictories but need not be so. The two parts of a dichotomy may be contraries in the sense of diametric difference, opposition in position or direction (antipodes), adversarial opposition (the political division of the world into East and West), or complementarity (Northern and Southern Hemispheres). Returning now to Gould's observations, we will appreciate our human propensity for dichotomizing in its twofold nature as a function of the usefulness of division in cerebration (*analysis* = ''break up'') and the impoverishment that comes from reading the opposite terms as always conflicting or contradictory.

Returning now to our pair of terms, the literal is a literary convention, as is the figurative or metaphoric. That they are indeed a pair of opposites, or a dichotomy, becomes clear when we examine the dictionary articles on them. Under *literal* we find, for example, ''in accordance with, involving or being the primary or strict meaning of the word or words; *not* figurative or *metaphorical.*'' Under *metaphor* we find ''the application of a word or phrase to an object or concept which it does *not literally* denote in order to suggest comparison with another object or concept'' (emphases mine). Since dichotomy is by definition a *division,* it follows that there is (in a logical or existential sense) a prior whole. This whole in regard to the literal and the metaphoric is the entire universe of linguistic expression. Words or phrases are intended in the one way or the other. And such is the pairing of, and the opposition between, the two that the very definition of each, as seen in the words which I have italicized, requires the negation of the other. The question, however, is whether the negation is merely the negation of opposition or also the negation of contradiction. A significant proportion of the debates on the literal or metaphoric intent of a given expression in a given place is due to the assumption that the negation there is one of contradiction, whereas it is in reality only one of opposition. And the futility of such debate is in large measure attributable to this erroneous assumption.

If we consider that metaphor suggests image, attribute, or symbol *carried over* (the denotation of *metaphor* in the original Greek) from its matrix to a comparable nexus, we shall not be inclined to resist the observation that language itself is essentially metaphoric. All words are symbols. Language is essentially speech. Written language is, therefore, a visual symbol for an oral symbol. Letters are written symbols of which written words are constituted. And the letters that represent vowel and consonantal sounds are, like those separate sounds themselves, intrinsically meaningless. The very word *literal* (= "according to the letter") is a metaphor and, when strictly construed, is essentially meaningless; for if the letter is essentially meaningless, then so must be that which is "according to the letter."[27] The word *literal* is, however, an elliptical expression; it stands for *literal sense* or *literal meaning,* by which we intend the narrowest or strictest sense construction of a word or phrase, in contrast to the metaphoric or figurative sense, by which we intend a broader or freer construction. The categories of literal and metaphoric are therefore not absolutes (hence, they cannot be contradictories of one another). Each instance of these two categories represents a spectrum, so to speak, of constructions with the constructions representing a range of more-or-less strict at one end and more-or-less free at the other.

On a purely linguistic or literal level we can examine a term and its range of usages (in linguistic terminology, its "distribution") without reference to metaliterary categories. By a process of reasoning based on assumed probabilities or with the help of a diachronic study of texts, we can often arrive at a meaning that we can characterize as original and denotative, namely, its narrowest and most specific sense; and then go on to locate ever-later stages of meaning, acquired by operation of association or metonymy, until we finish with the most recent and most broadly connotative sense. The beginning denotation may be characterized as the most literal sense and the final connotation as the most metaphorical sense. Let us take the term *table* as an example. Beginning with a formal definition, we might see its denotation as a flat expanse of surface (length and breadth) separated from its physical environment by its dimension in height or by its (vertical) distance from the ground. (This formal definition would be applicable to an artifact of lumber, stone, or any other material or a shape of rock occurring in nature, like a slab or an elevation—for example, a mesa.) Continuing with an extension of meaning in functional terms, we might then conceive of such a three-dimensional object as one upon which we prepare or arrange, or at which or from which we take, our food. Both the formal and the physical sense would disappear altogether in a further extended and figurative sense of *table* as "provision of food," and sharing one's own table and sharing another's as metaphors for the role of host and guest, respectively. On another axis of meaning, the formal denotation of *table* may be extended functionally to a small slab covered with wax to receive the "writing" impressed by a stylus. By metonymy, any ordering of data that might be listed on such an object might then be abstracted from the object and exist only as an arrangement of ideas, such as the multiplication table or the periodic table. In between the most concrete and the most abstract of these usages (respectively literal and figurative), we might locate yet another usage, closer to the denotative in respect to physical dimension yet closer to the connotative in the sense of storage places for data, and for all that purely figurative: *the tables of the heart.*

As with a single term, so with a longer expression, the metaphoric intent can sometimes be discerned and the literal denotation ruled out purely by reference to universal experience. An example of this is ready to hand in a biblical metaphor for oversatiation and a variation of it in English idiom. In Numbers 11:20 the Israelites, who had complained of their vegetarian diet, are told that the meat which God will supply them will be so plentiful that they will eat of it "until it comes out of your nose." In English, which has so many idioms taken from literal translations of the biblical Hebrew, the variation on this metaphor may be a clue to its originality. One eats and eats "until it comes out of one's ears." Both of these idioms are stamped as metaphor by the incongruity of the image with our experience. Forced feeding or gluttonous overeating, we have learned, frequently results in nausea, culminating in regurgitation. Regurgitation or vomiting is, however, always through the mouth. Both the Hebrew and English metaphors suggest that the food egested comes in such a flood that it must seek egress from other orifices. The point I would make with this instance of recognition of metaphor is in how it differs from such recognition by virtue of purely linguistic factors. Here the judgment in favor of the metaphoric over against the literal depends on an empiric factor: our witness that vomit always issues from the mouth. If this experience were less than universal, less than certain, it would be a *convention* of physiology rather than a *law*. And in regard to its being featured in a literary judgment, it would be a metaliterary convention. That is to say, if we were informed in a biblical narrative that a certain person ate and ate until the food came out of his nose, there would be those who would argue that infrequent though such regurgitation is, the egress through the nose here is to be taken literally. If a number of such instances were to appear in Scripture, scholars might even conclude that the Hebrew expression for vomit, inclusive of the mucous flow that attends violent vomiting, is what is responsible for the literal description. Or to extend our speculation further, if the idiom "(to eat) until it comes out of your nose" is found in the literature of other ancient societies, we might arrive at the convention that in antiquity the nature of the human physiology of vomiting was such that vomit issued both from mouth and nose. We might even find that the likelihood of such a phenomenon in antiquity is bolstered by reports of missionaries and anthropologists who have witnessed such extremes of regurgitation among the primitives of this or that dark corner of earth. As far-fetched as such speculation may seem, it is, I believe, a fair analogy to the role played in literary analysis by metaliterary conventions, particularly the metaliterary conventions that derive from putatively historical accounts and disclose themselves in the dichotomy of human behavior or capacities then and today.

As good an example as any of this last point is the **primitive–sophisticated** dichotomy. Given the assumption, for example, that the civilization that produced the Enuma Elish epic was, in a meaningful sense "primitive" (as compared to ours today or to that of Athens in Socrates' time) or (if *primitive* is too strong a term for comparison) that that era in Babylon was lacking in the sophistication we credit to the Western mind since the Hellenic Dawn and, further, that metaphor comes with sophistication while literalness is a hallmark of the naive mind (a common assumption), we shall readily accede to the characterization of this *Creation Epic* as a "Babylonian fairy tale." This characterization depends on the presumption that the text of Enuma Elish is generally to be taken on a literal level; however silly the text seems to us on a literal level, it was literally understood by its audience in antiquity and did not seem silly to them. The analogical experience in regard to Genesis 1 will not be lost on my readers.

The Interpretation of Ancient Literature

In the course of the interpretive essays that follow, I shall have occasion to expose the role played in biblical interpretation by an assortment of metaliterary conventions and, further, to challenge their validity or plausibility. Among these will be such constructs as the following:

1. Xenophobia is more characteristic of ancient societies than of modern ones.
2. Fertility rites are a salient feature in ancient religions.
3. Fertility cults, distinguishable from other kinds of religious institutions, are a feature of ancient religions.
4. Sacred prostitution was a widespread phenomenon in antiquity, both in connection with, and apart from, the fertility cults.
5. Sacrifice of animals to deity is a universal and ubiquitious feature of ancient, or primitive, religion.
6. The ancients believed that sacrifices to the gods, particularly meat offerings, were the nourishment on which the supernals depended.
7. Human sacrifice, too, was widespread in antiquity across the globe.
8. Child sacrifice, that is, the slaughtering of one's own children and burning them on altars as offerings to one's tutelary deities, was similarly widespread.
9. Cannibalism, often as a religious rite, was far more frequent then than today. Cannibalism is, indeed, a feature of primitive or backward or savage societies, so much so that the eating of one's own children might be freely confessed.
10. The personification of natural powers in pagan polytheism is not a poetic device but rather reflects a literal and ontological belief.
11. In like manner, the sculpted representations of these deities, in form human or demonic, were not artistic projections of divine power and grace but rather idols; that is, they were the gods materially.
12. The imagery of paganism expressed in mythology and in art finds its rejection in the biblical antipathy to mythology and its explicit proscription of much of the plastic arts.

This list of examples is not exhaustive. It serves as a rudimentary checklist against which we may anticipate the role of metaliterary conventions in affecting the critic's judgment in regard to the location of literary elements on the literal–metaphoric spectrum and the influence of this judgment on the interpretation of the literary composition as a whole. For example (referring to the above list), convention 1 may color our judgment of the Sodomites' reception of Lot's guests (Gen. 19), perhaps in designed contrast to Abraham's hospitality (Gen. 18), and the behavior of the Benjaminites in respect to the Levite from Judah (Judg. 19); conventions 2–4 our judgment of Scripture's Baalim, Judah's daughter-in-law (Gen. 38), and prophetic denunciations of Israelite apostasy in terms of whorishness; conventions 5–6 or judgment of the offerings of Cain and Abel and of Noah; conventions 7–8 our judgment of the Binding of Isaac (Gen. 22), the daughter of Jepthah (Judges 11), and the prince of Moab (2 Kings 3); convention 9 our judgment of the reneging cannibal (2 Kings 6:24–31); convention 10 our judgment of Genesis 1 and personifications of Sea, Ocean, and so on in poetic passages; and conventions 11–12 our judgment of the Golden Calf (Exod. 32–34).

The Debt to Literary Expression

To anticipate the nature of the challenge I shall offer to many of these metaliterary conventions, I shall not, for one thing, argue that no human parent ever sacrificed a child, that no hierodule ever played the harlot, that no rites were ever performed in the pursuit of fertility. What I *shall* argue is that the basis for such putative beliefs, practices, and institutions in historic societies is derived largely from literature— literature in which these themes are preponderantly figurative but have been misread as "historic accounts" or "literal events." Students of ancient literature have then borrowed these historical constructs from historians and anthropologists as facts about the past and, presuming these "facts" in the mind of the ancient authors and their audiences, have gone on to interpret the literary texts on the basis of such perceptions of beliefs held and practices observed by the ancients. Here, then, we have one interface between the purely literary convention and the metaliterary convention.[28]

An example of such an interface (chosen from several extrabiblical ones that I propose to treat extensively in another volume) is the definition of the genre epic and the presuppositions behind that definition in regard to the author of the *Iliad* and the audience for which he intended it. An epic is a narrative of considerable length celebrating, in the form of poetry, heroic characters at the center of legendary heroic events. The term *epic* in its narrow sense requires that the form be poetry. When used loosely, as in the phrase *prose epic,* the emphasis is on features of content (heroic characters, a series of events, and the legendary nature of both characters and events). It will be useful to quote my earlier discussion of legend:

> Legend, in our dictionary's definition, is at one remove from the fantasy element in myth: "any fictional story, sometimes involving the supernatural, and usually concerned with a real person, place or other subject." The supernatural need not be involved in legend; and the focus is on a person, place, or subject that is in some undefined sense "real." Legend, therefore, is essentially credible in terms of what may actually have occurred to real people. Its intent would seem to be to inform rather than to entertain or explain. Its claim on our own belief is greater than that of myth.

Now let us consider Homer's *Iliad.* The "realness" of the city of Troy, the Greeks who besieged it, and the Trojans who defended it (or at least the general perception of this "realness" on the part of scholars) is readily apparent in the (metaliterary) convention that Schliemann's excavations at Hissarlik uncovered the historic "Troy" for which the contestants of the *Iliad* warred. Many historians doubted, until Schliemann's ravaging of the site at Hissarlik, whether there ever was a Troy or an Ilium that was sacked by Greeks commanded by a king of Mycenae; and for all that the dig and particularly the discovery of treasure there somehow led most historians to relinquish their doubt, nothing discovered at that site really militates against it. The *Iliad* is read today almost universally as a legend containing many kernels of historic truth, and almost no one doubts that the epic was composed in high seriousness by a masterful artist to celebrate the feats of the legendary warriors who, in the traditions that had come down to him, had gained immortality before Ilium's walls. This despite evidence on every other page of the epic that the author believed not in the heroism of the humans (for the most craven warrior becomes lion-hearted when a god comes to his side, and the bravest turns tail upon the intervention of a god in support of his adversary) nor credited the gods (who ever-so-ineptly take part in, or interrupt, the story's action) with much mind, courage, motive, or resolve. Not a single character, divine or human, is developed with any consistency unless it be that they are consistently inconsistent, absent-minded, silly in argument, and tagged with arbitrary epithets that are never germane in the immediate context. Read as a serious work, the *Iliad* is unbelievably dull; the long rambling speeches in the thick of battle make the speakers look ridiculous; and the alternation in the protagonists of rashness, cowardice, peevishness, venality, prodigality, good common sense, and idiotic nonsequiturs and many more violations of poetical criteria for good art (not to speak of great art) have not led critics to question the genre to which the work is assigned. More myth than legend, more fairy tale than myth, poetry by a master craftsman who can spin similes of interminable length and nonsensical extension, is it possible that the epic is itself a satire upon both myth and legend of the heroic age, the warrior ethos, and the warrior lord–patrons of the bard who thus toasted them in their besotted carousals?

Although the above is a sketch of the argument I shall plead in detail, I am not pleading for it now. At this point I am pointing to a possible interpretation of the *Iliad*, both novel and radical, that is likely to elicit condescending smiles or contemptuous silence from academic establishments in classical and comparative literature. But if I could persuade my reader to suppose for a moment that a convincing argument could be made for this interpretation, the crucial question would be, Why has this view not been advanced before? Why have we failed to apply purely literary criteria to the epic? If we have likewise failed for other ancient classics as well, why have we failed? There would seem to be a twofold answer. Antiquarians—be they historians or humanists, classicists, Semitists, Sinologists, or biblicists—are for the most part philologians in a narrow sense and not well versed in the methods of literary criticism. Literary critics, for their part, are likely to be overly modest in regard to their own competence in the matter of a literature in a foreign and ancient tongue that has become the domain (virtually the monopoly) of the philological specialists; fearful of rushing in where angels fear to tread, they leave the field and its fallen to the paramedical technicians. Reinforcement of such praiseworthy, if excessive, modesty is provided by the ever-quickening pace of ever-narrowing specialization in modern times, and, to boot, by a number of melancholy examples of specialists celebrated in their own fields sallying forth to achieve ignominy in territories adjacent to their own. Under such conditions the inappropriate weight of metaliterary factors will be overlooked by the literary critic. If, for example, the historicity of the Trojan War becomes ever more a staple of classical scholarship and if, furthermore, archaeology is bolstered by the reading of Homer and historicity is a function of Homer's being read in the light of archaeology, how pretentious and arrogant will appear the modern critic who applies modern tools to the products of the ancient mind, who finds fault with the poetic creations of antiquity's greatest bard only to salvage the reputation of that bard by imputing to him a sophistication proper only to a twentieth-century postivist!

Figures of Speech

Figures of Speech and Translation

I have dealt with **metaphor** as a broad factor in the interpretation of **ancient literature** (that is to say, in the context of an entire conceptual complex) rather than in a word or group of words; specifically, it was in connection with an assessment of literal or figurative intent in a context where such an assessment is more a metaliterary than a literary factor. In regard to metaphor in a narrower focus, many **literary expressions,** such as metonyms and merisms, are metaphors whose figurative nature is quite often masked for the reader, especially one who encounters them in a language that is not native to him. And very often, to miss the figure (that is, to take the word or words in their most literal sense) is to distort the sense of the entire context. In my translations of the biblical text I shall sometimes render a figure into English by its semantic equivalent (whenever I can find one); at other times I shall follow the Hebrew closely and explain the figure in the commentary. In the former case, the reader who has missed the force of the figure will see me as taking unwarranted liberties with the text. In the latter case, he may be tempted to see my argument as cutting the cloth to tailor my suit. My purpose in discussing a few of these figures (which appear often in the Hebrew Bible and have proven perplexing to my students), is (1) to defend against such perceptions and (2) to preclude the necessity of demonstrating figurativeness each time that recognition of it is absent in standard translations and commentaries.

Idiomatic Expressions

This category often includes figures that have become so common in our everyday speech that it is only with a start of surprise that we are reminded that they are indeed figurative. Consider, for example, in ascending order from colloquial to literary usage, *shake a leg, foot the bill, last leg* (of a journey), and *drag one's feet.* Even an idiom in English that has been outmoded by change in fashion may be used to render a text from a bygone era. Thus, while skirts are today an item of female dress, it may be the famous kilt of the warlike Scot that keeps us from sensing incongruity in the expression, *"He picked up his skirts* and ran with the best of them." Yet when the Hebrew in 1 Kings 18:48, in telling of Elijah's marathon run, features an otherwise unattested verb (*šns*) with a noun that is probably the area of the human body from the

waist down past buttocks and upper thighs, the Authorized Version's wrong but not
seriously misleading "girt up his loins" has prevailed in English versions; while the
recent and bolder New English Bible dares to have him "tuck up his robe," an action
clearly conforming to the intent of the Hebrew except that *robe* for the prophet's attire
is at odds with the notice elsewhere that this particular wilderness-dwelling prophet
covers his nakedness with sheep or goatskin. Similarly, the Hebrew that has King
Saul entering a cave, literally "to cover his legs" (AV) would be meaningless to most
readers and justifies the rendering "to relieve himself" (RSV, NEB). The literal
Hebrew is not in itself a euphemism, it is a description of what disappears from sight
when a robed or kilted male squats to evacuate his bowels. (One marvels therefore,
that the new Jewish Publication Society version accepts the Revised Standard
Version's translation yet adds in a footnote that the meaning of the Hebrew is
uncertain.) What makes this expression an apparent and somewhat gratuitous
euphemism here is Saul's entering a cave where no one would see him at his royal
easing.[29] A related idiom is a biblical expression for the human male, comparable to
our *every last mother's son,* namely, "every one who urinates against a wall." This
designation is for males only, for women's anatomy does not permit so easy a voiding
of urine while preserving their modesty. The context of denigration, however, is
expressed in this idiom by an extended synecdoche of part for the whole, the male not
in any dignified configuration but in the vulnerability of his unprotected back as he
stands to pass water.

Another example of biblical idioms often misconstrued is the Hebrew verbs
employed for aspects, degrees, and places of locomotion. Hebrew will often focus on
topography in terms of going up or down terrain. Cities are built on elevations for
reasons of defense, fortification bespeaks a place valued and prestigious, and in both
English and Hebrew *ascent* is generally positive and *descent* is generally negative.
But consider the meanings of the verb *'lh* in the following contexts: *to go up to*
Jerusalem on pilgrimage, *to go up* YHWH's mount for audience, as against *to go up
against* a fortified city for battle. Exodus 1:9–10, where Pharaoh views with concern
the growing population of Israel (note the preposition *min,* "from, of, than [with a
comparative]" "The Israelites are *too many for* our own good" [or *"too strong for us*
to cope with," but not *"more numerous than* we are"]), culminates in his anticipation, "Should war come upon us they may join our enemies in battle and *go up from*
the land." This sense of the Hebrew bespeaks his reluctance to lose his supply of
corvée labor but is not in consonance with his desire to control the size of the Israelite
population; hence the clause *too many for* is a double entendre with the additional
sense of *to prove stronger than* or *have the upper hand over* Egypt.

The verb *bō'* "to come (into), enter, penetrate" also means, as in English or
French, "to arrive" in the sense of *to succeed, to achieve.* An antipodal verb is *yṣ'*
"to go (out)," "to go free," "to leave"; when this last sense is in regard to a city
beleaguered or under threat of impending siege it may refer to a sally against the
enemy, a surrender to him, or a strategic retreat. Thus, in the context of David's flight
from Jerusalem before the advance of Absalom's forces, Shimei ben Gera in his
mocking jeer (2 Sam. 16:7) employs the imperative of this verb in the sense of "Run,
run [with your tail between your legs]"; and in 2 Kings 2:23 where Elisha is making

his way *up country* to Bethel, the young lads of that city employ the imperative of '*lh* to bar him from entering their city: *Keep on moving, mister, don't stop here*. Yet in the former case scholars have interpreted the verb as spell-casting language, and in the latter case the failure to appreciate the idiomatic usage has contributed to the misunderstanding of a perplexing passage.[30] So, similarly, have scholars discerned the Israelites engaging in a magical incantation in Numbers 21:17, where the context of digging for water and the anticipation of God's fulfilling his promise makes it altogether clear that the verb '*lh* here stands for "*Well up/bubble up*, O spring water."[31]

Metonymy

Essentially a *metonym* is a metaphor in a single word. When the metonym is a noun, it frequently appears as synecdoche of part for the whole, whole for the part, effect for cause and cause for effect. While metonyms are frequently a feature of poetic diction, they are also frequent in commonplace speech and may therefore signal no artful intent on the author's part. Thus, for example, Hebrew's usage of '*āwōn* for both *crime* and *punishment* (as also *hēt'*) has an exact parallel in Akkadian *arnum*.

A particularly pernicious pitfall for the interpreter is the use of a term that has a specific denotation as well as an extensive metonymic range in his own language to render a term or a complex of terms in another language. For example, the word *idolatry*. Literally "idol worship," metonymically "harboring/pursuing false values," it is applied by translators to render a wide range of biblical Hebrew expressions, from *iconoplasm* (in YHWH worship) and *apostasy* (literal or figurative) to *mistaken ideals* or *proscribed ritual practices*. The danger is, of course, greater in the degree that the interpreter is unaware that he is letting himself be doubly tricked (by his own language and by the one he is translating).

Particularly important for the translation and interpretation of biblical texts is recognition of the contribution to metonymic extension made possible in the Hebrew by the flexible semantic functions of the stem conjugations: causative, elative, factitive, privative, durative, reflexive, reciprocal, and so on. A parade example is the use for acts of purification and palliation of a variety of terms that in their ground stems are related by synonymous overlap and antonymous exclusion: *thr, tm', ht', kpr, qdš.*

Wordplay

This category of rhetorical expression may overlap some of the others discussed here, as well as categories so easily and universally recognized as to require no discussion (such as hyperbole, litotes, aposiopesis). Notable examples of wordplay are puns, hypocoristics, alliteration, onomatopoeia, double entendre, and oxymoron. For example, the exploitation of the comparative and privative uses of the preposition *min* in Exodus 1:10 (see p. 38) and of the partitive and privative functions of that preposition in Genesis 27:28, 39.

What I would stress in regard to biblical instances of wordplay (as, indeed, to biblical Hebrew diction in general) is that these are never idle, never inappropriate to their contexts. This is not to say that they may not be expressive of humor, whimsy, and the like but rather that the humor or whimsy is in consonance with the tone and purport of its context. Again and again we shall have occasion to see that the problem of a seemingly crude or absurd event in Scripture may be resolved by attention to the diction or wordplay in the original Hebrew.

To offer a single example from a story that will receive detailed treatment in a following essay, the verb *bq'* featured in 2 Kings 2:24 is everywhere attested in the denotation "to split." Thus, such translations of the verb as "tore" (RSV), "mangled" (JPS), and "mauled" (NEB) are departures from the sense of the Hebrew. Curiously, the New English Bible translation, had it been the product of a translator into American English, would have been literally exact and have pointed to the intended metaphor. Alas, in the English of Britain, *to maul* (deriving from the noun *maul,* "a heavy hammer") has only the sense of "to handle roughly, paw, injure by rough treatment." Only in American English can the verb be used also in the sense "to split with a maul and a wedge, as a wooden rail" (*Random House Unabridged Dictionary*).

Hendiadys

This figure is the expression of a single idea, concept, or meaning by the use of two separate terms joined by a conjunction. Common in biblical Hebrew and in classical literature, it is rather rare in English; but a few examples not borrowed from either of the foregoing should help us to develop a feel for this figure where it is featured in interpretive cruxes. *Good and mad,* that is, "very angry" and *sick and tired,* that is, "weary to the point of nausea" are colloquial instances; whereas *tried and true,* that is, "proved true by trial" is, if not purely literary, at least an instance of high style. But we know this only because we are so at home in the English idiom. By analogy this should alert us to the danger of assuming that the very figurativeness of hendiadys (as also of many other figures) is in itself a token of high style or (a less likely mistake) of a colloquialism.

As each instance of hendiadys must be separately investigatd before we can conclude, if at all, the level of speech or style it represents, so, too, must the possible shades of meaning in any oft-encountered hendiadys or of such a given hendiadys in one context as against another. That the conjunction in a hendiadys is not a conjunction semantically ("and") but an instance of a multivalent copula (see on **parataxis,** especially note 19) is to be seen in the function of the two terms coupled by it. The order of the terms is of no significance: either may function as an adjectival or adverbial modifier of the other, even when the terms are virtually synonyms. For example, in the case of the divine decree that dooms Cain to be a *nā' wānād* on earth (Gen. 4:12, 14) the verbs of both participles have the basic sense "to move, to stir." The Revised Standard Version's "a fugitive and a wanderer" is clearly wrong in that neither verb has the sense of flight (from danger or pursuit). That sense is borrowed from the following context of Cain's complaint that the decree which makes him a *nā'* *wānād* also renders him rootless, homeless, a stranger to—and therefore a defenseless prey in—any society. The meaning of the hendiadys, "one ever on the go," in the immediate context of the decree is that the earth that he has force-fed on his brother's blood will not respond to his cultivation of it; he will have to revert to gathering his food from plants growing in the wild. The New English Bible's "a vagrant and a wanderer," while avoiding the traditional import of fugitiveness, betrays no sense of the hendiadys. The Jewish Publication Society's "a ceaseless wanderer" is, therefore, the best of the three renderings. Yet even this one makes the decree inconsonant with the notice in verses 16–17 that "Cain settled in the Territory of Movement (Nōd) over against/facing Eden" and "founded a settlement, which he named after his son Enoch." The sense of a settlement is not in itself a contradiction of the "moving about" if the latter is itself a metaphor for food gathering in an unfertile territory. This last image gains in poignancy in the contrast with the remembered fertility of the paradise lost.

Not often recognized (hence not reflected in translations) is that the coupling of two verbs, one in an auxiliary or adverbial function, is also an aspect of hendiadys in biblical Hebrew. In such instances as "Abraham [proceeded] early and saddled" (= "At first light Abraham saddled") or "He arose and went" (= "He promptly set out on the journey") in Genesis 22:3, the nuance of the hendiadys is subtle and of no great import. Far more serious is the failure to perceive the nuancing in so celebrated a passage as Deuteronomy 26:1–3: here the verbal hendiadys pivots on the extended sense of *bō'*, "arrive" (see on **Figures of speech**). To render the verbs in verse 1 in paratactic literalness ("When you come and possess and settle in the land . . .") makes the narrator out as simplemindedly childish; one cannot settle a land possessed by others nor take possession without entering it. The correct translation, "When you succeed in wresting possession of and populating the land that YHWH your God is granting you," prepares us to understand that the purport of the formula prescribed for all future generations (verse 3), "I have hereby declared to YHWH your God that I have entered into possession," is the acknowledgment that every future generation must make that not only has God fulfilled his promise to the fathers who crossed the Jordan but each generation succeeding in inheritance is a fresh recipient of God's benefice.

Whether a given hendiadys is an expression of everyday speech or an indicator of high style must (as in the case of **metonyms**) be determined by examination of the distribution of this particular hendiadys and in contrast with alternative usages. For example, biblical Hebrew has no single word corresponding to English *justice*. The basic meaning of the term *ṣdq* is *right* or *correct* or *accurate* in an ethical or moral sense; a basic meaning of *špt* is *to judge, to make a decision*. The concept of justice, (namely, *correct judgment*) is clearly expressed in the explicit (and rare) construct *mišpaṭ-ṣedeq* (Deut. 16:18). Otherwise it is expressed by the hendiadys *ṣedeq umišpāṭ* or *mišpāṭ uṣ^{e}dāqā*. In the case of this hendiadys, there can be no question that regardless of the order of the two terms, *mšpṭ* is the primary noun and *ṣdq* serves as the adjective. By contrast, in the hendiadys *ḥesed w$^{e'e}$met* we may have the double force, "enduringly faithful" or "faithfully true." Finally, we must be alert to such phenomena as separate negations of the two terms of a hendiadys (such as Hos. 4:1a–

b) or the distribution of the two terms in two hemistichs (such as Isa. 1:28, 5:16), which does not destroy the sense of the hendiadys even while it mocks the best efforts of the translator.

Merism

A *merism* is an expression for a totality consisting of two opposites joined by a conjunction. More common in English than hendiadys, many instances of merism in English correspond to biblical Hebrew equivalents: up and down, near and far, heaven and earth. Such common oppositions as these—like the nonbiblical *through thick and thin*, suggest roots in common speech. Others may have their origins in the picturesqueness of everyday speech and lend themselves to artfully crafted diction or to exploitation in the elegance of sober incantation. Consider, for example, *to play fast and loose, it was touch and go, for better or worse, in sickness and in health.*

It is a mistake, however (and not an uncommon one), to imprison this figure (as other figures also) in a cast of straitened meaning. For example, the English *by and large* is a merism even though it seems so incongruously to conjoin a preposition and an adjective. Both terms are, however, elliptical and adverbial (*nearby, at large*); and the sense of added opposites is not quite an entirety or totality but a related phenomenon, a generality. *By and large,* like the disjunctive merism *more or less,* means "generally speaking" or "on the whole." Such elasticity in linguistic figuration should be kept in mind when one addresses a semantic crux such as Hebrew *ṭōb wārā'* (Is it "good and *bad*" or "good and *evil*"?) in the interpretive crux of the Tree of Knowledge in the context of the larger puzzle of the Eden narrative.

An example of an inverted merism is present in Deuteronomy 29:18, where the recognition of this figure is crucial to a resolution of the problems besetting the exegete in the entire pericope of 29:9–28. *Thirstily dry* and *sopping wet* constitute a dichotomous division of the universe into opposites. Here the two opposites are not joined in a merism, but one term cancels out its opposite member.[32]

I shall conclude this discussion with an example of a virtuoso performance in the manipulation of merisms by one of Scripture's supreme poets, of which one of our better scholars could yet write, "The expression of the thought is unaccountably labored and obscure." Speaking of the New Jerusalem that God will create, the prophet says:

> Of that place there will not be an *infant*
> [surviving only] a *few days,*
> Nor an *old man* who *fills not out his days* [literally, *years*]
> For the *young* man will die *a hundred years old,*
> Anyone falling short of the age of a hundred
> will be [deemed] afflicted [indeed]. (Isa. 65:20)

The merism of *young and old* is explicit in the first stich, where infant mortality will be zero and rich old age the statistical norm. In the second stich the merism is made implicit (though most effectively abandoned) by defining the hundred-year-old as a short-lived youth and anyone failing to reach that age as one cut down in his prime (of course, by act of God).

Oblique Expression

In our discussion of **the literal and the metaphoric,** we saw that these two terms may (but need not necessarily) constitute contradictory phenomena; they may be (relatively) opposed members of a series or sections on a gamut, scale, range, or spectrum. The oppositional and complementary aspects of the literal–metaphoric continuum are inherent in the carrying over of a word, phrase, or image from its natural matrix to a different context by way of suggesting a likeness or analogy between matrix and context. Thus, from the matrix of agriculture where the plow cuts into the soil to the context of seafaring where the ship's prow cuts into the water, the similarity is in a sharp edge digging into a substance and, as it moves forward, lifting and displacing some of that substance; the difference is in the substance (solid versus liquid) and in the instrument, denotative in its matrix and connotative in its figurative context. In like manner is the carrying over of *wake,* the short-lived track of a ship in the water, for the long-enduring track of the plow's furrowing: *the wake of the plow* will suggest the dynamism of the hydraulic *process* as against the static *result* of the plowshare's progress. Thus, the literal is to the figurative as the denotative to the connotative; and similarity is an assertion of likeness (or synonymity) even as it is a denial of sameness (or an assertion of antonymity).[33]

Oblique expression or style is a term I employed some years ago to characterize certain instances of biblical idiom that are clearly figurative (in that the literal is ruled out by the context), yet for which the term *metaphoric* is not quite suitable.[34] In metaphor there is a stress on likeness, in obliqueness a stress on difference. Metaphor consists in the carrying over of a word or image from a denotative matrix to a connotative context; it is an elliptic simile with the *like, as,* or *as it were* understood or implicit. In oblique idiom the vector of transference seems just the reverse: the oblique idiom says one thing literally yet intends something quite different; the literal sense goes far beyond what the speaker actually intends. The implicit *as it were* of metaphor is absent and inappropriate in obliqueness. For example, in the metaphoric expression, "Carelessness is a crime," the *as it were* is implicit because crime requires intent and carelessness bespeaks the absence of intent. In a corresponding oblique expression the figure would say, "Your behavior is criminal," when the intent is, "Your carelessness is so injurious to me that you might as well be my enemy."

The last example reveals another feature of obliqueness that sets it apart from metaphor: the element of consciousness, intent, and will. Metaphor features an inanimate object or a brutish subject: a vessel plowing the sea, a bull in a china shop. Obliqueness always features a motivation or imputed motivation that, literally absurd, stamps the expression as figurative in the modality of obliqueness.

It is my suspicion that the biblical instances of oblique expression are always instances of high style or sophisticated thinking, hence (in this consideration) unlike metaphor, which is often rooted in the mundane, the prosaic, and the casual. This need not mean that obliqueness is artificial; rather, it involves an ellipsis of a chain of reasoning rather than of the simple *as it were* of metaphor. Scuh an extended ellipsis between what is intended and what is actually said often renders the oblique figure incongruous, grotesque, or even comic. Consider the unfactitious cry of the widow as her husband's coffin is lowered into the grave, "Why, oh why, Nathan, are you doing this to me?" The widow's words are an expression of pain, a testimony to the grievousness of her loss by her husband's death; yet what she actually expresses is that her husband's departure is willful, designed by him for the sole purpose of injuring her, who has deserved better at his hands. The *as it were* consists in reading (her) loss as injury, the injury traced to a blow, the blow to a hostile striker, the hostility to vengefulness. A less incongruous instance of obliqueness is an expression suggesting that we are our own worst enemies, that we will use our freedom of action to encompass our own downfall: "Give him enough rope and he will hang himself." The rope is both leash and hangman's knot, the length of the leash an image of freedom to act, and the knot symbolic of how little rope suffices for the suicidal noose. The obliqueness expresses a sardonic view of human nature, our capacity to turn freedom and power against our best interests. Perhaps a converse of this oblique expression is reflected in the metonymic chain that binds life to rationality, suicide to insanity, and self-destruction to a deprivation of the faculty of reason: "Whom the gods wish to destroy they first drive mad."

A fuller appreciation of the poetical function of obliqueness in Scripture must await our investigation of its various and varied occurrences. For the present I shall content myself with closing this discussion with an illustration of Scripture's poetical ingenuity, one that features at its heart the oblique figures we have just been considering, exploits poetic structure in a prose prescription, and has been assigned by modern scholarship as belonging to a legal code called H. The passage is Leviticus 19:17–18:

> a. Hate not your brother
> > in your heart:
> b. Verily, reprove your fellow
> > [on immorality bent],
> c. And suffer not punishment to befall him.
> d. Nurse not enduring vengefulness 'gainst [any of] your kinfolk:
> e. Rather love your fellow as [you do] yourself.

The clue to the meaning of this prescription lies in the seemingly superfluous phrase *in your heart*. Where else is emotion felt? This phrase points to the oblique expression of hatred for your brother, refraining from warning him of his act's consequences (on the grounds, we may guess, of nonmeddling and general prudence) in smug anticipation of the pending retribution. Lines *a* and *e* clearly constitute an *inclusio*; *b* is the antithetic explication of *a*; *c* and *d* point to the motivation that make the prescription in *a* and *b* necessary; *d* is a reprise of *a* neatly balancing *in your heart*—a grudge held over for so long that no one else can even remember the incident that occasioned it.

2

Preface to the
Exegetical Essays

Claims to Objectivity in Scholarship

In the Preface I made reference to "scientific Bible study" and to methodologies in modern biblical scholarship which have become synonymous with this expression. This expression is a claim on the part of biblicists that bespeaks our hope or prayer rather than a universal recognition of prudent methodology and objectivity of judgment. As far as the hope or prayer is concerned, I suggest that the pretension to the adjective *scientific* may be expressive of a desire to participate in the prestige of the empirical (or natural) sciences. If this be in any considerable measure true, it is an indictable offense to which we might plead a guilt palliated by being shared with the fields of investigation that style themselves the "social sciences." The guilt, if any, is of course in the measure that the choice of *science* owes, not to a preference for the Latin term for "knowledge" but to a claim of validation for objectivity that the natural sciences achieve today as a function of their ability to predict. It is noteworthy, however, that with the exception, perhaps, of history (which is sometimes listed with the social studies, sometimes with the humanities), none of the areas of study designated as "the humanities" has laid claim to the label *science* or the adjective *scientific*. Why the field of Bible study stands alone in this respect is hardly a mystery. The reason is anticipated in my alternative suggestion for the purport of this claim: it is a disclaimer of dogmatic prejudice—that is to say, in claiming to be scientific, modern biblicists are avowing that their approach to the text, however great their reverence for it and its inspiration, is not determined by the prejudices of one or another doctrinal interpretation. By implication, however, the interpretations of Scripture that are associated with religions and denominations for whom the literature is sacred authority are intrinsically not scientific. Very few of us are aware, I am afraid, how close we come in this to conveying a sense of contempt for the exegetes of synagogue and church (whose philological standing is not inferior to our own), how close we come in this to sanctifying the presumptions and conclusions of the genetic approaches to Scripture. If my reader finds this hyperbolic, I suggest he do a sample analysis of the articles in any issue of any journal devoted to Bible: how much space is devoted to the biblical text itself and how much to discussions of it in secondary sources; how often is an exegetical solution based on, or supported by, citation of a modern scholar who has shown, proven, or demonstrated that such and such is true.

Claims and Disclaimer for Biblical Poetics

In advocating a poetical approach to biblical literature as an alternative, not an accompaniment, to the various genetic and historical approaches, I feel it incumbent upon me to spell out the claims and disclaimers for this method. Poetical analysis is an art, not a science. It aims to discern the nature of a literary text and to arrive at the message that the author was trying to communicate in it. Its discourse, therefore, is in the nature of argument, not proof or demonstration; it aims to persuade the reader as to the reasonableness of the argument—to convince him that essential elements of the argument correspond to what it was in the author's mind to convey. The argument, further, eventuates in *an* interpretation, not *the* interpretation, of the text. When the text is Scripture, the interpretation proposed and the argument for it are often referred to as *exegesis*. This term, reading the meaning *out of* a text (as opposed to *eisegesis*, reading *into* a text meanings that the author never intended) will be used here as a descriptive term, not a self-serving value term, and not in exclusion of other interpretations. No poetical argument may ever be claimed as conclusive, for art is open-ended; it cannot be the last word. As such, each argument presents itself to each reader for his or her verdict, in respect for the autonomy of that judgment. It is a success when it opens windows on a vista, a self-confessed failure when it claims to be the vista itself captured, so to speak, on photographic film.

Format of the Interpretive Essays

In the interpretive essays on narrative that follow I shall attempt to disclose both certain vistas in the Bible—and actions played out on those vistas—as I see them, and how I have come to see them (or better, *read* them) as I do. This will involve in each case my own translation of the biblical Hebrew. On the question of how close to the literal sense of the original I deem these translations to be, I refer the reader first to my discussion of **the literal.** Beyond that, I confess my belief that almost inevitably "to translate is to traduce" in the sense of the Italian original, that every translation is by definition unfaithful to the original. If my translations strike the reader as more free than literal, I would ask him not to draw the further conclusion that this freer translation is any the less faithful to the text and texture of the Hebrew. A faithful translation is one that conveys what the original means, not what it supposedly *says* in the narrowest denotation of its terms.

Every translation is in itself a commentary—implicitly so. I shall try to supplement the commentary implicit in my translation by explicatory additions in square brackets and, where my translation differs radically from the standard ones, to justify them by discussing the broader philological issues in the following commentary, where I shall deal with issues of ancient Israelite concepts and larger questions of poetical analysis, and the narrower ones in the notes. In each case the commentary will entail, or conclude in, an interpretation of the text and an explication of its message. By this latter I mean essentially its preachment or (to borrow a term from

New Testament scholarship) its kerygma. For whatever else Scripture is, it is always a preachment: be its form poetry or prose, narrative or prescription, homily or history, the question that must ultimately be addressed to any portion of Scripture is, What is it doing there? How does it fit into its ideological setting?

At the end of some of the essays I shall append a poetical retrospect, that is, a review of the poetical elements featured in the interpretation, with reference (in boldface) to the discussion of these elements here in part 1. These reviews are not in the interest of pedantry but in the hope that I can help my reader locate the nexus or plexus where he must part company with me.

Contrasts in Poetical Interpretations: Conflicts and Complementarity

My characterization of poetical criticism as art rather than science is not to be taken as an elevation of this method to the level of the creative art of which it seeks a better appreciation nor as confession that the method makes no claim to objectivity. The subjective element in art criticism is in the matter of whether or to what extent the artist's product is pleasing. But such matters as palette, brush stroke, chiaroscuro, perspective, balance, trompe l'oeil, formal configuration if any, and the like may be clearly delimited and studied in terms of design and function, achievement of verisimilitude or illusion, and as elements of artistic originality or imitativeness. And even beyond the purely technical aspects of the art work there may be meaningful discourse, informed opinion, and degrees of agreement as to the presence of a statement—social, political, psychological, theological—in a depicted scene. So, too, in the case of the art of poetical criticism.

To illustrate my meaning, my claim for the validity of my interpretations, the bearing on this claim of different interpretations arrived at by the use of the same poetical–critical method, and the claim's bearing on them, I present the narrative in 1 Kings 3:16–28 (in my own translation), a condensed version of Meir Sternberg's analysis,[1] my own significantly different analysis, and certain conclusions about Scripture as art in part and in toto.

16. In yon time 'twas, came[2] two women, harlots, to the king. They stood arraigned before him. 17. The one of them said, " 'Struth,[3] my lord, I and that woman dwell under one roof. And I gave birth, alongside her, there within. 18. Two days after my bearing child that woman also gave birth—we together, no third party in the house with us, just the two of us in there. 19. That woman's son died in the night, because she lay atop him. 20. She made bold in the thick of night to take my son from my side—your subject deep in sleep—enfolding him in her bosom, while her son the dead one she left to my embrace. 21. When I bestirred myself in the morning to suckle my son, lo, [he's] dead! I looked closer in that morning light and lo, 'twas not my son, the one that I had born!" 22. The other woman said, "Not so! The living child is my son, your son is the dead one." The other one then saying, "Not so, your son's the dead one, my son's the live one." Thus did they wrangle before the king. 23. And the king mused, "This one says, 'This living child here is my son, your son's the dead one,' and that one says, 'Not so, your son's the dead one, mine is the live one.' "

24. The king then called, "Fetch me a blade." And they produced a blade at the king's behest. 25. The king pronounced, "Split the living child in two, give one half

to the one and one half to the other.'' 26.Then said the woman whose son was the
living one to the king, her innards churning with love for her child, she said, ''By my
troth,[4] m'lord, give her the living babe—[anything at all, but] just don't kill him!''
(the other one the while declaring, ''Neither mine nor yours will he be: slice away!'')

27.The king then spoke up [in command], '' 'Give her the living babe—
[anything at all, but] just don't kill him!' [testifies that] that one is his mother.''[5]

28.When Israel at large heard the judgment that the king had given, they stood in
awe of the king, for they realized that there was in him a divine wisdom for the
exercise of judgment. (1 Kings 3:16–28)

Meir Sternberg's interpretation of this narrative is in exemplification of the play of
perspective in Scripture's art, just one of his many brilliant contributions to the syntax
of biblical poetics. The following condensation of his argument consists mostly of
excerpts, a tribute to the spareness of his style.

> The effect of this tale depends on its blending of genres. Thematically it enacts a test
> and a riddle—a puzzle-solving exploit by which an untried youth makes a name for
> himself as a possessor of divine wisdom. Perspectivally, however, it operates with
> conventions (or rather inventions) that were to be codified only as late as the detective
> story. The most basic of these is the fair play rule, whereby the reader must be given
> the same data to make inferences from as the detective himself. So if the thematic
> materials foreground the puzzle-solution movement to the glory of Solomon, then
> the superimposed perspectival equality drives his achievements home by contriving a
> two-level riddle and a double test: it challenges us to match wits with Solomon and,
> indirectly, with his heavenly source of inspiration. . . .
>
> Neither Solomon nor the reader has had any previous acquaintance with the two
> whores. . . . Neither has had the slightest advance warning of the point at
> issue . . . which unfolds in and through the dialogue alone. And when the case
> does begin to unfold, neither could foresee what an impasse it will reach and how
> little hard evidence will emerge in the process. . . . For the defendant's strongest
> point is that she has no story to tell: she did nothing, saw nothing, suffered no loss,
> and would not care to advance any theories about the death of another's child. (PBN,
> 167–68)

Solomon's decision to divide the child, Sternberg continues, might strike the baffled
reader as a (ludicrous) variation on the cutting of the Gordian knot, a decision hardly
arguing for a sagacity of divine inspiration. But the appeal to the sword is revealed as
a trap not just for the reader, but ''a false solution put forward to trick the culprit into
self-betrayal.''

> We never find out for sure which of the harlots (''the one'' or ''the other'') is the
> mother. . . . All that matters is that he [Solomon] succeeded where everyone,
> given an equal opportunity, would fail—as we have reason to know. It is therefore
> both fair to the winner and comforting to the loser to deduce with all Israel, that
> ''God's wisdom was in him.'' From an illusion of equality to an admission of
> inferiority with a loophole for self-esteem: this is the route marked out for the reader
> by the dynamics of point of view. Having started out by putting us in Solomon's
> place, the tale concludes by putting us in our own.
>
> Like some other dialogic occasions, Solomon's Judgment occupies the exact
> middle of the scale leading from the reader-elevating to the character-elevating

position. The even-handed treatment establishes a parity in both the raw information
and the modes of processing, since the two detectives must weigh the same evidence
by the same lifelike standards. It is precisely this that rubs in the ultimate superiority
of divinely inspired over native wit. (PBN, 169)

Perhaps. And perhaps not. That this story has captivated generations of readers and of
young audiences who could not read for themselves is testimony to its gripping
qualities, to the capacity of the human mind to accept this example of virtuosity in
judgment as witness to a superhuman sagacity. And Sternberg's revelation of the
dynamic architectonics of perspectival positions in narrative art; of reader and
character at opposite ends of the seesaw with the narrator as fulcrum; of equilibrium
assumed, disturbed, restored—is a breathtaking performance in the art of poetical
analysis. But for all the logic, precision, and persuasiveness of Sternberg's analysis,
there may yet be a demurrer: all that you say is true—for some, many, perhaps most
readers. But not for every reader. For all the ingenuity of the ancient king and his
ancient historian, for all the ingeniousness of the modern poetical critic, there are
things about the story and the analysis that remain questionable. In many respects
both problem and solution, as well as triumphant QED, evoke the plaintive skepticism of many a tyro upon being introduced to the verbal pyrotechnics of Socrates in
Plato's *Dialogues*. The strawman is too often betrayed by the rents in his attire.
Socrates, for all his paternal endearments to his tuition-free students and protestations
of respect for the minds and rhetoric of his Sophistic opponents, is too glib, too quick
in analogical thinking, and too quick to shift his ground before his opponent has
caught him out, heading for a cul-de-sac. And his playrwright endows his hearers and
antagonists with a docility and tractability that render them less than equal as
adversarial foils or competent as juries, petit or grand.

In regard to the QED in which Sternberg's argument culminates, is this conclusion compelling or necessary by the canons of logic or of experience? For all those
who gape at prestidigitator's feats or in gullibility accept the thaumaturge's pretensions, how many will accept them as seats of the divine oracles compelling faith? And
what of Solomon's inspired test? Though the response of the true mother was
altogether predictable, was that of the thieving whore inevitable? Must a harlot be so
stupid and guileless as to expose her barbarous bloodthirstiness (not to mention her
total lack of maternal affection), especially—according to all translations other than
mine—*after* she has heard her rival surrender the prize to her alive! If, as Sternberg
has stressed, "the defendant's strongest point is that she has no story to tell: she did
nothing, saw nothing, suffered no loss, and would not care to advance any theories
about the death of another's child," how could this false claimant to the child, be she
the plaintiff or defendant, not at the least keep lip buttoned at this critical moment?
For that matter, is Sternberg correct that we have no clue indicating "for sure which
of the harlots [plaintiff or defendant] is the mother?" Is it even correct to suppose, as
Sternberg does at another point in his discussion, that the reader and the king are equal
in ignorance, in that "neither has had the slightest advance warning of the point at
issue . . . which unfolds in and through the dialogue alone?"

It is noteworthy that all of these questions, raised in challenge to an easy
acceptance of Sternberg's argument, fall into the category of metaliterary conven-

tions; they all reflect assumptions of a logical or psychological (or other experiential) nature that are factors in interpretation; not one is a focus of purely literary interpretation; not a one of the purely literary foci in his argument comes under challenge.

Two purely literary questions, neither of them raised by Sternberg, are what set me on the road to question the level on which the narrative is intended (that is, degree of literality or figurativeness) and to reexamine the metaliterary conventions that must be assumed if the story is meant primarily as as particular example, a single instance of a king's judgment in demonstration of "the ultimate superiority of divinely inspired over native wit." These are (1) the narrator's telling of the false claimant's response to the king's verdict only after the true mother's response has assured her of victory; (2) the earlier choice of the narrator to divide the king's verdict into two dialogic steps separated by an action. The first problem is so obvious and glaring that one is tempted to question it as possibly an artistic lapse; the second is ever so subtle, hence possibly moot: True, the call for a homicidal instrument is a sudden and dramatic eruption on the part of a king who, as Sternberg has insightfully pointed out, has just summed up the impasse by a repetition of the harlots' contradictory dialogue in the absence of additional witness. But why the call for the sword in dialogue, the action of its being brought at the king's behest,[6] and only then a resumption of dialogue in the command to split the child in twain? A sitting judge is not without a bailiff in attendance, how much less a king whose guard is never out of vocal reach. Why does the narrator not have the king bid a soldier, "Draw your blade," and continue with his command? Is there a deeper significance to the bared blade, summoned at the judge's order, lying before him in all its menace? And if there is, is that significance intended for the quick apprehension of the two litigants, or is it, as an instance of **free direct discourse** there primarily for the close reader to ponder?

Let us go back now to the story's beginning, to the formulation of the narrative's first verse (3:16). Compare my translation with Sternberg's, which is in keeping with his (and Alter's) preference for rendering the waw in parataxis by *and*): "Two women, harlots, came to the king and stood before him." Were this the purport of the Hebrew, normal and regular narrative style would have the first clause featuring a waw with the imperfect verb followed by the subject. Instead, we have the anomalous beginning with '*āz* followed by what looks like a verb in the imperfect but which cannot be rendered by any sense of that tense.[7] The temporal '*āz* (which Sternberg omits in translation) is in reference to the preceding verses which tell of Solomon's return to Jerusalem, where Solomon feasts his servants in grateful celebration of YHWH's promise, in the dream at Gibeon, to grant him the wisdom in judgment he had asked for (as well as the personal blessings he had refrained from asking). The first clause, governed by this adverb, is contextual and neutral in connotation. One may "come to" the king, as here, for any number of reasons but to come to court is not the same as to be granted audience. This latter sense is what is expressed in the second clause (the verb "to stand" with the preposition *lifney* "in the presence of" a king being a usage either for courtiers in attendance or litigants suing for justice, hence my translation "stood arraigned").

At this juncture we must ask a question in regard to a metaliterary convention: the appearance of two contestants before the throne for judgment. The king's obligation "to rule," that is, to serve as chief magistrate in the dispensation of justice, is a convention known to us from many passages in Scripture and from the literature of Israel's neighboring societies as well. But the king's justice is administered by lower and higher courts deputed of him; the royal person himself is the court of final and highest appeal, the court of last resort.[8] As such, it is or should be clear that supplicants will be admitted to the throne for judgment only after their case has been screened. Have the litigants had recourse to the magistrates in their locality, have these proved inadequate to the case in question, is there a presumption that the king may succeed where his deputized judges have failed? A further matter of standard operating procedure raises additional doubt that this opening of the biblical narrative is intended to be taken literally as the whole story without ellipsis (in Sternberg's felicitous coinage "gapping"), that is, omission of information that either remains ever the same or is to be bridged by the text later or by the reader's common sense. The two parties to a suit do not just appear in court of equal volition. Unless they agree amicably to abide by a third party's arbitration, it is the plaintiff who appears to lodge her complaint and the defendant who must then, if grounds for the suit are deemed adequate by the court or its deputies, be haled to court for a hearing.

In view of these considerations we must reconsider Sternberg's assumptions that the king had no advance warning of the point at issue between the two women, that the appeal to the sword was a sudden inspiration, that despite being "caught off guard," the king had yet "the presence of mind to transform a shriek of renunciation ('Give her the living child and by no means kill him') into a verdict of restoration ('Give her the living child and by no means kill him') by a simple shift in reference." Yet the point of the king's sagacity is in no way lessened in this understanding. If anything, the point of that sagacity is complemented and reinforced by the recognition that the king, having heard from his clerks of the nature of this case, must have been able to plan a strategy for judgment such as to warrant his hearing the suit; that is to say, the king must have been confident that his strategy was foolproof, else he would have exposed himself to ridicule in his august role as court of last resort. If we may borrow Sternberg's adducing of the detective story analogy, how would the great sleuth appear if, in the final and resolutional scene, having gathered all the suspects for a review of the circumstances of the crime, he springs the surprise question that must startle the perpetrator into a self-incriminating statement or act and she simply fails to rise to the bait?

This brings us back to the question of the author's competence in his planning of a trap into which only the silliest of women would have fallen, his covering up this weakness by the deus ex machina device of "Well, she fell for it, didn't she?" and then arranging for this self-incriminating burst to take place after the other claimant's renunciation has yielded her the prize, a burst so idiotic as to be incredible: "Now that I have won my child, go ahead and kill him." A close reading of the Hebrew text, however, discloses that our author is guilty of none of these faults. On the contrary, he has deliberately arranged for the at-first-blush incongruities to alert the careful reader to the need to read between the lines. If we may again compare translations, Sternberg renders the second part of verse 26 (the italics are mine): *"And the other one said,* He shall be neither mine nor thine. Divide him." Sternberg thus renders the Hebrew just as if the syntax were the normal narrative usage of waw-conversive with the imperfect in paratactic sequence, a usage that, he feels, is best interpreted by rendering the waw always as *and*. And in so doing, he overlooks the significance of the author's having departed here from normal paratactic narrative style. The waw here is (apparently, not semantically) conjunctive, attached to a pronoun, followed by a participial verb—in clear **hypotactical** construction. Such a construction frequently signals an interruption of sequence in narrative time, often best rendered for the reader in a modern language (including modern Hebrew) by bracketing in parenthesis: "(the other one the while declaring: 'Neither mine nor yours will he be: slice away!')."

What the narrator achieves by this parenthetic aside is, on the one hand, to contrast the second woman's response to that of the first, while allowing, in the regular narrative time sequence, the king's response to follow immediately and hard upon the first woman's response. The immediacy of this sequence is not only a logical requirement of the dialogic action, it is absolutely dictated by the narrator's ingenious use of double entendre in the king's rejoinder; for in his repetition of the true mother's plea, he is not so much commanding (note the parenthesis in my translation) as, in quoting her verbatim, declaring that her statement is the definitive proof of her motherhood—as if to say, her words, not mine, compel the verdict.

As delicious a touch as this is, it should not distract us from notice of two other considerations. First, the parenthetic break in narrative time sequence is not dictated by either the logic of the action or the dramatic touch of a quotation's becoming a verdict; for the false mother's response could, and normally would, precede the true mother's response in the telling. Second, there is something peculiar about the dialogue that the narrator puts into the false mother's mouth: her first address is not to the king but, in fishwife manner, a gloating taunt of her rival (which makes her acceptance of the king's verdict implicit); and her second address, the command to execute the child, is in the plural, addressed this time to the king's minions. This double failure to address the king—unexpected, impolitic, uncalled for, and (like the insistence on execution) hardly credible in the mouth of a woman who need be neither suicidal nor imbecilic—must be seen as an instance of free direct discourse. The purport of the thieving whore's answer is unambiguous, the incongruities in the formulation of that purport are intended for the reader. For one thing, the words constitute a metaphor—not for uncouthness (which they actually bespeak) but for affront to the majesty of justice as symbolized in the king. For another, taken together with their placement in the parenthetic aside, they are a signal to the reader to look beneath the simple, straightforwardness of the narrative, to read between the lines and check the vagaries of a storyteller against the canons of verisimilitude.

To respond to this signal we must go back to the beginning of the narrative and not only retrace the steps telescoped in the narrative's march but expand on the mental processes that must have preceded each step. I have already argued for the necessity of assuming that the king decided to have the defendant arraigned before his throne to confront and respond to the plaintiff's charge. For him to have come to this decision he must have first determined on a strategy for the trial that would elicit the truth of the matter. What was that strategy? The one thing it was not is what the story says it was, to trap the false mother into calling for the execution of the child.

At first blush, the circumstances of this case seem indeed to constitute an extraordinarily difficult dilemma for any one called on to resolve it with justice. In contrast with the Gordian knot, the perplexity of the challenge is due not to the complexity of the problem but to its simplicity. One child. Two claimants. No witnesses. One must be lying. One must be telling the truth. An arbitrary guess has a fifty–fifty chance of hitting on the truth. But the judge is not allowed to gamble, especially with other people's money in a game at/from which he has nothing to lose or gain (except his judicial reputation, which he has already put at stake by volunteering to hear the case). And let us remember the loneliness of the royal judge in this case; he is unassisted by prosecutor, paraclete, or jury. He alone must fill all these roles responsibly and alone bear responsibility for verdict, sentence, and execution of it. Such is his responsibility. And his quandary seems to be even more formidable: he is like the detective in a mystery story, peering into a hermetically sealed room and pondering a corpse therein—victim of an impossible murder.

So it seems at first blush. On second thought, however, the case before Solomon is far from exceptional. Any forensic buff will testify that the circumstances of this case have more, rather than less, in common with most cases tried today, especially those tried without a jury before a single magistrate. Be it a suit in petty claims as to liability for a garment damaged during cleaning and the value of the garment before damage or a suit for divorce and property settlement hinging on whether a mate did or did not witness an act of conjugal infidelity, the testimony of corroborating witnesses is often moot. Yet the magistrate will again and again manage, in what will seem to the inexperienced layman a virtuoso performance, to render a compelling verdict, one based on lines of questioning of the two parties that elicit traits of consistency, character, and credibility.

So now Solomon, as he ponders the question, To hear the case or not to hear? The plaintiff is clearly not in possession of the child, else there would be no complaint. Assuming that the woman in possession will categorically deny her rival's claim, is there any presumption in favor of the plaintiff? After all, possession is nine points of the law, and there is only the unsubstantiated word of the plaintiff: Why not dismiss her suit as frivolous? Yes, there is a presumption in the plaintiff's favor: the very absurdity of a lying woman's coming to the king in an appeal to wrest, without support of evidence or testimony, a child from a woman she knows that woman has mothered. But if there is a presumption in favor of the plaintiff and the accused has no way to substantiate her motherhood, why not rule now in favor of the plaintiff? The answer, of course, is that presumption of sincerity on the part of the plaintiff is warrant for trial, not for a verdict. In our case, the accused has possession in her favor, but further, if the absurdity of a litigant's bringing an action without supporting evidence were once accepted as the basis for a verdict in that litigant's favor, it would also be the last time it could serve to create such a presumption: the verdict would have destroyed the presumption of absurdity for all future time. Hence, we have reason to hear the case, to subject both women to examination and cross-examination in the hope of the guilty one's becoming so flustered as to somehow betray herself.

Is there anything the king can do—any pressure he can bring to bear on the litigants—to heighten the tension that the false mother must surely feel, to increase the chances of rattling her, short of devising some sort of ordeal? Yes, there is. The appeal to the sword! One of the offenses proscribed in the Decalogue is theft, interpreted by the rabbis as theft not of property but of a person, kidnapping. Support, perhaps the very basis for this interpretation, is the wording of Deuteronomy 24:7: "Should there ever be an instance of a man who steals a person from among his brethren, from among the Israelites, whether he make him his peon or sell him off, that felon–thief is to die: [thus] you are to purge evil from your midst."[9] The call for the executioner's sword, its blade bared in menace, is metaphoric expression of the king's inspiration. The harlots may think theirs is a wrangle over property. No, the king has raised the seriousness of the issue and the stakes for the litigants. The charge now is not tort but felony—and a capital crime at that. The woman in possession of the child, if she is not left in possession of the child, is adjudged to be a kidnapper. The plaintiff, if the judge rules against her, has borne false witness in a capital charge and is subject to the same penalty her testimony would have brought down on the accused.[10] For either woman now, what is at stake is not just possession of the child but retention of her head!

So much for the meaning of the bared blade. What of the preposterous proposal to split the child down the middle from head to toe and to give each woman, mother and child stealer alike, an equal half? Could even the most naive of bystanders in any courtroom of ancient Israel's credit (or is the proper word *debit?*) the royal magistrate with literal intent? And if not literal, what is the metaphor intended here? Sternberg, discussing the testimony of the plaintiff, writes, "Even the frankness of the speaker's reference to her uninterrupted sleep does not count in her favor, since the very logic of her tale forces this admission. How else could she explain her failure to resist the switch?" A point well taken. But what of the next question: If she could not resist the theft for having slept through it, why, upon discovering the switch of babes, did she make no attempt to retrieve her own? The answer to this is clear, and it is foreshadowed (or, to coin a neologism, *postshadowed*) in the option presented to the contestants by the king. For the plaintiff to have attempted to wrest her child from the grip of the thief would have rendered the child not the prize of battle but the field of battle itself, the rope in a tug-of-war. So here we have another presumption that the plaintiff is the true mother, based on her refusal to harm the child, while the woman in possession by her readiness to risk the child's being torn apart in such a tug-of-war has already, so to speak, given her answer to the king's proposal. This answer, let us remember, is one that the narrator put into a parenthesis, following in the narrative telling—but not in narrative time—the true mother's answer. This narrative device thus suggests that the examination of the litigants took place in one another's absence, while the free direct discourse of the thief's answer, addressed not to the king but to the mother, spells out for the reader the full baseness of her motive, which was not so much to hold a warm child to her bosom as to deny to her neighbor the blessing she herself has been denied, the blessing (if we are to credit the speculation of the true mother) of which she has deprived herself by her besotted smothering of the newborn infant.

At the risk of provoking my reader into a protest that I am exceeding the limits of interpretation (if indeed he has not already come to that conclusion), I shall revert to the interpretation of the bared sword metaphor. If to yield her claim to her child is for the plaintiff equivalent to a confession of false witness and self-exposure to capital punishment, then her mother love is equal to that sacrifice. The thief, on the other hand, would not consider for a moment confessing to a capital crime in order to save the life of a babe she was ready to tear apart in a tug-of-war—hence the force of her hyperbolic bidding, in free direct discourse, to the executioners: $G^e z\bar{o}r\bar{u}$, "Slice away!"

The Cumulative Factor in the Interpretation of Biblical Narrative

How far the metaphor? The individual reader must judge this for himself. From the extreme of literalness with which children of six years might take this story (should we risk the nightmares it might occasion them), through Sternberg's sophisticated exposition of the tale as an example of the Bible's mighty rhetorical art, to the extreme of metaphoric extension in which my own exegesis culminates, every reader may stake out his own interpretive position. One important factor in anyone's choice of position will be the metaliterary one of sophistication, a factor that may operate in several different loci, namely, the degree of sophistication (1) that the reader himself has reached; (2) that he is prepared to attribute to the author and to the most advanced audience in the author's society; (3) that he is led to believe the author presumed in the audience for which he intended his story.

In regard to locus 1, neither average six-year-old readers nor, perhaps, a good part of today's adult population probably own a sophistication equal to the task of following Sternberg's argument, even if his diction were simplified and each step in his argument slowly rehearsed. In regard to locus 2, the convention of the king's sitting in judgment and being in his own person available to every plaintiff without preliminary screening, while hard to credit in a kingdom with a population numbering a million or more, can be plausibly assumed perhaps in a city–state with a population of fifty thousand; but even an author in the larger kingdom might, for the purposes of his tale, assume the circumstances of the king on a barnstorming tour holding court spontaneously for all comers. In the latter case, the author will, in regard to locus 3, be presuming an audience ranging from the most simple- and literal-minded to the cognoscenti who are well aware of the uses of metaphor.

But there is yet another locus in which estimates of sophistication may be a crucial factor for interpretation: the constitutive conventions of an ancient society—for example, its legal institutions and presumptions as to legal rights. An assumption that I made in my exegesis of Solomon's Judgment—one open to challenge—was that the legal principle of possession counting as nine points of the law would have been operative in ancient Israel. Were I to defend that assumption, I might adduce the talmudic equivalent of this principle: *hammōṣî' mihavērō, 'ālāw har^e'āyā*, "the one who would extract [something] from [the possession of] his fellow, upon him [falls the burden of] the proof." While this would bring the assumed principle closer to Solomon's domain geographically, ethnologically, and chronologically, a chronological gap of considerable size would yet remain to be bridged.

I shall not now try to bridge that gap. But I shall present four considerations in support of the plausibility of ancient Israel's legal thinking's having advanced to the "sophisticated" concept of possession's creating a strong presumption in favor of the possessor. First, the notion that this concept is advanced or sophisticated is itself a presumption in both senses of the word: it is an assumption, and it has very little to recommend it. If we consider the alternative to this presumption in favor of the possessor we will quickly realize that this "legal principle" is hardly more than simple common sense. Second, if this were not the case—if it represented more than just common sense—the existence of the principle in the legal systems of other societies, closer in space and time than our own to the biblical setting would, in the absence of a contrary principle, place the burden of proof on the critic who would question the reasonableness of assuming this principle for Israelitish law. Third, if the citations from, say, Roman, talmudic, or our own law are judged as being reflective of an advanced state of juridical development and the question before us is whether Israelite culture had reached that high level, the presumption for an affirmative verdict would certainly be strengthened by an accumulation of arguments demonstrating the advanced state of Israel's literary art; for except for the biblical literature, we have no information at all about the state or level of ancient Israel's culture. In connection with this last point, if there is no fallacy of logic in this argument—if there is indeed a correspondence between a high level of rhetorical art and a high level of intellectual attainment in other areas of humankind's pursuit of the good and the beautiful, of the true and the just, of the abiding constancy of certain principles in a world that seems more characterized by change than by stability, or of the regulatory dynamics of a world that mutatis mutandis, seems ever to revert to equilibrium or stasis—then Sternberg's claims for the poetics of biblical narrative has raised the proper assessment of ancient Israel's intellectual achievement as at least comparable to that of Plato's Athens and more likely to our own today.

This argument may tempt the reader to ask me whether in my pleading for a higher level of sophistication and abstraction than even Sternberg discerns in Solomon's Judgment, I am not, in effect, inviting the inference that the intellectual achievement evidenced by the Bible transcends our own today. The implicit self-contradiction in such an inference should preclude its being drawn. But the question that I suspect will continue to lurk in the reader's mind will have to do with the incongruity between my acceptance of Sternberg's poetics and its implications and my somehow going beyond him in my application of a similar, if not quite the same, system or method of poetical analytics. My suspicion is that the answer, as regards the narrative of Solomon's Judgment, lies in our differing estimate of the level of the audience that the biblical author had in mind in constructing his narrative. Clearly, Sternberg's estimate is neither insulting to the biblical author's target audience nor denigrating of its sophistication; nor is it so in regard to the modern critical reader. But I would argue that the art of the narrative is such that it is meaningfully read on quite a few levels. Thus, the most naive and literal-minded have delighted in this story and its several lessons about human nature and the glory of justice done for long centuries before Sternberg's analysis of its rhetorical use of perspectives appeared to enhance our appreciation of its technique and its message. Many readers who, for a number of reasons, will find themselves unable or unwilling to follow his analysis will continue to find the story admirable. By the same token, to read the story, as I have done, on the basis of both several purely literary features of the Hebrew that have hitherto gone unremarked, and a differing metaliterary assessment of the author's appreciation of juridical procedure, as well as of his assessment of the level of his most sophisticated readers, is to extend our grasp of its depths yet not to vitiate Sternberg's analysis.

There is always, for all this, the danger of overinterpretation, a danger that increases as the argument becomes ever more subtle and the discourse of the narrative is understood on an ever-higher level of abstraction. The more subtle the art and the more sophisticated the message we discern in a given narrative, the more we would look elsewhere in that literature for support of the presumption for such high levels of discourse. To put it somewhat differently, we seek to add weight to the weighty interpretation of a particular narrative by adducing the similar weight of companion narratives. This I propose to do now in regard to Solomon's Judgment by recourse to several narratives from elsewhere in Scripture that also deal with juridical issues. For the present I shall refrain from formulating the kerygma of the Solomon narrative that issues from my interpretation of it.

Judging the King's Judgment and Justice

It will not seem strange if the narrative most germane to Solomon's Judgment is one we might title David's Judgment. As the former celebrates the wisdom of a king newly seated on his throne, so the latter features a king in the declining years of his reign, Solomon's father.

We are told that David, early in his consolidated reign over the United Kingdom, had sought out a surviving scion of Saul's line, Mephibosheth son of Jonathan. To him the king had turned over all the personal estate of Saul, to be administered for him by Ziba, a former retainer of Saul's, while Mephibosheth himself was to reside in Jerusalem that he might dine regularly at the king's table. We are also informed that Mephibosheth had as a child sustained a fall from his nurse's arms that rendered him a paraplegic. The first part of our narrative is 2 Sam. 16:1–4. David, in his flight from Jerusalem before the advance of his rebel son Absalom's forces, is met by Ziba, who presents to the king two asses laden with food and drink. When the king asks about the whereabouts of "your master's [Saul's] son," Ziba replies, "Lo, he stays on in Jerusalem; his thought is, 'Now is the time that the House of Israel will restore to me my father's sovereignty.'" To this the king responds with a grant that Ziba accepts with alacrity, "Yours now is all the property of Mephibosheth's!"

The second part of the narrative is 2 Sam. 19:25–31. Here we are first told that Mephibosheth had denied himself pedicure, barbering, or change into clean clothes from the time David abandoned his capital until the day he returned there in triumph. Although Mephibosheth made his way down to the Jordan's ford to greet the returning king, the latter took no notice of him until they were back in Jerusalem.

> The king asked him, "Why did you not accompany me [in my retreat], Mephibosheth?" He replied, "My lord the king, my servant tricked me. Verily, your subject proposed, 'I'll saddle me an ass upon which to mount and accompany the king'—your servant being a cripple. Now he has treacherously slandered your subject to my lord the king. Yet is my lord the king like an angel of God's, so do as you deem proper. Truly, there was not a single one of my father's line but was justly slated by my lord the king for execution. Yet did you place your subject among those

dining at your table. Have I then any claim yet to merit that I might plead before the
king?'' The king then said to him, ''Why must you carry on with your prattle? I have
decided: You and Ziba are to divide the estate!'' Said Mephibosheth to the king, ''Let
him take it all—now that my lord the king has arrived safely home.'' (2 Sam. 19:25–
31)

Talk of a preposterous division! Talk of dialectics! Can there be any doubt that the
inclusion of this story is by design or that the design is for it to be equipoise to
Solomon's Judgment? The young Solomon is no party to the contest before him, has
nothing to lose or gain but a reputation for wisdom. He takes that risk to hear the case.
He properly escalates the suit from property claim to a capital charge. He lays bare the
psychology of homicide in the crime of kidnap, and achieves justice for all by his
mockingly incongruous proposal to split the child and divide him between the
plaintiff and the defendant, the defendant whose possession of the child spoke against
her even as it constituted a legal presumption in her favor.

Now old David. No suit is laid before him. There are no litigants and no *chose* but
of his own making. Two assloads of provender purchase his favor and goodwill. He
invites the servant to comment on the failure of his master, a cripple, to present
himself as companion to the king on the retreat into the desert. A steward, long in
service to grandson and grandfather, who would betray his master's line has no claim
on anyone's trust. Yet does the king, on the basis of unsupported speculation and
uncalled-for denunciation, without waiting to hear a defense, deprive his ever-loyal
friend Jonathan's son of the estate in possession of which he has himself confirmed
him and bestows the estate on his treasonous lackey. Surely, we might expect that
even in those benighted days, when confiscation and distribution of a subject's
property were loyal prerogatives, a king would not so act except in accordance with
due process of one sort or another. And when the king's fortunes have once more
taken the turn, for which he prayed to God, when loyalties can be rewarded and old
scores settled, he is confronted by this embarrassment of a cripple he has judged a
traitor, bearing on his person all the marks of a loyal courtier long in grieving over his
suzerain's troubles. The king asks a question he knows to be ridiculous. He receives
an answer he should have himself anticipated. To top it all off, the son of his old friend
does not remonstrate but, rather, expresses gratitude for the favor shown him by a
king who might well have extirpated the line of the former dynast. Now this evokes
another consideration. It is no longer a question of reconsidering a decision to credit a
denunciation, to confiscate from one and give title to another. If Mephibosheth is
lying, he has been treasonous to his gracious king and deserves to die. If innocent, he
must be restored to full estate. If Ziba has lied, borne false witness in the capital crime
of treason, he should be put on trial for his life; in any case he has no claim to the estate
into possession of which he was placed by the rash decree of a king in flight. But it is
some sense of that rashness of his—that intemperate rush to an unjust judgment—that
now troubles the conscience of the king. Has the king learned from his troubles so
newly overcome? The king, himself so recently restored to his estate, who so
mourned the death of the son who had usurped him, who saw the judgment of God
both in his initial defeat and then in his vindication, is now torn between self-reproach
and the unpleasantness of having to repudiate his earlier verdict. Irritated by the very
refusal of Mephibosheth to plead for such a reversal, he cuts him off in the very course

of exonerating the king of fault whatever be his decision. Accusing the unimportunate prince of talking too much, he announces his decision, one that sidesteps the capital elements of the case, gives no verdict of guilt or innocence, and divides an uncontested babe in two!

At least one rabbi understood why this story is told of David:

> R. Judah cited a pronouncement of Rav's: when David said to Mephibosheth, "I have decided: you and Ziba are to divide the estate," a whisper [from Heaven] declared to him, "Reheboam and Jereboam will divide the kingdom!" [The point of] Rav as quoted by R. Judah: Had David not given credence to slander the kingdom of David would not have been divided. (Bab. Tal. *Shabbat* 56b).

What makes the above passage of special interest is the seemingly gratuitous explanation of Rav's saying. The Talmud, however, is taking pains to explain that the oracle giving judgment on David's judgment is not to be taken literally. The decree announced to David, if taken literally, would be in contradiction of the biblical narrative that has that same oracle delivered only a generation later, to Solomon (1 Kings 11:11–13). The decree to David (like the decree to Solomon) is a metaphor for one of the most frequent kerygmas in Scripture: throne and dynasty are founded upon justice; the failure to rule justly, or indeed to function as God's deputy in the administration of justice, undermines the throne's foundation—so, in poetic formulation, of earthly thrones in Proverbs 16:12, 20:28, 29:14 and the Heavenly Throne in Psalm 89:15. This metaphor and this kerygma, more than any rhetorical point that is the particular property of Solomon's Judgment, is the deeper purport of this narrative.

The Unity of Scripture

The metaphor and kerygma of this narrative is certainly one of the central ones in the Hebrew Scriptures, on earth and in heaven, from Genesis to Chronicles, from the election of Abraham to Scripture's concluding verse, namely, the proclamation of the new Solomon, Cyrus of Persia, that sovereignty of all earth has been granted to him by YHWH so that he might restore in Jerusalem the earthly judgment seat of YHWH, God of *heaven* (not heaven and earth; 2 Chron. 36:23).

Our first immediate concern here, however, is to support our interpretation of the narrative that is the focus of our discussion, Solomon's Judgment. Thus, Sternberg notes at the beginning of his discussion that this narrative follows "God's promise to grant him supreme wisdom" and at its conclusion that "it is therefore both fair to the winner [Solomon] and comforting to the loser [the reader] to deduce, with all Israel, that "God's wisdom was in him." In each instance, correct though he may be as far as he goes, he does not exhaust the full significance of the verse in question. In regard to the latter observation, for example, the concluding verse does not stop with the presence in him of God's wisdom but goes on to specify the specific area and function of that wisdom: "to exercise judgment." Furthermore, the verse concludes with this deduction; it does not begin with it. It begins with the effect of this deduction on all Israel: "When all Israel heard the judgment which the king had rendered *they were afraid of the king*." Why does the omniscient narrator choose to tell us of this fear, why not of their rejoicing to have so wise a ruler? The answer, I would suggest, is twofold. On the one hand, in regard to this particular judgment, there was the king's awful insight which led him to escalate the level of the trial from that of tort to that of felony (and a capital crime at that), in keeping with Deuteronomy 24:7. On the other hand, the fear of the people resonates with the rubric with which Deuteronomy 24:7 concludes: "Thus shall you purge evil from your midst." This rubric appears in Deuteronomy nine more times, in connection with the mandating of capital punishment for certain offenses; in four of these instances it appears along with the supplementary motive for the mandating of so severe a punishment: ". . . and all the people will hear and be afraid." The force of this expression in Hebrew is echoed in the English idiom "to put the fear of God into someone." Of these ten instances,[11] the context of one is of particular interest for its bearing on our understanding of the kerygma of Solomon's Judgment. Deuteronomy 17:8 anticipates occasions in Israel's future on the promised land when the local judges will find themselves faced by an impasse in legal contests where the stakes may vary from payment for injury (to person or property) to capital punishment. It requires the local magistracy to repair to the capital city (which YHWH will choose for his judgment seat) and put the case to either the Levitical priests or the Ruler (Haššofet) for resolution; whatever guidance line or verdict they prescribe is to be followed exactly, and anyone defying the authority of priest or Ruler is to be purged from the community.

As for Sternberg's observation that the story of Solomon's Judgment follows "God's promise to grant him wisdom," the full context of the passage preceding this narrative must be cited, not the promise alone. In this passage, 1 Kings 3:5–14, YHWH reveals himself to Solomon in an incubation dream and invites him to make of him any request he will. Solomon's response, expressing gratitude for his succession to the throne, a succession not owing to his own merit but to God's rewarding David for loyalty to him, stressing his own inexperience and his responsibility for the countless multitude of his chosen people, asks for "a sensitive intelligence to rule [govern, judge] Your people, to discriminate between right and wrong—can anyone [human] rule this numerous people of yours!" God's answer, granting Solomon his request as well as the wealth, prestige, triumphs, and longevity that he had not asked for, points up Solomon's sense of responsibility as over concern for self. He thus already has the widsom he seeks. Another, more recent version of this metaphor is God's response to Blaise Pascal: 'You would not be seeking me if you had not already found me."

The citation of so many passages from Deuteronomy in connection with an interpretation of a narrative in Kings will not occasion comment from scholars brought up on the hermeneutics of source criticism; for most of the two books of Kings are attributed by this school to *the Deuteronomistic historians.* And the presence in ancient Israel (as where not?) of a local magistracy of one kind or another (assumed in Deuteronomy 17:8 ff., which was discussed above) is prescribed (as if it could not be taken for granted) by God in Deuteronomy 16:18: "Empowered judges are you to provide yourself in all the chief cities that your God YHWH grants you, in [all] your tribal districts, that they may judge the populace with justice." The point of having God command what was already standard practice in Israel as well as in neighboring societies is (as I have argued elsewhere in regard to such prohibitions as murder and adultery) that what makes an act right or wrong is not the fact that society has chosen to make it legal or illegal (that is, a pragmatic recognition on the part of a society) but that God has ordained it so.[12] We have in any case further strengthened two of the assumptions or conventions accepted in our interpretation of Solomon's Judgment: (1) that lower courts or officials would normally hear a complaint or a case before it came to the king and (2) that the kerygma is the obligation of the king to serve as God's vicar for judgment on earth and God's granting this human deputy the wisdom necessary for the task.

A general conclusion for biblical poetics that I would, however, derive from this discussion is occasioned by the presence in Exodus of a passage to the same effect as Deuteronomy 16:18, with one significant difference. In the Exodus passage (18:13–26), the appointing of lower and higher courts with a supreme magistrate as a last court of appeal, as the mediator of God's law to the people, is not an ordination by Deity but a pragmatic recommendation to Moses by his pagan father-in-law, Jethro. This difference alone might well be seized upon by source critics as pointing to opposing points of view reflecting separate and differing authorships. I would argue to the contrary that the similarities in the two passages are crucial and that doublets of this sort (particularly in the Pentateuch) merely stress different faces of the matter.

It is indeed the very incongruities of this Exodus passage in its context that must (like the syntactic parenthesis about the false mother, in Solomon's Judgment) be examined for the message between the lines. The central motif of Exodus' beginning is the injustice of the servitude forced upon Israel. The liberation from that injustice is effected by the God of Justice acting through a human intermediary and culminates in the revelation at the Mount of God, where he is revealed in his demand for recognition of his absolute sovereignty, a recognition to be expressed in the acceptance and execution of his laws. The liberation, however, is not a quickly achieved event, nor one owing to the faith and merit of the beneficiaries; it is a drawn-out tale of repeated miraculous afflictions on behalf of an Israel that cannot sustain her faith in her redeemer. In this context of miracles, faith and faithlessness, injustice and the demands of the God of Justice, we are suddenly, anticlimactically, brought down to earth, to a society where normalcy reigns. As if already settled on its own territories and not at the very beginning of its transitional desert experience, apparently oblivious of the so-recent escapes from Pharaoh and Amalek, of their dependency in this wilderness waste on the providential manna and quail, equally unaware of their impending rendezvous with God, this people has settled down to engage in the mundane bickerings and litigious actions that tax the energies of the one leader and judge, Moses. A Moses who speaks to, hears from, and acts on, the orders of God but must get from his pagan father-in-law the practical advice on courts that, in the Book of Deuteronomy, he will be depicted as requiring of Israel in the name of God.

The unity of Scripture is not (as Renan said biblical monotheism was) a reflection of the monotony of the desert. It is a unity that emerges somehow, dimly in some parts, brilliantly in others, out of the rhythms of birth and death, of the tension of pendulum swing, of the dynamics of conflict and reconciliation, of sinners groping for visions of salvation and the righteous falling into error. More a library than a book, poetry and prose, narrative and prescription, deriving from preachers and writers over a span of centuries, it can yet lay claim to one authorial voice. Seemingly gratuitous repetitions turn out to be subtle variations on a theme (the same phenomenon seen from different perspectives and shifting phenomena from a single perspective) while contrary accounts, opposite versions, and even seemingly contradictory dogmas can be resolved without deprecatory comment on the editors or the editorial process that made for the encompassing unity (called Scripture in English and Torah in Hebrew) out of the scriptures that relate the time-bound to the timeless and aim for a separation of the holy from the profane even as they wed immanence and transcendence. The kerygmas are sometimes formulated in propositions so simple as to be clear to the youngest child, sometimes they are encoded (as, for example, Solomon's Judgment) in a tale whose simplicity is a mask for the intricate traceries of its art. Scripture's prose is as *supple* as its teachings are (so often) *subtle*. When the diction suddenly becomes heavy, the formulation labored, the preaching wearisome, those are the times when the reader must shake off the smugness of the sophisticate snug in his critic's chair and look all the more humbly for the motivation of poets who can, and usually do, write like Plato and choose to resort to the diction of an Aristotle or Hegel.

Scripture, so rich itself in metaphor, defies attempts to encapsulate it in a metaphor or simile. I should like, nevertheless, to recall the insight of my teacher, Henry Slominsky, into the form of many a classical midrash: in commenting on an abstruse verse from, let us say, Leviticus, the midrash will juxtapose a verse from the Song of Songs, and out of this confrontation derive a meaning of dazzling richness. This method, he explained, is neither test nor boast of the expounder's ingenuity nor a capricious exploitation of the preacher's freedom to take license with the text. It is, rather, a serious game often playfully indulged in, based on Scripture as the literal words of the all-creating Artist, which therefore requires us to see every word, even every letter, in all Scripture as blades of grass above ground, all rooted beneath the surface where in the depths every rhizome is joined to every other.

By analogy with this hyperbolic yet meaningful characterization of midrashic presumption, I would propose a more modest simile for Scripture's poetical unity (one that falls short by analogizing Scripture to a vista spread on a flat surface rather than one with the twisting contours of a three-dimensional map or the topological paradox of an Escher stairway): it is like a gigantic jigsaw puzzle made up of smaller jigsaw puzzles, the adjoining rectangles of which are all the more difficult to piece together for the difficulty in seeing how the upper right-hand corner of the macrovista relates to the upper left-hand corner, which is located in another room. The clues to any given pericope in Scripture are often to be found in another section of Scripture rather than (where it is our wont to look) in the disciplines of sociology or history, anthropology or comparative religion, or even in poetics borrowed from an extra-biblical literature. To be sure, we face the perplexity of missing the forest for the trees or the trees for the forest. That is why in examining every tree or copse we must be ever ready to take to the air and scan the forest's complex topology and then, with the expansion of our yet-limited field of vision, return to this tree and that copse for a better grasp of how the pieces and parts join to form the whole(s).

This sketch of Scripture's unity, inadequate and impressionistic as it surely is, is put forward as a hypothesis to be tested, not as a conclusion demonstrated. I make no claims to have solved more than a few of the puzzles nor to have laid bare the larger number of Scripture's poetical principles—hence the title *Toward a Grammar*. It is my hope that such successes as the critical reader will allow in the interpretive essays that follow will be enough in their cumulative weight to recommend the poetical hermeneutics and the assumption of Scripture's unity, in art and in kerygma, to students privileged to delve into the mysteries of this inexhaustible book.

II

Exegetical Essays: Tales of the Prophets

3

"And Much Cattle": YHWH's Last Words to a Reluctant Prophet

Few of the books which constitute the Hebrew Scriptures lend themselves so admirably to a poetical analysis and interpretation as does the Book of Jonah. A freestanding narrative of four chapters with a single plot line and a limited cast of characters, it is comparatively (and refreshingly) unencumbered in modern scholarship by the methodologies of source criticism, tradition history, and the like; a condition reflected in the words of Brevard Childs: "In sum, the basic unity of the book has been strongly maintained by modern critical scholarship."[1]

For all this, the extent of disagreement as to the literary genre of the Book of Jonah may be discerned in the following characterizations of it: "fable, didactic novel, prophetic legend and parable . . . a midrash . . . an allegory . . . narrated dogmatics . . . mixed genre with the presence of many eclectic elements."[2] Almost as diverse as the proposed genres of this narrative are the interpretations of its major purpose and the proposals for the date of its composition.[3] And for all of Childs's apparent agreement as to the book's basic unity, the emphasis of his brief treatment of it is the failure of the major interpretations "to do justice to the full range of exegetical problems in the biblical text" (p. 421). This leads him to the function of the book in its "canonical context," to its "canonical shape . . . [as] example of an editorial process which retained intact elements of an earlier interpretation" (p. 426). Similarly, Landes, who is alert to many of the literary elements that point to the book's artistic unity and argues for the appositeness of Jonah's psalm in its context,[4] concurs in the consensus as to the psalm's provenience in an "earlier . . . epoch" (i.e. pre-Exilic) while "we should look toward a post-sixth-century date for the writing down of the narrative portions of the book" (pp. 8–9); and at that, he makes bold to ascribe the borrowing of the psalm to "the prose writer" rather than "some later redactor."

Given the presumptions of modern critical Bible scholarship, it is not surprising that even readers sensitive to many of the nuances of biblical Hebrew and appreciative of purely artistic techniques and effects should nevertheless discern in this book good reasons to deny it a consistent congruity or esthetic unity. Many of the details in these four chapters seem, in contrast with the normal terseness of biblical narrative, superfluous; many more are incongruous in themselves and even more so in their immediate context, while other details necessary for a discernment of development of character or plot are conspicuous by their absence (an oxymoronic phenomenon that Sternberg dubs "gapping"). Characters act in ways that seem completely out of character; motivations are murky or imponderable; dialogue is disjointed or disordered, with questions left unanswered or asked only after the answer has been given; and time flow is distorted, and narrative sequence confused.

Every one of these literary problematics will yield to the poetical elements as discussed in chapter 1. My argument is that the Book of Jonah is from beginning to end, in form and content, in diction, phraseology, and style, a masterpiece of rhetoric. It is the work of a single artist, free from editorial comment or gloss; every word is in place, and every sentence. As an esthetic achievement the marvel of its creation is surpassed, if anything, by the marvel of its pristine preservation and transmission over a period of twenty-five centuries and more.

The Story of Jonah

> 1.The word of YHWH came to Jonah ben-Amittai to this effect: 2."Up! Go to Nineveh—that great city—and proclaim her condemnation. Yes, the indictment against them has come up on my docket." 3.Up got Jonah—to decamp to Tarshish— in flight from YHWH. He went down to Jaffa, found a ship Tarshish-bound. He paid the fare and went aboard—with the others Tarshish-bound—in flight from YHWH. (1:1–3)

One would be hard put to imagine a narrative beginning better designed to strike an ancient Israelite audience as discordant, incongruous, absurd. A monarch charges a deputy, trusted and long in his service, with a mission which, altogether in the line of his duty, will take him to one end of his lord's far-flung empire. Without a word of demurral, without a suggestion of motive, the deputy proceeds—and ever so casually—to head in the opposite direction. He does not—as a number of his fellow courtiers have been known to do—plead unworthiness. He does not plead superannuation and beg leave to spend his retirement years on his ancestral estate. He abandons all in the hope of achieving a goal that to his fellows was a fate worse than death, that is, to become a nameless outcast somewhere at world's end.[5]

Scholars have attempted to mitigate the absurdity of Jonah's response to the call of his God by suggesting that at the time of the book's composition there prevailed a notion that as is the case with other national deities, YHWH's power was confined to the neighborhood of his people's domain.[6] Support for the existence of this notion is adduced from a few passages that express an altogether different, and well-attested, Scriptural teaching: worship proper of Israel's God, that is, offering of animal sacrifices, might be conducted on the sacred soil of Israel and there alone.[7] More to the point, however, is the fact that even among Israel's pagan neighbors there is no notion that a national god is impotent beyond his frontiers. Quite to the contrary, such tutelary divinities as Marduk of Babylon or Asshur of Assyria are seen as cosmic gods, responsible for the creation of heaven and earth and for the imperial extension of their proteges' domains.

The flight of Jonah must stand out, as the author intended, in all its existential absurdity. By the peremptoriness of the divine command and the immediate and contrary response of the prophet; by the threefold repetition of *Tarshish-bound* (the westernmost shores of the Mediterranean, whereas Nineveh lies overland at the uttermost east) and the repetition of *in flight from* yhwh; by such details—normally eschewed in terse biblical style—as finding the right vessel, purchasing a ticket, and boarding "with the others" (just another commercial traveler crossing the ocean on the first tramper outwardbound); and, finally, by continuing his tale with (in the Hebrew) a jerky inversion of subject and verb so that the first word in verse 4 will be the same as the last word in verse 3—the author achieves this intent.

> 4.yhwh, now, hurled a mighty wind upon the sea. A great gale churned the sea—the ship herself figured she was about to break up—5.and the sailors, terrified, shrieked each to his gods while they hurled the ship's cargo into the sea hoping to diminish the threat pending o'er them. Jonah, however, had gone below to the deepest hold where he now lay fast asleep. 6.The skipper, coming upon him, cried, "What in the blazes! Sound asleep? Up, up! Call to your gods. Maybe heaven will yet give us a thought . . . so that we shan't founder." 7.The crewmen agreed to a suggestion, "Let's cast lots to discover on whose account this disaster has befallen us." They cast lots and the short end fell to Jonah. (1:4–7)

The jerky effect of the inversion of subject–verb order appears twice more in these few verses. Once in the attribution of thought to the vessel[8]—"I'm a goner," she thinks—and again in the aside of Jonah's descent into the hold. The perplexing obtuseness of a prophet who undertakes flight from his God is again conveyed to us by his capacity for untroubled slumber while the ship heaves, crashes, and shudders in the hurricane's grip, while a nuance of judgment suggests itself in the repetition of the verb *go down,* which was first, and rather anomalously, used in verse 3 for Jonah's going aboard ship. The increasingly desperate plight of ship and crew is reflected first in the shipmaster's hope that one more prayer might yet turn the tide and then by the seamen's recourse to the casting of lots. Knowing, as all men do, that God helps those who help themselves, they have done all they could while invoking the powers that govern the elements. Heaven has turned a deaf ear to their efforts and prayers alike. What now? To the modern reader the transition to the drawing of straws is a step smooth and logical. Not so to an audience in antiquity. True, human misfortune was for them, pagan and Israelite alike, a visitation from above; visitation was retribution for wrongdoing—by someone, sometime, somewhere; and examples are abundant of recourse to oracles to pinpoint the malefactor. But when an entire community suffers defeat, drought, or plague no one family casts lots to determine which of its members is the guilty party. No single ship's crew is likely to regard itself as the only vessel caught up in the gale, not even when a squall strikes without warning shortly after they have set sail on a tranquil sea. Some such consideration must have prompted the rabbinic tradition that only the one ship suffered the tempest: its crew could look past the sheets of rain on all sides and spy other vessels sailing serenely under a blue and sunny sky.

> 8."Now, then, Mr. On-Whose-Account-This-Misfortune-Has-Come-Upon-Us,[9] tell us," they said, "what's your line of work, what brings you here, to what land are

you native, which folk call you kin?'' 9.He answered ''I am an *'ibrī;* and it is YHWH I
worship—God of Heaven, he who created sea and shore.''

10.Terror-stricken, they cried, ''How could you? To do such a thing!'' This was
their response to the enormity he disclosed to them: that it was from YHWH's Presence
he was trying to escape. 11.''What now,'' they asked, ''are we to do with you so that
the sea may ease up on us?'' The gale was yet rising in force. 12.''Pick me up,'' he
said, ''hurl me into the sea, the sea will ease up on you. Full well I know on my
account alone does this tempest rage against you.'' 13.Yet the crew turned to their
oars, rowing strenuously for the shore. In vain, no headway at all, so fierce the
boiling waves against them. (1:8–13)

Free direct discourse characterizes an oft-recurring peculiarity of biblical narrative
style. Dialogue, explicitly marked off as a direct quote, represents not the exact words
of the speaker but the gist of what he intended to say. This gist, even as it omits details
of what must have been said, will include what the speaker thought but did not
actually express; and frequently will include an intrusion on the part of the narrative's
author. So, for example, in verse 2, God's address to Jonah, the characterization of
Nineveh as ''that great city'' is by the narrator for his audience and not by God for
Jonah. The narrator knows that the normal sequence of questions to Jonah would have
been: Who are you? Where do you come from? What brings you aboard our ship? He
cunningly reverses the order to achieve a subtle effect. Passengers on a train striking
up an acquaintance may get around to discussing their occupations. The ''brash
American'' is remarked in Europe for beginning his conversation with the intrusive,
''What's your line?'' (preparatory, to be sure, to the inquiry, ''How much do you
earn?''). An author who gave priority to such a question from the mouths of seamen at
the point of shipwreck would normally be stamped as inept, if not idiotic. Our author
is neither. The Hebrew word for ''work'' here ($m^e l\bar{a}$'$k\bar{a}$) also has the sense of ''task,
role, vocation, mission.'' And the performer of such work, *mal'ak,* is rendered in
varying contexts as ''agent, deputy, legate, ambassador'' and (from the Greek word
for messenger) ''angel.'' The question thus expresses the assumption of the crew that
their passenger must be vested with a high office indeed. No minor functionary could
draw upon himself so violent an attention from so powerful a god. The audience, of
course, hardly needs to be reminded of the dignity of the prophetic office, chosen
minister of the King of Kings.

Jonah's response is further indication of the narrator's single-minded focus on the
office of his protagonist, an office so exalted that its very naming by the one who is
attempting to escape it makes for an overpowering irony. For Jonah's reply ignores all
the questions except the first: he names neither his land nor his people. The
appearance of the word *'ibrī* here and its translation as ''Hebrew'' is what accounts
for the mistaking of the author's intention by generations of exegetes.[10] The original
connotation of this term was membership in a heterogeneous group of immigrants
whose origins or lineage were of little concern or interest to the established natives of
the territories they were infiltrating. The applicability of this term to the members of
the various Israelite tribes in the land of Canaan ceased with their conquest of this
territory and with the consolidation of this conquest with the beginnings of monarchical rule in Israel. From that time on the term *'ibrī* stands for a person, such as an
indentured servant, who is neither slave nor free. Jonah's answer therefore is first, ''I

am a vassal.'' He then goes on to name his liege lord, not a mortal but YHWH—a name unknown to the pagan seamen—whom he proclaims God of Heaven, Creator of sea and land, a double merism that effectively excludes any other divinity. The artistry of the narrator is revealed in what immediately follows: the shock and incredulity of the seamen that a vassal so great to a lord so mighty could act so suicidally. Only then does the author remark that this reaction, to be sure, followed the disclosure that the vassal was indeed fleeing from a Power he had himself implied could not be escaped.[11]

The dramatic irony continues to build. The seamen accept Jonah at his word. Being privy to the will of the Deity, the prophet will know the remedy for their plight. They ask, they receive the answer, and proceed to ignore it. The prophet does not offer to jump into the sea; and they for their part, pagans though they are, will not lay hands on so sacred a person. The ship's sails having been lowered to ride out the storm (or ripped from the masts by its sudden onset), the sailors have only one recourse— the galley's oars. *To return to the shore:* these few words confirm the rabbinic perception as to the miraculous nature of a storm that rages against one ship alone and within sight, at that, of the shore it has just left. The ancient Israelite audience that knew not the compass, that knew that a stormy sky would black out celestial aids to navigation, could only conclude that the mariners could row for the shore because they could see it.

> 14.Only then did they call upon YHWH: "Please, O YHWH, please, let us not perish in retribution for this man's life. Do not charge us with the murder of an innocent. You alone, YHWH, have willed all this to pass!" 15.With this they picked up Jonah and hurled him into the sea. Instantly the raging sea was calmed. 16.The crewmen were gripped by an awesome fear of YHWH. They offered sacrifice to YHWH. And they made vows [to bring more in the future]. (1:14–16)

One of the safer generalizations one can make about the mind of the ancients is that it did not give a place to atheism. Skepticism about divinity centered upon its (or their) nature, not on its existence. Nor did the ancients doubt that the divine communicated with humankind—hence the ubiquity of revelation, through dreams and divination, oracles and prophets. The problem of a prophet with an incredulous audience lay not in his hearers' disbelief in deity (although this is how the prophets prefer to formulate it) but in their doubt as to the prophet's credentials or as to the accuracy of his reception of a message.

Jonah's audience have no grounds for doubt. The casting of lots is an oft-attested oracular device. It is not a recourse to haphazard chance or to a magical rite. It is based on the assumption that the Power responsible for the storm will determine the cast. That the Power has, indeed, done so is confirmed by Jonah himself. No wonder, then, that the pagans now acknowledge that Power by the name they have hitherto not known or acknowledged. Previously, they had called each to his own god, the captain had spoken of gods. . . . Heaven''; now they address YHWH.

The address itself is masterful rhetoric. In three terse sentences the pagan sailors reveal a theological good sense and a moral scrupulousness that provides a sardonic contrast to that of YHWH's prophet. Their refusal to save themselves at the cost of heaving Jonah overboard shows a consideration for the preciousness of a single life;

the contrast of this sensitivity with Jonah's own stance will only appear to us in chapter 4. But their sense that they have no right to act on their own recognizance expresses an exquisite sensibility. Jonah himself has confessed that his irresponsibility has endangered them, they who are innocent. In referring to Jonah's death as the death of "an innocent," they are not exculpating him of a crime against his God; they are rather acknowledging that his offense against them does not warrant their appointing themselves his executioners. YHWH is forcing their hand. In heaving him overboard, they will act as YHWH's agents.

The instantaneous calming of the sea is the conclusive demonstration of YHWH's power, of his immanence in human affairs. For all that, there is something strange in the concluding notice that these pagans expressed their awe (conversion, perhaps?) by offering sacrifice. Where and when? On board the ship? On return to the shore? Or upon arrival at their first destination? And stranger yet, their vows to bring future sacrifices. The text continues,

> 1. YHWH commissioned a huge fish to swallow Jonah. Jonah spent three days and three nights in the belly of that fish. 2. Jonah prayed to his God, YHWH, from the belly of that fish.

> 3. In my trouble I called to YHWH—
> He answered me.
> From the belly of Netherworld I cried out—
> You heeded my plea.
> 4. When you cast me into the depths,
> Into the heart of the sea,
> Currents swirling 'round me,
> Your breakers and waves crashing over me—
> 5. My thought was; I am driven out,
>   out of Your sight.
> Yet . . . I shall again gaze
>   upon Your holy Temple.
> 6. Water smothering my life breath,
> The deep swirling 'round me,
> Seaweed 'round my head entwined,
> 7. To the base of mountains sinking,
> Earth's bars forever 'gainst me fastened—
> Then up from the pit you raised my life,
> O YHWH, my God!
> 8. As my life force ebbed away,
> 'Twas YHWH I invoked.
> And my prayer reached your notice
> In your sacred Temple shrine.
> 9. Those who make will-o'-the-wisps their guardians
> Play treason to their own interests;
> 10. But I now, voicing thanks,
> Will offer sacrifice to you
> What I vowed pay in full.
> Deliverance is from YHWH only! (2:1–10)

Most modern scholars are agreed that this psalm is not from the pen of the story's narrator. It was, they feel confident, borrowed from another context and interpolated by an anonymous editor. The reasons for this consensus are that the imagery of the psalm is more apposite to a drowning man than to one snug in the belly of a whale; that on the contrary, the psalm is one of thanksgiving for deliverance and not when one expects a petition for help; and that the repeated reference to YHWH's sacred temple in Jerusalem is as incongruous in this oceanic ambience as it is anomalous in the mouth of a prophet whose sole activity (as told in 2 Kings 14) was confined to the northern kingdom of Israel, whose shrines were in bitter competition with the prestigious center of the Judean kingdom to the south.

The incongruities are clear. The reasoning that they have stimulated is absurd. The imagery is indeed that of a drowning person, an oft-employed metaphor in the Book of Psalms (and hyperbolic at that) for a person at death's door. But its literal appositeness would be most incongruous in the mouth of a person whose plight it actually described. A drowning man does not recite psalms, describe ocean's canyons, or complain that he has been wreathed in a turban of seaweed. And to be sure, he is too busy praying for help to bribe the Deity with a hymn of praise recounting past beneficence or with vows pledging future sacrifices. The psalm of Jonah does not appear elsewhere in the Bible. To suppose that so inapposite a hymn was borrowed or composed for insertion is simply to solve the conundrum of a narrator's idiocy by attributing that idiocy to a supposed editor.

Let us consider first the featuring of the Jerusalem Temple in the psalm and the prophet–psalmist's offering of sacrifice and paying of vows. The choice of this theme, which appears often in other psalms, on the part of an interpolating editor makes no sense at all. But as the choice of the narrator himself—assuming for the moment that he has apropriated an existing psalm—it makes excellent sense, for it continues the ironic theme of chapter 1: the ironic contrast between the incongruous behavior of YHWH's chosen messenger and that of the pagan seamen. The voice from the belly of the fish acknowledges Jerusalem's shrine as the primary earthly terminal for a line of communication that stretches to YHWH's heavenly court. That is the sense of verse 8—the prophet's prayer reaching God via his earthly seat.[12] But this seat is also, according to Judean theology, the only place on earth where sacrifice may be offered to YHWH. At an earlier stage in Israel's history such worship could be offered anywhere on Israel's sacred soil (but not, for example, on Israelite territory east of the Jordan, which is not sacred).[13] In 2 Kings 5, we are told of Naaman, an Aramean general who is cured of a dread skin disease through the agency of YHWH's prophet, Elisha. Naaman vows henceforth to offer sacrifice to no god but YHWH and to that end asks for two muleloads of the sacred soil to be taken back with him to his Syrian homeland. There is in this account a bit of gentle humor at the expense of a pagan who, converted to the worship of YHWH, thinks he can so easily circumvent the restriction of YHWH worship to his sacred soil. In the case of the (converted?) seamen, at a time when the restriction of such worship has been narrowed down to the Jerusalem Temple, the humor is directed not at them but rather at the prophet they honored. Jonah talks of offering sacrifice and paying vows in the Jerusalem Temple—this, when he is fleeing from his God! The seamen, however, offer sacrifice to YHWH immediately after the calming of the storm. They cannot make their way into the mountains of Judea, they have a voyage to complete. But, as if they had already been instructed in the niceties of YHWH's worship, they make vows to offer sacrifices on their safe return—this time, of course, in pilgrimage to YHWH's exclusive shrine in Jerusalem.

Once we discern how fitting the psalm is in its context, there is no longer any reason to deny its authorship to the narrator.[14] The diction and style of this psalm is impeccable. It might hold a prideful place in the Psalter. Yet it contains not a single word or image which is original. It is a poetic mosaic, skillfully fitted together, each stone copied from one or another placement in the Psalter. And the composition as a whole and in detail is artfully contrived to serve the narrator's purpose: to bring into bold relief the absurdity of the prophet's plight and to contrast his behavior with that of the pagan seamen.

Our narrator introduces the poem with the normal Hebrew rubric for prayer in the sense of petition and then proceeds to detail an address to God that is devoid (explicitly, at least) of supplication. And as it is devoid of petition, so is it lacking in other elements common to psalms of petition: there is no mention of any charge against the petitioner, no protestation of innocence or of extenuating circumstance, no confession of wrongdoing, no hint of penitence or penance. And the reason should be obvious. How could Jonah possibly pray to the One Power he acknowledges and Whose service he has repudiated? Jonah's actions that precede the psalm call for regret or contrition on Jonah's part. His actions that follow it do not.

The psalm marks the hinge, the point of change or reversal in the drama's action, what the Greeks called the peripeteia. The psalm serves not only to retard the dramatic action, it holds it suspended between the last notice of the pagan's sailor's reaction to the events and the resumption of Jonah's travels. This time in the reverse direction of his flight, in the direction which marks obedience to his Lord, but—as we shall soon learn—a forced obedience, an obedience uninformed by change of heart or spirit.

The last verse of chapter 2 reads, "YHWH gave a command to the fish, whereupon it spat Jonah out onto the shore."

1.Once again came the command of YHWH to Jonah, to this effect: 2."Up! Go! To Nineveh—that great city—and deliver the proclamation I speak to you." 3.Up got Jonah and proceeded to Nineveh in obedience to YHWH's command. Nineveh, now, was a city of awesome size, [its avenues] a three-days' walk. 4.Jonah began his mission, traversing the city['s streets] one full day, the gist of his proclamation, "Forty days more and Nineveh will lie in ruins!" 5.The magistrates of Nineveh, giving credence to heaven, proclaimed a fast; sackcloth was donned by great and small alike.

[This is how it came about:] 6.When the word reached the king of Nineveh he left his throne, threw off his royal robe, donned sackcloth, sat himself down covered with ashes. 7.Throughout Nineveh he made the heralds' cry resound: "By decree of the king and his councilors: O humans and cattle—herds and flocks—not a one is to taste anything, to nibble a blade, to take a sip of water! 8.All are to don sackcloth— humans and cattle alike—all are to call upon heaven with might and main! Each and all are to turn back from evildoing, from whatever lawlessness they are engaged in! 9.Who knows but that heaven will turn about and relent of its purpose, turn back from fury's course—and we not perish."

10. God took notice of their doings, of their turnabout from their wicked ways. So God, relenting of the punishment he had threatened upon them, did nothing. (3:1–10)

Nineveh in its heyday, was—for its time—a huge city indeed. Covering some eighteen thousand acres, it had only two or three peers. Scholars, however, have been misled by the description of its size here to assume that our narrator has wildly exaggerated its dimensions. Translating the Hebrew by "a three-days' walk across," they assume a north–south or east–west dimension of anywhere from five miles (a leisurely stroll for a tourist) to twenty, resulting in an assumed area of 25–400 square miles! Our own translation discloses the narrator's intention. Walking its avenues from one end to the other, back and forth on parallel streets, would have required three days for the prophet to cover the city with his proclamation. Remarkably enough, the response to the prophet's call was belief and acceptance. A pagan city that is the capital of the world's greatest empire and boasts a population of 120,000, would know itself favored of the gods and credit itself with some merit for that favor. Yet it accepts the indictment of an alien prophet from a small and far distant hill-town just beyond the imperial borders, a prophet representing a deity whom it has reason neither to know nor to honor, and this without the performance of a single prodigy to support the judgment and threat and, more remarkable yet, on the very first day of the prophet's call. Guilt is accepted and repentance is immediate, total, and absolute. An individual's survival of three days in the belly of a fish pales beside such a wonder of human behavior.

And this wondrous behavior—a pagan city brought to instantaneous acceptance of God's word—stands in parallel and antithetic plot movement to the prophet's instantaneous rejection of God's command at the beginning of our narrative. The question is moot as to which of the wonders is greater. For in the one case an entire population must accept one man's word for the authentic voice of Deity, while in the other the mediator must reject the voice that comes to him directly. And in both cases nothing intervenes, in the telling, between the call and the response. This "gapping" on the narrator's part is, of course, a **dramatic technique**.[15] For the bridging of the gap in regard to the prophet's response we shall have to wait until the last chapter. The bridging of the gap in regard to Nineveh's response follows immediately upon that response in verse 5. Here, then, we have our first instance of the **synoptic/resumptive-expansive** narrative technique.[16] To read verses 5 and 6 in simple paratactic (hence chronological) order is to commit the incongruity of having a fast proclaimed before the king who proclaims it gets word of the occasion that will prompt his action. Our author is not guilty of such ineptitude. On the contrary, having compressed challenge and response and arranged for the reader to be pulled up short by the impact of the "bottom line," only then does the narrator back up in time to describe in detail just how that denouement was arrived at.

That denouement—the synoptic conclusion of Jonah's first day of preaching—is expressed in verse 5. In this one verse the elements are the subject (the city authorities), their faith ("in the gods," which means in Jonah's credentials and in the judgment upon them), their response (proclamation of the fast), and the response of the city's inhabitants' all ("great and small alike") assuming sackcloth in sign of penitence.

The expansion of the one-verse synoptic conclusion requires five times that space. The king himself sets the example. The decree proclaimed in the city is not by his authority alone but by that of "his great ones" who share governorship with him. There is both similarity to, and contrast with, the scene at sea. The sailors who witnessed the prodigy of the storm did not know it to be an indictment against them, prayed to their gods, and jettisoned cargo, while the captain expressed the desperate hope that prayer might yet save them. Nineveh's citizens witness no prodigies yet hear the indictment against them and accept it as just; they cast off their pride and abandon their sinful ways; and the prayer that concludes the royal decree parallels the prayer of the ship captain.

If that were all! But no, the dread note of the prophet's denunciation and threat, the remarkable turnabout of Nineveh's populace and their contrition, the sombre tone of decree and prayer, is all set to naught (or nearly so) by a comic inspiration. The decree is addressed to human and cattle alike! (What, by the way, are herds and flocks doing within the city walls?) Lest this point be overlooked, the decreeing of the fast specifies no grazing; and the call to don sackcloth emphasizes that the animals, too, are to so bedeck themselves! All alike are to invoke the gods (every low and bleat counts); and since the crimes of cattle are bound to differ in nature from those of humans, the call for turnabout specifies that each creature abandon whatever wrongdoing it has been engaged in committing![17]

Our narrator, to judge from this chapter's concluding verse, manages to maintain a poker face. He reports that God took notice of the universal repentance, relented of his resolve to punish (how not?), and did nothing. The question we must ask ourselves (since we cannot question the narrator) is rather awkward to express: When? When, would the narrator have us understand, did God do nothing? At the end of the first day? Or at the end of the third? Or at the end of the forty days? Or perhaps at the end of the day (how many years later?) when the author was telling this story? This question of narrative time is crucial for the interpretation of the concluding chapter.

> 1.Jonah was terribly displeased, not a little upset.[18] 2.He prayed to YHWH, in this vein: "Now then, O YHWH, was this not my prediction back then in my native land? My reason for taking flight toward Tarshish? How well I know that you are a God gracious and compassionate, slow to anger, exceedingly kind and relenting when it comes to punishment! 3.So then, YHWH, take now my life—I would far rather die than live." 4.YHWH said, "My, you really are upset, aren't you!" (4:1–4)

What is the occasion and what the cause for Jonah's disgruntlement? The answer to this question must lie in the context of what immediately precedes and what immediately follows. The occasion is given in the preceding verse: Jonah's knowledge of the repentance on the part of Nineveh and of God's decision to do nothing. Nineveh has been spared destruction. The reason for his unhappiness may lie, however, more narrowly, either in God's decision or in Nineveh's contrition, which eventuated in God's decision. Jonah's words make it clear that it is not the latter. For the first time we are given a clue as to what prompted Jonah's initial disobedience: knowledge that God's attributes of patience, pity, and compassion would inevitably result (as it now has turned out) in the sparing of Nineveh. For these attributes of God to come into play, however, Nineveh must first repent. Strangely enough, then,

Jonah, contrary to the experience of all prophets, had to have assumed that his alien status would not militate against him, that his credentials as a representative of the supernal powers would not be questioned, that the threat would be credible, that guilt would be acknowledged, and that penance would follow contrition!

As absurd as this set of assumptions may appear, it does pinpoint Jonah's complaint: not the response of the sinners but the response of God to their response. God accepts the repentance! By clear implication, then, Jonah, in God's place, would not have done so. He is then questioning God's judgment. God, he says, is a softy.

A study of the section on Jonah's anticipation (see pp. 83–85) will reveal that for a close reader of the biblical texts the narrator could not have made clearer the crux of Jonah's disgruntlement. His God is, in our colloquial idiom, a softy, a pushover. He knew it when he first received his marching orders; what he knew then has now been confirmed. And *disgruntlement* is a poor word to describe his emotional state. This attribute of his God has brought him to despair. Now, finally, he submits his resignation—in hyperbolic formulation. He wants to die, he wants to get out of the game, or (in our own gambling metaphor) he is ready to cash in his chips.

God's answer to Jonah is as sardonic as it is terse. The answer, three words in the Hebrew, is rendered in the older standard translations as a straight question: "Doest thou well to be angry?" For the absurdity of such painful pedantry our narrator is not to be held responsible. The more recent translations recognize that the question is rhetorical—hence to be rendered along the lines of "Are you so angry?" (NEB) or "Are you that deeply grieved?"[19] (JPS). But they still fall short of the narrator's intent. Those three words are a statement of scorn. Jonah is addressed as if a child in a tantrum, "My, we are upset, aren't we!" This reply of God's is the bottom line of the entire narrative: God says, "I couldn't care less."

If this bottom line were also the last line of the narrative, perhaps the sense of it ~vould have been perceived earlier. But it is not the last line, and we shall have to go on to complete the chapter. Before we do, however, we should note, in addition to the absurdity of Jonah's implicit claim to foreknowledge of Nineveh's repentance, the absurdity of a prophet's being reduced to despair by the success of his mission. The notion often put forth, that the repentance, precluding the destruction, discredits Jonah and makes him out to be a false prophet is untenable. For all that the gist of the prophet's proclamation is given in one sentence, "Forty days from now Nineveh will lie in ruins," we have here another example of free direct discourse. The office of the prophet is to call for repentance, with threat as the stick and hope as the carrot. This is to say that virtually every prophecy of doom is conditional—whether in respect to the inevitability of the punishment or in respect to its proximity. No prophet would regard the "conversion" of his audience as a sign of his failure. Yet Jonah, apparently, does so regard the denouement of his mission.

5. When Jonah came out from the city, he sat himself down on the city's eastern side, improvising for himself a lean-to under whose shade he sat, waiting to learn what would come to pass in the city. 6. YHWH–God commissioned a broad-leafed vine that climbed Jonah['s lean-to] overhead, to serve him as a shade-cover for his head to provide a relief from his misery. Jonah's delight in the vine-plant was ecstatic. 7. God commissioned a worm-pest toward daybreak of the next day; it attacked the vine-plant so that it shriveled. 8. Then at sunrise God commissioned an east wind of

numbing effect. The sun beat down on Jonah's head; he languished so that he told himself to give up the ghost:[20] "Better death than to live so," he thought. 9.God said to Jonah, "You are really upset over the vine['s death] aren't you?" He answered, "Grievously upset—I long for death." 10.YHWH then said, "You, now, have been brought to such concern for the vine-plant, over which you expended no labor, in whose growth you played no part, which in the space of a night came to be and in the space of a night perished. 11.And I—I am to show no concern for Nineveh, this city so great, within which there is a human populace of twelve times ten thousand humans and more—which cannot tell right hand from left—to say nothing of cattle, so much?" (4:5–11)

Once again, our narrator uses with telling effect the peculiarly biblical narrative technique of synoptic-conclusive followed by resumptive-expansive. Here, too, as in chapter 3, the narrator backtracks in order to flesh out, to supplement and comple-ment, the bottom line that was reached in the synoptic-conclusive. It is clear from the end of verse 5, ". . . waiting to learn what would transpire in the city," that we are now in an earlier time frame; Jonah does not yet know that the city has taken the track of penance. (Nor, in this version, does he claim to have known all along that it would repent.) The notion of Jonah's upset in the first (the synoptic) version—and the upset itself—is owing to the decision to spare the city; while God's three-word sardonic reply is the bottom line. In the resumptive version the same three-word reply of God's both introduces the notion of his upset and traces the upset to another cause altogether. In the resumptive version no reason is given for Jonah's initial reluctance to accept the mission. The prophet's misery is unrelated to the success or failure of that mission, it is rather a function of his self-centeredness.

The God who blew up a storm to thwart one man now commissions a three-step scenario to drive home to us just why the bottom line was, "I couldn't care less." The hapless prophet makes it easy for God by choosing the east side of the city: with his back to the wall Jonah will suffer the assault of the sun from early morning on. God provides Jonah's lean-to with a providential layer of natural insulation. The destruc-tion of this insulating element by the most unprovidential God-sent worm increases the misery of the prophet under his now denuded trellis. The third step is the sirocco or hamsin, a wind from the eastern desert that is unbearable not for its directional force but for the hot, numbing pressure from above. As in the first version, Jonah wishes for death; but this time the wish constitutes no resignation from service—it is his physical wretchedness that leads him to wish it were all over. As his answer to God's question about his upset reveals, he would not feel so desperate were the vine-plant insulation still in place. The question that leads to this answer, the repetition of the three-word "Upset, are you?" must be our cynosure. For in adding a phrase—"on the plant's account"—our narrator gives it a masterful twist. The question seems to suggest that the upset is of the nature of sorrow or grief and that the grief derives from a disinterested pity for the poor plant that was cut down in its prime. And Jonah's answer suggests that he is altogether content to let this construction of his sorrow stand.

A hodiernal analogy would be an ecological concern for the preservation of an unspoiled wilderness out of a principled presumption that wildernesses have an intrinsic right to exist rather than out of a concern to preserve for ourselves a pleasurable resource. Jonah's grief—if that is the correct nuance of the upset—is not for the plant but for himself. His concern for his comfort and his compassion are only self-pity. God's final response contrasts the infinite dimension of the noumenal with the transiency of the phenomenal; affirms the priority of the animal over the vegetable kingdom by purpose of the Power that creates both and not by reason of one animal species' self-arrogation of dignity; and contrasts again the long-term interests of God with the selfish, self-centered, and self-serving interests of the prophet, who purports to be God's tool. And by concluding with the consideration of God's interest in his cattle (incidentally characterizing the mass of humanity as such, in their bovine lack of moral discrimination), the narrator pulls us up short on an abrupt note of bathos, that classic device of satire. God's jeer at Jonah's values, which concluded the synoptic version with "We are upset, aren't we?" concludes the resumptive version with this figure of descent from the sublime to the ridiculous. If Jonah, God's reply suggests, has no feeling for immature humanity, perhaps he can be moved to pity for the "much cattle."

The detailed analysis of our text is now complete.[21] We may now address ourselves to the synthetic question, What is the central message of the book? The two concluding versions of chapter 4, where the answer must appear, are clearly complementary.[22] Jonah finds fault with God's exercise of compassion. A compassion that foreordained (as he sees it) the outcome of his mission. God rebukes Jonah for his lack of compassion, accusing him of egocentricity and selfishness. But to say this is to leave unexplained the absurdities of a prophet seeking to escape his calling, of an unprecedentedly immediate and absolute conversion to repentance of an entire populace, and of a prophet's despair over the success of his mission.

In answer to these questions, the consensus of modern scholarship focuses on Nineveh as a non-Israelite city, as the capital of the Assyrian empire, which had destroyed the Northern Kingdom, Israel, and thus set the pattern for its neo-Babylonian successors, who destroyed the Southern Kingdom, Judah; deported its inhabitants; and put YHWH's Jerusalem Temple to the torch. These disasters resulted in a xenophobic hatred for all peoples not Jewish and an exclusivist ethnic, racial, religious, and nationalistic self-centeredness that is to be discerned in the books of Ezra and Nehemiah, written after Persia's restoration of the Judean exiles to the city–state, Jerusalem. The Book of Jonah is a protest against this prevailing mood. Jonah is a symbol of the Jewish people, chosen as God's prophet-people to the nations, which is now being chastised for its unbecoming hatred for the gentiles and called back to its true vocation as prophetic witness to God's love for all humanity, including those who had smitten it on both cheeks.

There is nothing inherently implausible in a single prophet standing, in a parable, for collective Israel. But plausibility falls short of probability. Against this interpretation of the Book of Jonah are the following considerations:

1. The dating of scriptural books or chapters on the basis of supposed stances of tolerance or intolerance, universalism or particularism, or of judgments as to primitivism or sophistication of theological or moral values, although still widely practiced, is, in general, coming under ever-increasing attack.

2. The intolerance of mixed marriages is not a novelty of the Restoration period;

it appears in admittedly early strata of the Pentateuch. As for xenophobia or hatred of persecutors, the literature of the Restoration period is mild when compared to the anguished cries of prophets and psalmists in pre-Exilic times.

3. A protest against Israelitic or Judean xenophobia would have been directed against the conquering armies of Babylon, not against the Assyrian city that those armies had razed and of whose empire they had become the new masters.

4. There is not a hint of hatred or hostility on the part of Jonah either for gentiles in general or Assyrians in particular.

The explanation for the absurdities must be found within the Book of Jonah itself. And the crux of the absurdities lies in Jonah's rejection of God's judgment, a judgment that he anticipated and rejected the moment he received his commission. And that judgment hinges on the repentance of the Ninevites. God judges that repentance acceptable. Jonah disagrees. Assuming that people are lacking in faith and wanting in obedience to God, assuming that the prophet is the authorized spokesman of God, who threatens dire punishment unless they change (tešūvā, the Hebrew word for *repentance* means "moral change"), how much change warrants God's annulment or suspension of the sentence? Here is the problem that elicits the anguished perplexity of prophet after prophet. The prophet threatens, sometimes to be heeded but more often not. Punishment often comes, but never in the degree the prophet deems sufficient. For behold, the people backslide once more, the prophet is once again faced by self-righteous mockers, even by enemies who seek to deprive him of his life—sometimes even of his livelihood! Jeremiah is the most eloquent of the prophets in expressing his agony: God commissioned him in the womb, promised him invulnerability—but he is accused of treason in face of the besieging enemy. He curses the day he was born, he accuses God of having seduced him into his service, he wishes he could resign—but alas, the compulsion to preach, planted in him by his Creator, is like a sword burning in its scabbard. God's answer to the prophet is in the Book of Jonah: I never promised that it would be easy or the journey short.

There is a snare for the prophet in his office. A snare whose components are pride, self-righteousness, and impatience. As the agent of Deity he may forget that he is only the *arm* of God, not his mind or heart. As the spokesman of God to the people, high and low, he is with him and against them: "It's you and me, Lord, against them; and sometimes, Lord, I wonder about you." This is the snare into which Jonah falls. As hyperbolic as the mechanisms of this parable are—Jonah's flight, freak storm, pious pagan seamen, big fish, Nineveh's repentance—one element of hyperbole would yet ring false. An impatience for vindication on the part of the prophet so great that it will settle for nothing short of the doom of his listeners. For the prophet—Jeremiah is again the best example—is the most tenderhearted of men, the most loving of his people, weeping over travails they have survived: "Would that my head were water, my eye a fount of tears, that I might weep day and night for the slain of my beloved people!" Hence the choice of Nineveh—capital of a legendary empire, no longer cruel mistress of the world—as the object of the prophet's callous indifference. Not that God's concern for all humanity is not expressed here: it is so here, as everywhere throughout Scripture. But the choice of Nineveh is to make credible a prophetic impatience carried to such an extreme.

Yet the very difficulty of judging between the extreme (the incorrect and unacceptable) and the moderate (correct and acceptable) must have led to the invention of that story form we call *parable*. Whether it appear in the guise of a fable, a short story or a *piéce á thése,* the effectiveness of the parable is likely to be in proportion to the cunning with which its intent is concealed. The plot will be at a significant remove from the situation in which the listeners are likely to imagine themselves; hence, they can remain objective and uninvolved. The characters, too, will not lure them into self-identification; no suspicion that they themselves may be the objects of the parody will arouse anxiety and distract their attention. And when the pieces of the puzzle all fall together, it is too late to walk out. To the narrator's last line a disembodied voice adds the mocking coda: *De te fabula narratur,* "It is you the story is about."

The reader is invited to reread the Book of Jonah, without interruption; to judge for himself the cunning of its author's art. As far as the interpretation of Scripture is concerned, the lesson of this exercise in interpretation is threefold. For the literary critic, that the sophistication of the biblical authors must be the presumption with which he begins. For the biblical scholar, that any section of Scripture is best understood in the reference frame of the Bible itself. For the layman who would not surrender the sacred writings to dogmatists or academicians of any stripe or denomination, that the approach to the Bible as literature—far from diminishing its dignity or sanctity—offers perhaps the most promising access to its theological and moral affirmations.

The story of Jonah would seem to be an intraprophetic phenomenon. If there were schools for prophets, as is widely presumed, one can imagine a distinguished visiting alumnus—perhaps a Jeremiah speaking ruefully out of his own experience, or gratefully out of a sense of having skirted "the last impediment of [a] noble mind"— lecturing the young seminarians: spinning this homily to warn them of the greatest pitfall facing them in their vocation. Rabbis, priests, and ministers will testify that little has changed over the centuries. The perennial question of the ministry is, "Does preaching do any good?" The perennial complaint is against the worship-attending flock, which listens week after week—to what effect? Somehow, the laity does not come up to the preacher's standards. And as for those who shun church and synagogue altogether, is it not a wonder of God's grace that they do not call down his wrath to the ruin of us all?

But the Bible was written for all men, not for prophets alone. And which of us do not, at some time in our lives, play the prophet (certain of our standards of right and wrong, pronouncing judgment on our stubborn adversaries who will not acknowledge the eternal verities as we see them, and even, after events have proved us to have been in the right, suffering disappointment because our discredited opponents have somehow managed to escape annihilation)? Scripture is preachment. And the preachment of this book is to the Jonah in each one of us.

Wordplay, Repetition,
and the Role of Humor

We have seen the artistry of the author in his ordering of action and dialogue in such a way as to compel the reader to deduce for himself the logic of the narrative's formal flow and the absurd grammar of the events and dialogue in the light of his own experience of existential reality. Another of the narrator's bags of tricks is discernible in his wordplay. At time he uses a simple word repeatedly when his lexicon afforded him a choice of synonyms more appropriate to the contexts. For example, the word *big* as characterizing the city of Nineveh at the beginning of chapters 1 and 3 and in the narrative's final line in chapter 4; similarly in Jonah 1:4, the wind is "big," the storm is "big"; in verse 10 the sailors' fear is "big"; in verse 12 the tempest is "big"; and in verse 16 the sailor's fear is again "big." Despite the availability of terms in the Hebrew Bible for huge sea monsters (Leviathan, *tannīn*), Jonah is swallowed by a "big fish." In 4:1 he suffers a "big" displeasure, while in 4:6 his joy is "big."[23]

Another example, the word *hurl*. YHWH "hurls" the big wind (1:4), the sailors "hurl" the cargo overboard (1:5), Jonah suggests that he be "hurled" into the sea (1:12), and the sailors do so "hurl" him (1:16).

The word *bad* appears in verse 2 for (the charge of) wickedness against Nineveh. The same word, adjective or noun, stands for misfortune (1:7), immorality (3:8, 10), punishment (3:10, 4:3), displeasure (4:1), and misery (4:6).

An extremely rare word appears to describe God's action in four places. Normally rendered "prepare" or "provide," this verb is common in later Hebrew in the sense of "designate, appoint." I have rendered it by "commission." Thus, YHWH "commissions" the fish to swallow Jonah, "commissions" the broad-leafed plant to shade Jonah from the sun (4:6), "commissions" the worm-pest to kill the plant (4:7), and "commissions" the paralyzing sirocco to torture Jonah (4:8). The image of God using flora, fauna, and natural phenomena to convey his commands and achieve his purposes—oft repeated in Scripture—is encapsulated in Psalm 104:4, "He makes winds His agents / messengers [Hebrew *mal'ak*], flashing firebolts his servitors." The choice of this verb, then, is to underscore the vein of irony in our narrative: the human commissioned by God for his service rebels, while the forces and creatures of nature, gross and fine, animate and inanimate, perform their missions like clockwork.

I suggested earlier that the use of the verb *go down* in 1:3 for Jonah's boarding (going up a ramp to) the ship is as inappropriate there as it is appropriate in 1:5 for his descent to the hold. The verb appears again in the psalm (2:7) where Jonah goes down to the base of the mountains on the ocean floor. In view of the richness of the material for interpretation, it is particularly unwise to risk overinterpretation. Yet the repetition of this verb and the images of "going down" suggest that the author may have been pointing to the moral descent of the prophet. As the fish's belly and the ocean floor are equated with Sheol, the biblical Hades, and the ship's hold is where Jonah hides from the God he is fleeing, so Nineveh figures as a moral sinkhole, reeking with corruption and a smell or foretaste of death. If that be so, we have another ironic twist by the author. Jonah spent three days in the belly of the fish; his mission in Nineveh, it is anticipated, will require three days for its completion—though as it turns out, one day suffices.

The modern reader is entitled to one more bit of information that has been suggested as accounting in part for the author's choice of Nineveh, rather than any of Assyria's other capital cities, for the locale of Jonah's mission.[24] Wordplays on the names of cities are not rare in the cuneiform scribal tradition of Mesopotamia. The syllable *nun* in Akkadian (the Semitic tongue of Babylon and Assyria) or *nin* in Aramaic (Hebrew's sister-language) means "fish." And the logogram or word-picture for the city Ninua (or Ninwe in its Hebrew pronunciation) is the frame of the conventional sign for place or dwelling and, enclosed within that sign, the sign for a fish! We can only surmise about our narrator's familiarity with the language or script of Assyria or about the likelihood that some, at least, in the narrator's audience may have shared such knowledge. The known and close connections between Assyria and Israel (the Northern Kingdom had surely been conquered by Assyria at the time of our book's composition) incline me to believe that the spoken pun in itself indicates that the association of the fish story with the "fish city" is neither coincidence nor accident. Jonah's three-day descent in the fish, the consequence of attempting to escape a three-day descent into the belly of the fish city, culminates in a climactic irony: the latter required only a third of that time.

I have expressed dissatisfaction with the category of *Leitwort* when presented in itself as an adequate explanation for a word's **repetition** in a literary text. The foregoing examples of **wordplay** in the Book of Jonah will serve to illustrate my contention that the biblical author's indulgence in whimsy or (even broad) humor (see also note 17) is never a matter of sheer whimsy engaged in as an authorial self-indulgence. In the case of the Book of Jonah, the playful humanization of animals and anthropomorphization of natural phenomena is part of the strategy whereby the author spins a lampoon and signals this intent even while concealing till the end the object or butt of the lampoon. For example, the (sly) invocation of Jonah's pity for a plant's brief life is not of a different order of humor or rhetorical purpose than the cartoonlike picturing of a ruminant's hoof as though it had the prehensility of a monkey's foot or the endowing of a ship with a face that can show the thought in her mind, made explicit, as it were, in the dialogue bubble "I'm a goner."

Although considerations of time and space preclude a detailed treatment of Numbers 22, one of the most fascinating of the tales of prophets, it is of importance to note that even within the Pentateuch there is room for whimsy, fantasy, and humor: a pagan prophet who acknowledges YHWH by name and a God who addresses his non-Israelite spokesman via a numen intermediary (*'elōhīm*) and an angel and yet resorts to empowering an ass with speech in order to teach a moral. Incidentally, all the discrepancies in the narrative will yield to an application of the synoptic/resumptive analytic tool.

Jonah's Anticipation of God's Forgiveness

The narrator of the Book of Jonah moves from a laconic account in 1:3 of Jonah's refusal to accept YHWH's charge to preach in Nineveh to an equally laconic description of his acceptance of that charge in 3:3 . While a reluctance to undergo another spell in a piscine host is obvious as the motive for this obedience, it leaves unexplained the motive for the disobedience to the initial call. The motive for this remains a mystery until Jonah himself clears it up for us in 4:2. And 1 remarked that Jonah's certainty that God would accept the contrition of Jonah's targeted audience requires an assumption on Jonah's part that his preachment would inspire both credence and faith. As contrary as this assumption is to the actual experience of God's prophets, the absurdity of it in Jonah's case is supported not by a different biographical experience unique to him but rather on the basis of Jonah's rooting his prescience in a canonical text, one pronounced by YHWH himself.

This text I reserved for discussion here to avoid an interruption of the Jonah narrative. Before we address this text we must examine a related text, which, like this one, deals with the question of how YHWH acts in terms of meting out or withholding of punishment. And neither of these texts can be fully appreciated without the seminal insight of Jochanan Muffs into the meaning of two antonymic idioms with God as subject: *nś' 'āwōn* and *pqd 'āwōn*.[25]

The prohibitions, in the Decalogue, of worship of other gods or the employment of images in (YHWH's) worship is followed by this expansion:

> For I, YHWH, your God,
>   am a Deity most exacting [punctilious/scrupulous],
> *Calling to account* for ancestral iniquity
>   the third and fourth generation of those who reject me;
> But graciously extending benevolence (*ḥesed*) even to the thousandth
>   generation of those who, paying me homage, observe my commands.
>   (Exod. 20:5–6)

Thus does the oft-supposed "jealous God of the Old Testament" contrast the shortness of his memory when it comes to punishment and its lengthiness when it comes to reward. The biblical writers knew from experience, as we know, that the offenses of one generation have inevitable consequences for later generations. Hence the *calling to account* of grandchildren for the crimes of their forebears. The Hebrew *nś' 'wn/ḥṭ'* is normally, and wrongly, translated as "to forgive sin," whereas the correct sense is that God forbears to act (carries the debit item on the books, refrains from foreclosure).[26] As we shall see in the following chapter, Moses asks God *to bear the offense* of the idolatrous Israelites, that is, not to write finis to their career. God accepts this plea but adds the warning, "At the time that I settle accounts (*ubᵉyōm poqdī*), I shall call them to account for this sin of theirs" (Exod. 32:34). By analogy with the phenomenon of radioactivity, good deeds have an extremely long half-life, while evil deeds have a very short half-life. If nothing is added to the latter, they die out in a century or less. But when, over a period of three or four generations, offenses build up to a critical mass, the generation that (to mix the metaphor) adds the last straw suffers the cataclysmic explosion. We may now turn to the text that our narrator has Jonah citing as source or confirmation of his implicit charge that YHWH is not as punctilious in exacting retribution as he should be.

> YHWH—YHWH is a Deity compassionate and gracious,
> Slow to anger [that is, long-suffering, patient],

Abundant in graciousness and truth,
Persevering in graciousness to the thousandth generation,
Carrying on the books iniquity, crime, sin—
But the slate he wipes not clean:
He calls to account for ancestral sins
Children and grandchildren in the third generation
And the fourth. (Exod. 34:6–7)

In this text, it is YHWH himself speaking, proclaiming his attributes as he expounds, so to speak, the name YHWH. In both this proclamation and in Exodus 20:5–6 there is the same asymmetrical balance of reward and punishment, yet the order is different. In Exodus 20, the pretext is admonitory so the punitive aspect comes first; the compensatory aspect of grace and reward, not crucial in this context, is underscored both by the length of God's memory of merit and by its being there at all. In the Exodus 34 pericope, the pretext is God's promise to parade his benevolence before Moses (Exod. 34:19), so it begins with that attribute; the balancing aspect of judgment and punishment is not crucial in this context, yet it is there and thereby underscored.

It is this second text that the narrator has Jonah citing, yet with some differences that manage to express what is underscored in the first. First, he alerts the reader by reversing the well-known citation of the attributes *compassionate and gracious*. This is a minor switch, for the two adjectives are complementary, a hendiadys meaning "compassionately gracious" or "graciously compassionate," whatever their order.[27] But this minor switch (after all, if one quotes, why not quote precisely?) signals a more important departure: Jonah breaks off the citation, leaves out the word *truth*, fails to cite God's rewarding of grandchildren for their ancestors' merit (for what merit would have accrued to Nineveh from her Assyrian forebears?), and totally ignores the concluding note of retribution! No—he does not, to be exact, ignore it: he substitutes for it its opposite! His conclusion is ". . . and relenting when it comes to punishment."

Poetical Review

Repetition (words). As against the Buber–Rosenzweig view of the function of verbatim repetition and in respect to reappearances of key words we have here five different functions of such repetition:

1. the repetitions of *big* and *bad* in artful narrative simplicity, suggesting a childlike narrator or a narrator addressing a childish audience;
2. the use of *hurl* to convey
   a. the power of YHWH,
   b. the vain reaction of the seamen to what that power produces,
   c. the awareness of Jonah of what is the only proper reaction to that power, and
   d. the execution of this appropriate reaction;

3. the repetitions of the unusual verb for "commission" to highlight the sovereign Deity's use of fauna, flora, insects, and meteorological phenomena as agents of his will—in implicit contrasts to the human agent whom he could do without and who considers himself indispensable;

4. the repetitions of the verb *to descend* to suggest the moral falling-off of Jonah in flight, the metaphorical equivalencies of (insensible slumber in) the hold of the ship and analogies to it (the fish's cavernous belly, the belly of the Underworld and—implicitly—the iniquitous fish city);

5. this last analogizing borne out in the three-days sojourn in the fish's belly and the three-day task of traversing the avenues of the fish city.

Repetition (incident). Two instances of the **synoptic/resumptive** technique: 3:4–5/3:6–9 and 3:10–4:4/4:5–11. These narrative strategies to be compared and contrasted with the parenthetic flashback **treatment of time** in 1:10 and 1:4b.

Dialogue. Never a matter of happenstance: 1:6–12, 14; 2:3–10; 3:1, 4b, 7a–9; 4:2b–3, 5, 9–11. In addition to these the **free direct discourse** in 2:1 and 3:2 and the free indirect discourse in 4:8a followed by the explanatory gloss in direct discourse, verse 8. The proportion of dialogue **(showing)** to narrative **(telling)** is almost fifty–fifty in a narrative consisting of some seven hundred words, a remarkable achievement. It is in dialogue that we are shown the irony of the pagans' trust in gods/God in contrast with Jonah's lack of confidence in him and their capability to do penance vis-à-vis his self-centered blindness to fault within himself. The very contents of Jonah's psalm (dialogue) is eloquent for what he fails to express: there is in his "prayer" no sense of any wrongdoing on his part, hence no penitence, and—except for the vague allusion to those so foolish as not to acknowledge the One God—almost no sense that there is anyone but himself and God in the universe. Were we told this rather than shown it by the prophet's own words, we should find it difficult to credit such solipsism.

Metaliterary conventions. Metaliterary conventions whose appositeness to the thought-world of biblical **ancient literature** was discussed in this essay included recourse to oracles for fixing blame for visitations from on high, navigability without a compass, the limitation of deities' power to geographical areas, and the limitation of sacrifices to YHWH to his sacred land or city. The particular metaliterary convention of the ancient's **naïveté versus modern sophistication** is a metaliterary convention affecting us in our judgment as to the author's capacity for personification and **metaphoric expressions** in such matters as attributing consciousness to a ship; a fish belly as submarine lodging; and animals in anthropomorphic and anthropopathic roles in dress, moral (or immoral) behavior, repentance, and atonement.

Characters. One would not expect round, full-fledged characters in a parable or fable, where the main personae are, virtually by definition, stock characters or types. Where (as in this book) the parable is a parody, this is all the more so. Particularly apposite to Berlin's category of *agents* are the personae, in this parody, of ship, fish, flocks, and herds, broad-leafed plant, worm-pest, *hamsin,* and gale.

Genre. This book, in its exploitation of a profession and a supposedly historical exemplar of that profession (see the next paragraph) as central character sharing center stage with the invisible God, is in itself an argument that the narratives of

Scripture, blending so many elements of narrative form, require a nomenclature of their own.

History and fiction. It is more of a general assumption than a consensus of the scholarship that views the Book of Jonah as a parable that the author of this book chose the prophet named in 2 Kings 14:25 for his protagonist. It is more likely, I believe, that verse 25b was inserted in 2 Kings 14 to lend a surface verisimilitude to the protagonist in the parable.[28] The putative historical prophet relates not at all to Assyria, only to Syrian territory restored to the Northern Kingdom before the tide of Assyrian imperialism began to reach for Israel's borders; while the Jonah of our story is pointedly focused on Jerusalem, and the Nineveh of his time is pictured as a legendary city–state with its own king, in no way tied to a larger national polity.

Poetic integrity and genetic provenance. The resolution of every poetical problem in the Book of Jonah on the basis of principles discussed in chapter 1 should lead even a Brevard Childs to reconsider the weight of the heritage of documentary provenance and redaction that led him to discuss this book in terms of its "canonical" development, context, and shape. (See also the Preface and pp. 11, 449, n. 13; and 451, n. 17).

4

The Worship of the
Golden Calf: Exemplar
of Biblical Idolatry

The reading of chapters 32–34 of the Book of Exodus that follows eventuates in the conclusion that this narrative constitutes a philosophical fable or myth—one that enlightens us as to the Bible's own understanding of idolatry and "iconoplasm" in the context of worship. Inasmuch as fable or myth are not categories of history, they must—if only by default—bespeak fiction. Most scholars read this story as history, real or intended. And it behooves us to note that a certain cluster of antonymous terms—descriptive and valuational—attend these poles of literary dichotomy.

History is commonly understood as a narrative of events that have actually taken place. The stuff of history is facts, that is, events to which we may apply the attributes of existence or of reality. Fiction, by contrast, is a term for a narrative whose events are inventions, imaginative fashionings. The events of fictional narrative are therefore denied the attributes of existence or reality. Fiction, indeed, is the most common antonym for fact. This is why religionists uphold the historicity of biblical narrative and generally bridle at the suggestion that such narrative may be fiction. Interestingly enough, adversaries of such religionists who themselves regard biblical narrative as being fiction for the most part are prone to assume that the authors of those narratives regarded themselves as presenting fact, not fiction.

A characteristic (perhaps a peculiar characteristic) of the modern mature (or adult) mind is its disposition to distinguish between the objective and the subjective, assigning the attribute of truth or reality to the former and denying it to the latter altogether or conceding to it a lesser degree of (whatever it is we mean by) the true or the real. Things are, or happen, objectively. The things that are or that happen are experienced, needless to say, subjectively. The experience of things or happenings external to us is somehow a warrant of their existence, their reality, their truth. But purely internal experiences (say, of dreams or the wakeful imagination) that are independent of external reality, private, not public, and hence publicly unverifiable are not really real (really true, truly real).

These habits of mind, assigning truth, reality, and objectivity to the factual (that is, history, or—more exactly—accurate historiography) and untruth, unreality, and subjectivity to the fictive are as ubiquitous as they are nonsensical. There are different orders of truth, reality, and subjectivity, as there are different orders of propositions, statements, and facts. A literary fact is as true and real and objective as is a physical fact. And a promise, for all that its fulfillment may yet lie in the future, may be as true and real as an asseveration about a past action. But the ubiquitous may not be ignored; and however much I could wish the protestation were unnecessary, I must avow my conviction that the theological lesson conveyed in Exodus 32–34 renders to this composition an order of value surpassing the value it would have if it were indeed a record of actual events.

And this brings us to reflect on another often misleading division of academic enterprise. The Greek word *onta,* "the things that exist," has given rise to the philosophic category *ontology,* "the science of being or reality." Another Greek word, *axios,* "worthy, valuable," has similarly eventuated in another category *axiology,* "the science of values." Whatever the defensibility of keeping these two as separate and independent categories in technical philosophy, it is absurd in the world of normal everyday discourse. For the very words *real* and *true* are inescapably terms of value, that is, human judgments on the worthiness of the things or occurrences that are supposedly the *onta.*

Language, the medium for communication of human experience, reveals our essential incapacity to divorce our sense of what is from some attachment of value. For example, the word *thing* applies both to an object, a fact, or artifact on the one hand (all things that are) and to an idea, a thought, or a project on the other (which may be or may yet be but are not yet, except in the mind). When a number of things are brought together in a span of time and related in some kind of activity, we have a phenomenon for which a number of synonyms are available: *occurrence, incident, happening, event, episode.* Synonyms, however, are almost never absolute, which is to say that they have not the same, but rather a similar, essential meaning. And in the case of our series of synonyms it is clear that what distinguishes them from one another is the degree of value, importance, or significance inhering in them. Closest to being neutral in value is *occurrence,* a something taking place. Greater import attaches to *incident.* An incident is an occurrence worthy of only brief remark, because its significance is slight in the larger context. A *happening,* until the last decade or so, was indistinguishable from an occurrence. Now it is used for a planned activity, one worth attending because it promises edification or delectation. How far it has been promoted from its older senses of happenstance or circumstance! It has indeed achieved the value of an *event.* The value inhering in this last word can be gauged by considering how even a minor event overshadows a major incident.

The focus of our interest in this essay is the word *episode.* Like *event,* it may be minor or major in context. A single act of adultery may be just an episode in the life of a faithful spouse; one of many events in a war may turn out to have been the fateful episode. Such uses of *episode* illustrate the third meaning given this word in *Webster's Collegiate Dictionary.* "A set of events that stand out or apart from others as of a particular moment." What distinguishes this word from its synonyms is its origin in artistic context. It is a term borrowed from classical Greek drama, where it is a label for the part of a tragedy between two choric songs. *Episode,* like *chapter,* is essentially a literary term, marking off a section of a narrative—hence Webster's second definition: "A separate but not unrelated incident introduced in narration, for variety or artistic effect; also, a similar digression in a musical composition."

History and fiction have this in common: both are narratives, both feature a selection of characters who play more or less significant roles in a sequence of events. Where the two differ is in the arrangement of events. In history, the events are arranged in chronological order. This is because history has causality (or the revelation of it) as a prime (perhaps *the* prime) purpose of its enterprise; and an event cannot precede either an event that caused it or a character trait that led up to it. The fictive enterprise, by contrast, may have a wide range of objectives, of which causality may be the least or even altogether absent. And the wide range of fiction's objectives allows for narrative strategies.

For example, the novel normally features a plot, which is developed in succeeding chapters. In a melodramatic novel, plot is central and characterization minimal. In a psychological novel, plot is subservient to characterization. A philosophical novel may advance (or deny) an underlying thesis, such as, *Character is fate.* The modern short story, by contrast, may dispense with plot altogether; it may constitute an isolated episode, altogether without incident, evocative merely of a mood or a change of mood. Time flow, the sequence of events, is the hallmark of the early novel. In the modern short story time flow may, even when events are featured, be of little or no importance. And with the increasing incorporation of short story techniques in novel writing, time flow may be deliberately blurred, distorted, or involuted, just as, in William Faulkner's *As I Lay Dying,* the narrator (not, be it noted, the author!) changes identities, flitting or shading like a dybbuk from one host mind to another; so that the sequence in time of the narrators or of their memories is (in our idiom for irrelevance) neither here nor there.

The foregoing is by way of introduction to my interpretation of that celebrated narrative in the Book of Exodus, the worship of the golden calf and its aftermath. Even in the most accurate and felicitous of translations the narrative seems to make little sense, both as to form and as to content. In terms of content, supposedly responsible adults behave like idiots; one hero is cast as villainous and escapes scot-free, other villains are exterminated only to leave associate criminals in the dock; these surviving accused are alternately punished, acquitted, punished again, and again acquitted. In terms of form, time flow is involuted at best; dialogues are repetitious and contradictory; and a drama—which for all its confusion reaches peaks of power—is interrupted by a self-contradictory notice of cult procedure and dribbles off into a finale which is a masterpiece of nonsequitur.

Little wonder, then, that modern scholars have discerned the hands of two or three different authors, fragments of whose creativity have been pieced together by an editor determined on leaving to posterity a single whole, constructed out of the pieces remaining from different jigsaw puzzles.[1]

It is my contention in this essay that chapters 32–34 of Exodus make up a carefully crafted narrative in the service of a single theme, that every discrepancy is deliberate, that a single author made use of an episodic narrative technique to weave a tapestrylike presentation of a theological principle. And in the process of interpreting,

I shall expose a deployment of sophisticated literary devices that will seem to have resurfaced in modern literary art after floating, barely submerged and unrecognized, for some two-and-a-half millennia.

The Story of the Calf Worship and Aftermath: Translation and Exegesis

Episode A

1. When the people realized how long overdue Moses was in coming down from the mountain, the people ganged up on Aaron. They said to him, "Come, make us a god, one that will go ahead of us. For that Moses—the Man who led us up from the land of Egypt—we know not what has become of him." 2. Aaron said to them, "Snap off the gold rings that are on the ears of your wives, your sons, and your daughters, and bring them to me." 3. All the people snapped off the gold rings that were on their ears and brought them to Aaron. 4. He took [this] from them and engraved it with a stylus. Thus he made it into a casting of a bull. They then exclaimed, "This now is your God, O Israel, which led you up from the land of Egypt. 5. When Aaron saw, he set up an altar before it. Aaron proclaimed, "Tomorrow—a feast to YHWH!" 6. Promptly on the morrow they offered up burnt offerings, presented sacrifices of well-being. They sat down to dine and wine; then they arose to make merry. (Exod. 32:1–6)

The Bible everywhere and tirelessly proscribes idolatry. Just what did the biblical writers understand by this term? A thoroughgoing address to this question must wait upon analysis of this narrative in all its tortuous complexities. For the present let us note that the people's demand of Aaron is that he make them an indeterminate god (Hebrew *'elōhīm*), the necessity of which is somehow associated with the disappearance of Moses, the man.[2] I have capitalized this last word in my translation to convey the force of the Hebrew word *'īsh*. Here, as in other contexts, it distinguishes neither male as opposed to female nor the one as opposed to the many but rather illustrious or paramount status. The function of this manufactured god is clear: it is to go in the lead of the Israelite host. But its form is unspecified. Neither the people who make the requisition nor Aaron, who complies, seem to have any idea (if they did, we are not told) of the shape that this god is to take. And the silence of the text on this point is only less baffling than the silence of Aaron on receiving the request. He makes no attempt to dissuade or remonstrate. He does not appear to remember—he certainly does not remind his petitioners or challengers—that the second command of the Decalogue proscribes the making of images, be they cast or sculpted, for use in worship. A single hint of reproof may be present in the repeated notice that the gold for the idol is to come from rings which adorn not the fingers, wrists, necks, or noses but the ears—the oft-attested symbol for obedience.

The production of the god is shrouded in mystery. The account of its making is— in six words in the Hebrew—both specific and elliptical. Literally rendered, verse 4 reads, "He took from their hand, engraved it with a stylus and made it a cast bull."[3] What is "it"? The verb *took* has no object and the object of *engraved* is clearly the gold in some form but not in the original form of earrings. We are left to infer that the rings have been melted down into a single ingot. What did Aaron engrave on the gold ingot? A word? A phrase? An image? We are not told. And what was the point of any engraving if the ingot was then to be melted down and poured into the mold of a bull?[4]

The response of the people to the completion of the bull is perplexing on two counts: (1) Are we really asked to believe that mature adults would hail as their liberator from Egypt a man-made image that had not come into existence until that very moment? (2) Would they have greeted the newly produced artifact with such confidence in its authenticity if Aaron had poured the molten gold into the mold, say, of an ass or an ape? Aaron's reaction is equally problematic. He proclaims the appearance of the bull as an occasion for a festival to YHWH, the invisible God of Israel, and this "when he saw." Saw what? The flawlessness of the cast image whose mold he had himself designed?

Episode B

> 7. YHWH addressed Moses, "Leave! Go down! How corrupt the behavior of this people of yours whom you led up from the land of Egypt! 8.They have lost no time in departing from the path I charged them with. They have made themselves a cast bull, payed it worship, offered sacrifices to it, proclaiming, 'This is your god, O Israel, who led you up from the land of Egypt!'" (Exod. 32:7–8)

This two-verse episode is limited to YHWH's address to Moses. Like the people in Episode A who first identified Moses as the man who led them up from Egypt and then credited with the very same role the hitherto-nonexistent bull image, YHWH assigns the historic role of leadership to Moses and then notes that the idolaters assign the credit that role to the bull. Unlike Aaron, he censures the manufacture of the bull as contrary to a prohibition that he has made explicit to Israel. As in Episode A, there is no condemnation of Aaron for his central role in the incident. Not only this, Aaron's role is totally ignored. Here it is not Aaron but the people who have made themselves a cast bull. And whereas Aaron in Episode A proclaimed that the building of the altar before the bull was for the purpose of celebrating a feast to YHWH, and the sacrifices offered upon it are not characterized in that episode as offerings to the idol, YHWH declares, that it is the idol which is worshiped and that it is to the idol that the sacrifices are made.

The points of agreement in the two episodes require no comment. Congruity is what we expect of an author. What about the discrepancies? Unless we are to challenge the author's competence, we must credit these, too, to his intention and search out what lies behind that intention. That intention is made clear if we compare the two episodes in terms of point of view. In Episode A, the narrator fades from our consciousness as he lets the facts speak for themselves and a judgmental note—if it is present at all—is subtly placed by him in the mouth of Aaron. Aaron says and does what he says and does, as do the people. In Episode B, we are given YHWH's interpretation of the acts and his judgment upon them. The act is perverse and in direct contravention of YHWH's earlier command. The responsibility for the act is charged to the people who commissioned it and not to the agent who effected it. The discrepancies between Episode B (a monologue) and Episode C (a dialogue) are more troublesome—so troublesome, indeed, that modern scholarship has assigned it to a second author.

Episode C

> 9.YHWH said to Moses, "I have reached a conclusion about that people. It is a people incorrigibly stiff-necked. 10.Now, then, stand aside, so that my anger may blaze against them and I annihilate them. But I will make of you a great nation." 11.Moses implored his God, YHWH, "How, O YHWH, can you let your anger blaze against your people—whom you delivered from the land of Egypt with such great force and mighty hand! 12.How let the Egyptians say, 'With malevolence did he deliver them—to kill them off in the mountains, to annihilate them from the face of the earth!' Turn away from your blazing anger, relent of this disastrous purpose against your people. 13.Remember for the sake of your servants Abraham, Isaac, and Israel how you swore to them by your Self, promising them, 'I will make your offspring as numerous as the stars of heaven, and all this land which I have designated will I give to your offspring to possess forever.'" 14.And YHWH did relent of the disastrous purpose he had proposed to wreak upon his people. (Exod. 32:9–14)

The discrepancy between this episode and the preceding one is blatant. In Episode B, YHWH's characterization of the Israelite offense is a pronouncement of his displeasure. But for all this note of judgment, there is no pronouncement of sentence. To the contrary, the bottom line of this episode appears in the first two words of YHWH's command to Moses: "Go down" (to your people at the base of the mount) implies that some remedial action may yet be taken. In clear contradiction to this is YHWH's opening of Episode C with his regretful conclusion that the people is "stiff-necked"—obstinate, incorrigible—and that his only recourse is to write finis to its career and to make a fresh start with Moses and his descendants.[5] Moses' reply to this proposal elicits no speech from YHWH. Instead, we are given the bottom line of this episode, YHWH's renunciation of his proposal, which is altogether congruent with the bottom line of Episode B. What we have before us here, then, is the first instance in this narrative of the synoptic/resumptive technique. A single incident is told in two episodic versions. The first episode (Episode B) is brief and synoptic. It relates the circumstance and the denouement. The second episode is resumptive and expansive; it doubles back on its tracks—goes back in time—and tells how the bottom line was arrived at.

Another feature of biblical narrative technique is to be noted in YHWH's address to Moses: free direct discourse. Words that are explicitly marked off as a quote—as a direct address—may represent what the speaker had in mind rather than what he actually said. This technique allows the author to intrude himself in the dialogue and thereby subtly and ever-so-concisely alert the reader to a key element in the story. The words we translated "stand aside," (more literally, "let me be") are pregnant with meaning. God does not require the permission of his servant to work his will. These words anticipate, and alert the reader to, a critical aspect of the prophet's role. Even as the prophet represents God to man, so must he represent man to God. God tells his servant what it is that his people deserves, so that his servant may fulfill his role as intercessor. This Moses does, convincing God (as it were) with a three-point argument. (The qualification *as it were* is to remind ourselves that the entire story is told to us, not to God; that it is not God who needs convincing but we to whom the message is addressed; and that message is the core of Scripture's teaching: God's relationship to humanity is one of benevolence. He is present in history, he acts with force in the interest of human freedom—and this out of his grace, for man rarely uses his freedom to earn God's beneficence.)

The first point in Moses' argument is that God has already invested considerable energy in liberating the Israelites from Egyptian subjugation.[6] What a waste of his effort if it is now to come to nothing! The second point is that the demonstration of God's benevolence to the Egyptians will be undone, for the Egyptians will not interpret the doom of the Israelites as a consequence of their disobedience in the wilderness; they will read the Israelites' very liberation from Egypt as a sign of Deity's malevolence. And the third point is not that God's promise to the patriarchs would be broken by the punishment he proposes; for Moses, too, is their descendant. The promise to the patriarchs is a subtle reminder that the liberation itself owes to the merit of the ancestors and not of the slave generation in Egypt. That same ancestral merit is now invoked on behalf of the generation that has again demonstrated its unworthiness.

Episode D

> 15.Moses turned and went down from the mountain, in his hand the two tablets of the covenant stipulations, tablets inscribed on both their surfaces, yes, inscribed on one side and the other. 16.(The tablets themselves were of God's making, the handwriting of God incised upon the tablets.) 17.When Joshua heard the sounds of the people's noisemaking, he said to Moses, "There's a sound of battle in the camp!" 18.He replied, "It is not the sound of the victor's cry, nor the sound of the vanquished. The sound I hear I cannot make out."[7] 19.When he came close to the camp, made out the bull and the dancing, Moses' anger flared up. He flung the tablets from his hands, shattering them there at the base of the mountain. 20.He seized the bull they had made, consigned it to fire, then ground it to powder, sprinkling this on some water which he forced the Israelites to drink. (Exod. 32:15–20)

In chapter 24 of Exodus, of which our narrative is a continuation, we are informed (in episodic narrative form) that Moses had transmitted the Decalogue and supplementary divine commands to the Israelites orally. The people accept these unanimously and a formal covenant ceremony between Israel and God ensues. When Moses reascends the mountain peak to receive the stone tablets, the divinely fashioned symbol of the covenant, he is accompanied by his attendant Joshua, who is left behind at some distance from the peak.

Episode D, harking back to chapter 24, has Moses, upon the conclusion of his audience with God, rejoining Joshua to descend with him to the camp. The episodic technique makes it possible for the author to proceed with the story line from Episode A, as if Episodes B and C had not intervened.[8] The lessons of these two episodes—in regard to the attributes of God, Moses, and the people—have been impressed upon us by the author's rhetorical skill. The story line can continue now in a more naturalistic vein, as though Moses makes his descent without any foreknowledge of what has been going on down there.

The first awareness that something unusual is taking place in the camp is Joshua's dim reception of the riotous revelry emanating from it. Like the first sounds from a spectator-packed stadium, the trumpeting of voices and instruments may be witness of athletes vying in sport or of gladiators dueling to the death. Joshua's first guess is the latter. Moses says *no*—the sounds may be ambiguous, but they are not those of war. At first sight of the idol and the dancing about it, Moses reacts with anger and despair. The anger is proportional to the shock of first discovery and directed against the revelers. The tablets, the handiwork of God himself, their words readable from any direction and their truth guaranteed by God's own script, symbol of sovereign YHWH's gracious pact with his servant people, are flung from Moses' hands in despair. "At the base of the mountain." Where else? Not geography is intended here, rather moral responsibility. Israel has shattered the covenant so newly entered with unanimous acclaim. Verse 20 concludes the episode with a synoptic summation of what Moses went on to do. He reduces the bull to formless gold; pulverizes that into gold dust; and, adding it to water, administers the potion to the people. Why? To what end? The answer must lie in the resumptive-expansive episodes that follow.

Episode E

> 21.Moses said to Aaron, "What was it that this people did to you that you brought upon them so great an offense?" 22.Aaron said, "Let my lord not be so intemperate in rage. You know this people from your own experience—that it is determined on evil-doing. 23.They said to me, 'Make us a god, one that will go ahead of us. For that Moses—the Man who led us up from the land of Egypt—we know not what has become of him.' 24.So I said to them, 'Anyone having gold, snap it off.' They gave it to me, I flung it into the fire—and out came that bull." (Exod. 32:21–24)

It is clear that Episode E opens at a point in time anterior to the destruction of the bull image by Moses. In his answer to Moses' question Aaron's use of the pronominal adjective *this/that* is tantamount to his pointing a finger at it. (Normally the pronoun *zē* is rendered as "this." When it is rendered as "that," it reflects a sense of scorn, the speaker distancing himself from the object, as in "that man in the White House.") Two features in Moses' question require comment. One is the phrase literally rendered, "brought upon them so great an offense." It is good Hebrew but unfelicitous English. For in our English expression it is the perpetrator who brings the offense upon himself. The Hebrew idiom derives from a regular **metonymic** (part for the whole, cause for effect) use of the terms for crime, sin, and guilt to stand for their consequences: punishment, expiation, and indemnity, respectively. The second feature is the implication that Aaron, in making the bull, was acting out of malice toward the people: he must have suffered some great wrong at their hands to go along with their proposal, a proposal which he must have known would bring down upon them dire retribution. Such **oblique expression**—the attribution of hostility to a person who acts mistakenly but with no malice prepense is a common form of hyperbole in Scripture. Here, the hyperbole lies chiefly in the implication that Aaron complied with the people's request willingly, not out of compulsion. That this implication is hyperbolic is guaranteed by the absurdity of attributing malice against the people to Aaron; for had Aaron acted other than out of compulsion, he would have brought a full measure of punishment upon himself as well.

Aaron's answer is directed to what both he and Moses know to be hyperbole, hyperbole expressing Moses' intemperate anger which rises out of the heat of provocation. Aaron first addresses Moses in the respectful third person, calling his brother "my lord," suggesting politely that he should cool down. He then switches to the intimacy of second person address, recalls to Moses his own experience with this people—it is hell-bent on self-destruction—perhaps hinting that Moses might not have handled the situation any better. He then quotes verbatim (from Episode A) the formulation of the people's request for a god to go in their lead, yes in the place of the leader Moses, whose failure to return in reasonable time suggests that this leader too must shoulder some small element of responsibility. But his description of his own response to the people's demand is even more terse and perplexing than the puzzling description of it by the narrator in Episode A. He does not plead in his own defense. He neglects to report calling the people's attention to their ears, an implied reproof. He does not mention his enigmatic inscription of the gold ingot or, perhaps, his tying the rings into a bundle. He does not recall that his proclamation of a celebration was to honor YHWH. In keeping with this account in Episode A, he makes no mention of the making of a mold into which the gold was to be formed. No, he says that he flung (the same verb used to describe Moses' hurling of the stone tablets) the gold into the fire; and the result was, not a shapeless blob, but—*mirabile visu*—the bull image now confronting Moses! The response to this incredible tale is silence. Silence gives consent. Moses accepts the story.

And we have no choice but to accept it also, that is, accept Aaron's story as factual; for there were many witnesses who might have given Aaron the lie, and did not. There is complete congruence as between Episode A and Episode E in respect to the initial responsibility of the people and the secondary responsibility of Aaron for the production of the golden image. And this congruence resolves a problem that has baffled generations of exegetes, orthodox religionists, and scientific modernists alike. How account for the central role of Aaron in the sinful production of the bull idol? The Aaron who is to be chosen by YHWH as the first high priest of Israel and the founder of the dynastic priestly line that is to monopolize the sacerdotal office for as long as Jerusalem's sacred Temple endures? (And let us remember, this narrative is the filling of a sandwich. It lies between God's instructions and the execution of his instructions concerning the making of the Tabernacle and the ordination of Aaron and his sons to its priesthood.) The logic of Aaron's role in history, and the logic of this narrative's placement within the account of how he attained that role require that Aaron emerge from this narrative as its hero, or at least blameless. And it is as such that the narrator contrives to present him.

The very artfulness of a literary creation often results in the subversion of its message, for example, the endurance of *Gulliver's Travels* as an adventurous fantasy for children or the attacks made during his lifetime on the same author, the leading Irish clergyman of his generation, for suggesting in *A Modest Proposal* that the English would both benefit the economy of Ireland and enhance the succulence of their own diet by substituting the flesh of Irish babies for suckling pig. In the case of the **ancient literature** before us, the determined insistence on making its author literal-minded has blinded us to his **sophisticated** intent.

The puzzling questions raised by the formulations in Episode A receive their resolution in Episode E. And how much greater the impact on the reader than if his attention had not been piqued in the first episode, or had been piqued only to be immediately rewarded! Now we know why the people were so confident of the bull image's authenticity as a representation of the God who had delivered them from Egypt. Now we can guess what Aaron may have incised on the ingot: " To YHWH"— or something along that line.[9] Now we know what it was that "Aaron saw" of a sudden and that prompted him to proclaim a festival to YHWH. He flung an ingot into the fire, he waited confidently for its reduction to a liquid blob, ready to turn on the image-demanding instigators with this proof of their folly. Instead, a miracle occurred. A miracle which could have been worked only by the will of YHWH!

This last conclusion—so absurd to our mentality—has been a contributing factor to our perception of the story as an enigma. But it would pose neither absurdity nor enigma to the biblical mind. Examples abound in Scripture of an oblique expression that has been labeled the *opus alienum,* the ascription to the working of YHWH of a *strange event,* an event by definition contrary to his will, an event that alienates from him the obedience of his creatures. A single example, chosen for its context of prophecy, will suffice. In Deuteronomy 13, Moses is addressing Israel:

> 2.Should there arise among you a prophet or one who receives dream oracles who presents you with a portentous sign, 3.even if that portentous sign comes to pass (which he predicted in [support of] a proposal to the effect, "Let us follow other gods"—gods of whom you have had no experience—"and pay them homage") 4.pay no heed to the proposal of that prophet or dream oracle recipient. It is only your God, YHWH, putting you to the test, to ascertain whether you are indeed loyal to YHWH, your God, heart and soul. 5.It is only YHWH, your God, you are to follow, him alone you are to revere, his call alone to obey, to him alone be steadfast in homage. 6.As for that prophet or dream oracle receiver, he is to be put to death. . . . (Deut. 13:2–6)

The purport of this passage could hardly be clearer. YHWH is the ultimate author of all truth and reality, of all events and experience. When any call to action proclaimed in the name of Deity is by its nature contradictory to the expressed will of YHWH, such as a call to apostasy, its herald must be put to death for his treason. Even if heaven itself seems to support his call by providing a predicted miracle. As for the obvious questions, "How can Deity collaborate with the false prophet, how permit the miraculous event to take place?"—the answer is, or should be, equally obvious. Miracles cannot override faith. YHWH, indeed, deploys miracles to put our faith to the test. In the case of the golden calf, the people who asked for a "god to lead them" had already failed the test. Aaron had relied on natural law to expose their faithlessness as absurd. And YHWH went one step further. He suspended natural law, performed a miracle to demonstrate the dire consequences of faithlessness, to demonstrate that faithlessness to his will is unreasonable even when reason itself is called into question by the occurrence of a miracle.

Episode F

> 25. Moses saw that the people was deranged, deranged by Aaron's act to the point of helplessness against any who would stand up against them.[10] 26. Moses took his stand in the gateway of the encampment and proclaimed, "Whoever is for YHWH, come to my side!" The Levites in full number rallied to his side. 27. To them he said, "Thus commands YHWH, the God of Israel: 'Belt on, every one, sword on thigh. Cross the camp back and forth from gateway to gateway. And kill, each one of you, your fellow—be he friend or kinsman.'" 28. The Levites went into action, obedient to Moses' charge. There fell of the people that day some three thousand men. 29. [In his charge to the Levites] Moses said, "Dedicate your hands to YHWH—for it is each of you against son and brother—that he may grant you immunity on this fateful day. (Exod. 32:25–29)

Episode F features the retribution for the sin of idolatry. Episode E, as we have seen is the expansion of the interval between Moses' despairing anger at first sight of the bull and his disposition of that golden monstrosity in verse 20, the last notice of Episode D. Thus, Episode F begins with Moses' evaluation of the condition of "the people." This word, *people,* is ambiguous; it can stand for the entire folk, for a council representing the folk, or for any group of leaders arrogating to itself the authority of such a council. In our narrative, the intent is unquestionably the last category, that is, the instigators or the ringleaders. When Moses issues the call for volunteers to stand up for YHWH it is only the Levites who respond: the Levites, a tribal caste from among whom the family of our hero, Aaron, is to emerge with a monopoly of the Temple priesthood. The other Levitical families are also destined for hierarchical privilege; they are destined to assist the Aaronides in their cult service; one of their chief functions will be to stand armed guard over the Temple's sacred property: its housing, furniture, and appurtenances. As Aaron's loyalty to YHWH, expressed in his attempt to thwart the idolaters, earns for his descendants the prerogatives of priesthood, so the Levitical minutemen earn for their descendants a sacred role second only to that of their Aaronide kinsmen.[11]

Two items call for comment. The first is the three thousand ringleaders who fall to the Levitical swordsmen. Who were they? Or, to put the question more precisely, how did the Levites know whom to single out for execution? The clue to the answer lies in chapter 5 of the Book of Numbers. There God ordains for the administration of a potion to be administered by the priest to a woman charged with adultery. The effects of the potions are oracular, they point to the woman's guilt or innocence. If guilty, she suffers a serious and undisguiseable affliction; if innocent, she is immune, suffering no effects from the potion.[12] Now the meaning of Moses' action in verse 20 becomes clear. He reduced the golden bull to a lump, pulverized it, sprinkled the dust upon the water, and administered the potion to "all the Israelites."[13] The presumption is clear that only the ringleaders were affected by the drink. We are left to guess the specific effects on the guilty—yellow jaundice, perhaps?—but their guilt was in some way made clear. And weakened by the ingestion of the golden god whose making they had instigated, they fell easy prey to YHWH's servants.

The second item requiring comment is verse 29. All the standard translations render Moses' reason for calling on the Levites to dedicate themselves to YHWH as describing an action in the past tense, for example, "for each of you has been against son and brother" (JPS) and "because you have turned each against his own son and brother" (NEB)—this, despite the fact that the Hebrew has no verb at all. Hebrew idiom regularly expresses the present–future of the verb *to be* by omitting it altogether. Thus, our translation is the only one faithful to the Hebrew original. The context for the statement, too, confirms this. For the Hebrew words *fill the hand*, an idiom for "dedicate, ordain," makes sense in this context only before the Levites launch themselves on their bloody mission. The bracketed words in our translation solve the problem which has caused translators to distort the original Hebrew. The bottom line of this episode is verse 28, the execution of the instigators. Verse 29, in the synoptic/resumptive technique with which we have now become familiar, fills in a detail of Moses' charge to the Levites. Had the executioners acted on their own recognizance, they would have been guilty not just of murder but of fratricide, a crime calling for capital punishment, to be exacted by the victims' next of kin. In dedicating their hands to YHWH, making themselves the instruments of YHWH, the Levites are guaranteed YHWH's *favor* and *protection*. Both these words are covered by the Hebrew word for "blessing."[14] The sense of favor is surely there, pointing to the preferred position which the Levites will be accorded in YHWH's Temple service. But protection—immunity from the retribution of blood vengeance—is the immediate focus of Moses' intent.[15]

Episode G

30.On the morrow, Moses addressed the people: "You, now, have been guilty of a great sin. I, then, will go up to YHWH—perhaps I may yet propitiate for this sin of yours." 31.Thus it was that Moses went back to YHWH. He said, "By Your leave! This people is guilty of committing a great sin—making themselves a god of gold. 32.Now then, if you will forbear in regard to this sin of theirs—but if not, erase me from the program you have written!" 33.YHWH said to Moses, "Whoever has sinned against me, him only will I erase from my program. 34.Now then, go! Lead the people to the destination which I have told you about—lo, my angel shall go ahead of you—but at that time when I call for an accounting, I shall call them to account for this sin of theirs." 35.[A time did come when] YHWH afflicted the people for making the bull, (the bull) which Aaron produced. (Exod. 32:30–35)

In our discussion of Episode F we anticipated, when we referred to the victims of the Levitical swordsmen as the ringleaders or instigators, this Episode G. For only now does it become clear that there were two classes of offenders constituting "the people." Those executed by the Levites were responsible for the commission of the sin. The rest—save Aaron and the Levites—were guilty of an act of omission. Their nonresistance of the offense rendered them, so to speak, accessories to the fact. As in Episode B where God attributes the making of the bull to the whole people, indicating thereby that he holds them all responsible, so here Moses charges the people with their responsibility for the event. And in confessing their guilt to God he uses the same terms that God used, "making themselves a god of gold." He then harks back to the theme of Episode C, the role of the prophet as intercessor for his sinful people. Indeed he takes up YHWH's threat in Episode C to annihilate the people and make a fresh start with Moses, as though he had not, in that episode, won a reprieve for his people. But in this episode, he does not argue with YHWH to persuade him that he should relent. He merely affirms his solidarity with his sinful people. His tone in pleading with YHWH is respectful, even abject; but it is firm nonetheless: it is all of us or none of us. If you forbear, well and good. If not, count me out; I shall share my people's fate.

The Hebrew term for Moses' plea is *bear the offense,* everywhere rendered "forgive their sin." Hebrew has other terms for forgiveness or pardon. This term, however, does not mean to forgive or to pardon, that is, to wipe the slate clean. It means to withhold punishment, to carry the debit on the books, to refrain from foreclosing.[16] Moses is only asking that his people be granted another chance. YHWH replies first in rejection, then in acceptance. He rejects Moses' formulation: go easy with them or count me out. YHWH's response that he will decide whom to count out is a rather gentle way of telling Moses not to present him with ultimatums. He does, however, accept the substance of Moses' plea. His command of dismissal to Moses makes this explicit. Resume your role as Israel's leader, resume the march toward the promised land. And then he reverts to his rejection of Moses' ultimatum. He reminds us that a reprieve has been asked and granted but not absolution. When Israel sins yet again, he says, providing him with cause to review the record, he will punish them for this offense as well.

The concluding notice of this episode has baffled translators and commentators alike. It seems to be a contradiction of the reprieve that YHWH has promised. But it only seems so if we forget that someone is telling a story. The narrator, at this point, returns (or jumps ahead in time, if you will) to the audience he is addressing. And tells them that at some later point in the ongoing narrative of Israel's history, YHWH was as good as his word. He did—in some unspecified way—punish Israel for their part in the sin of the golden calf. At first blush, one might wonder why this notice was included at all. And why, in telling us that the people were indeed punished, repeat the hyperbolic statement "for making the bull" and then almost pedantically underline the hyperbole by adding a reminder that they did not actually make the bull, it was the bull that Aaron had, so to speak, conjured up. The answer lies in the narrator's consciousness of his audience and their own immediate concerns. The audience are Israelites. They have been taught that God graciously credits ancestral merit to descendants in the thousandth generation; but also that he exacts payment for ancestral sins down to the fourth generation. In this concluding notice of the golden calf narrative, the narrator offers his audience a measure of reassurance. The offense of complicity in that early idolatry was expiated. The threat of punishment for that ancestral sin no longer hangs over them. If the generation listening to the story is ever to suffer for the sin of idolatry, it will be for an idolatry that is their own doing.

Episode G, concluding the narrower event of the golden calf, winds up on a note of retribution and, if our interpretation is valid, on a note of warning by the narrator to his audience. Our own exercise in literary criticism would be seriously deficient if we did not pause to review what our analysis has disclosed: a supreme achievement of narrative art. In one chapter, the biblical author has spun a tale of crime and punishment, sin, retribution, and expiation. He has depicted the sin as springing from anxiety; the varying degrees of guilt in the participants—active and passive; the acts and motives of the loyalist protagonists. He has maintained the suspense of the action throughout, never permitting his audience to forget that their own innocence or guilt, felicity or misfortune, are also at stake. He has unfolded the role of the prophet as the leader betrayed, angry judge, and self-sacrificial intercessor; the role of the priest, loyal to his God, temporizing with his flock, appealing to reason, and subverted in his efforts by the very God whose command he is defending; the attributes of God, alternately gracious in his dealings with an undeserving generation, fiercely exacting in sentence and execution, and again graciously relenting (to give human mortals another chance) and thereby protecting his own investment: the scenario he has authored for the creation he has staged and the actors he has chosen. All this the biblical author has achieved in seven episodes, packed into thirty-five verses— episodes intricately tied together by the use of varicolored threads, appearing and reappearing in separate scenes on two-dimensional tapestry—a tapestry that for all its flatness somehow succeeds in putting flesh on an incorporeal God, portraying man in three-dimensional corporeality, and mapping the protagonists and the events in the elusive fourth dimension of time.

Yet this perhaps-unequaled achievement of art is wrought by an author content to remain anonymous, striving for neither pay nor applause, perhaps even contemptuous of aesthetic values. He is a preacher with one aim in mind: the greater glory of his God. And the chapter we have just concluded is only the introduction of his theme, the proem to his message. Two more chapters remain to his composition, in which actions are minimal, and dialogue almost interminable. He will continue to employ the episodic technique, but his material will not lend itself to the dramatic. With voices predominating over actions, a musical analogy will prove more appropriate to the narrative technique. The development of our author's theme will more resemble the fugal than the tapestry art.

The link between these two sections of the narrative is present in verse 34. In bidding Moses to lead the Israelites toward their next destination, YHWH inserts a parenthetical remark: "My angel [or, messenger] will go ahead of you."

Episode H

1.YHWH addressed Moses, "Move on, you and the people which you led up from Egypt, to the land of which I swore to Abraham, Isaac, and Jacob, 'To your offspring will I give it.' (2.I will send an angel to advance before you and drive out the Canaanites, Amorites, Hittites, Perizzites, Hivites, and Jebusites), 3.a land oozing with milk and honey. That is [to say], I Myself will not travel in your company, lest—you being so stiffnecked a people—I annihilate you in midcourse." 4.When the people heard this ominous word, they reacted with grief. No one donned any finery.

5.[This came about in this way:] YHWH said to Moses, "Say to the Israelites [in My name], 'So stiffnecked a people are you, at any moment—were I to travel in your company—I might annihilate you! For the present, then, doff all finery, while I consider how I am to deal with you.'" 6.So it was that the Israelites went stripped of their finery from Mount Horeb on. (Exod. 33:1–6)

This episode resumes with God's bidding Moses, as in Episode G, to resume the trek toward the promised land. Again he treats Moses and his people as a single identity— "you and the people you led up from Egypt"—and, still addressing Moses (in verse

3), treats him as standing for the entire people: "You are an incorrigible folk." In qualifying the promised land, as promised not to this generation but to the patriarchs, he harks back to Episode C where Moses raised the plea of ancestral merit and thereby stresses that the permission to move on to their goal owes nothing to their own deserts. He again advises Moses (as in Episode G) that he will send his agent to go in the van but this time the role of "the angel" is to clear the promised territory of the populations whom Israel must displace from the land. Yet another reminder that it is by YHWH's undeserved grace, surely not by dint of their own prowess, that the land will become theirs. And in connection with the dispatch of this powerful agent, he underlines the double significance of this agent. YHWH in all his awesome power cannot permit himself to be part of their company. He is giving them another chance, but he is not sanguine about Israel's changing its nature so soon. By keeping a distance between them and himself, he will not be tempted (as it were) to fly off the handle and wreak the doom that they have just now so narrowly escaped. The angel thus represents God's grace (his power is deployed on their behalf) and His lack of confidence that Israel will not again put that grace to the test.

Verse 4 tells of the reaction of the people when they got word of God's intent. They react not to the fact of grace, but to—what? Chagrin that YHWH in all his dread power will not be present among them? Hardly likely! In 20:18–21 we are told that Israel knew enough to fear his proximity. Witnessing the thunder and lightning bolts, the trumpeting sounds and the smoking peak of the mount (emblems of YHWH's presence) they kept their distance, proposed to Moses that he represent them in dialogue with Deity: "You speak to us—we shall obey—but let not God speak to us, lest we perish!" It would appear then that the grief can only be due to God's scolding of them, his expression of no confidence. But is that a reason for not wearing jewelry? And then, after their reaction, comes verse 5 with the apparently repetitious characterization of their stiff-neckedness and the command of God to doff the jewelry that we have just been told they had not put on in the first place!

All this points again to the synoptic/resumptive narrative technique. Verse 4 gives the bottom line of the episode. Verses 5 and 6 go back in time, expand the synopsis, and, telling how the bottom line was arrived at, also explain its meaning. The offense of complicity in idolatry is still on the books. Even a minor sin would bring the atomic pile of God's anger to a critical mass. That is why God must absent himself. A prisoner in the dock does not dress as if for a festivity. God, in forbidding the wearing of finery, is employing a metaphor to remind Israel that it is still standing under judgment. And that judgment, we are told in verse 6, hung over Israel, like a Damoclean sword, from Horeb onward. Until when? Until that unspecified time of punishment and expiation related in the last verse of Chapter 32, the bottom line of Episode G.

Digression A

7.Now Moses would take a certain tent and pitch it for himself beyond the encampment's limit—a good distance beyond. It was called the Tent of Encounter. Anyone with a matter to lay before YHWH would go out to that Tent of Encounter, beyond the camp's limits. 8.When Moses, however, went out to the Tent of

Encounter, the entire people would stand to attention, each at his own tent-opening, their attention fixed on Moses until he arrived at the tent. 9.Once Moses arrived at the tent,[17] the cloud pillar would descend, stand stationary at the tent's entrance, and [he] would speak with Moses. 10.The moment all the people saw the cloud pillar stationed at the tent entrance, the people in its entirety dropped in obeisance, each at his own tent entrance. 11.YHWH would speak to Moses face to face, as one man speaks to another. Then, [the audience over], he would return to the camp—his aide Joshua bin-Nun, however, as duty officer, would not stir from out the tent. (Exod. 33:7–11)

We have labeled this passage a *digression,* a synonym for *episode,* connoting "a deviation, often at the expense of unity of effect, from the main subject of a discourse" (*Webster's Collegiate Dictionary*). Both its content and context have perplexed scholars. Tent of Encounter is a name for the Tabernacle that has yet to be built, the movable shrine whose design and furniture God was dictating to Moses on the mount until the moment of Israel's idolatry. Once this Tent of Meeting is constructed it will be pitched in an area considered outside the encampment proper, although ringed on all four sides by the twelve tribes. Can this passage, as some scholars insist, have reference to that yet-to-be-constructed Tabernacle? Yes, but only if it is by the hand of another author, interpolated here by an editor. But why, assuming such a snippet of tradition existed and found its way to an editor, would that editor choose to insert it into the narrative at this point, interrupting the narrative's flow? The job of an editor, after all, is to improve an author's work, not to garble it.

A number of considerations point to the conclusion that this passage, digression though it is, is an integral part of the story. And the Tent of Encounter here is not the ornate Tabernacle that has yet—as the author well knows—to be constructed. What the narrator is telling us is that prior to Moses' receipt of the instructions for the Tabernacle at Horeb and his execution of them, Moses had to improvise a place where God and man might meet (or *rendezdous,* the basic meaning of the word rendered "Meeting" or "Encounter"). Once God's instructions have been executed, the pitching and dismantling of the Tent of Encounter will be the business of the Levites, and the role of intermediary will be filled by the ordained priests—Aaron and his sons. But up to that point, it was Moses who pitched the tent "for himself"; and it was Moses' lieutenant, Joshua, who did service at the tent between Moses' comings and goings. Joshua was on duty at all times, ready to assist any Israelite who came with an appeal to YHWH. And at such times there was no visible manifestation of YHWH's presence—any more than we witness today at any synagogue, church, or mosque. Joshua served as a lay minister, so to speak; and the individual Israelite would, as we do today, answer for himself whether God had responded and how. In the case of Moses, we are told, when he came to the tent he did so as the representative of all Israel—as indicated by the formal stance of attention and homage that is assumed by the individual Israelites at their own tent entrances. And at such times, YHWH's presence was manifested in the marvelous descent of the cloud pillar. And the response that Moses received from the Presence in the cloud was not the kind of oracle normally received by ordinary people (vague, enigmatic, ambiguous), it was as clear and direct as one receives from a fellow human, with whom one speaks "face to face."

Why does the narrator introduce this digression here? For one thing, he must tell us where Moses received his direct communications from YHWH on other than those two or three rare occasions when he ascended Horeb–Sinai. For example in Episode G, when Moses "goes up" to intercede with YHWH, he does not again ascend the mount (as he will do in an episode-to-come in order to receive a duplicate set of tablets). The narrator is stressing that whether on the mount or at the tent, God's Presence was manifest for all to witness; and the message of God to Moses—as also Moses' retailing of that message to Israel—was clear, and not subject to doubt or question. For another, the digression deals with the issue that has just been raised in Episode H and will be the focus of Episode I: the Presence in person of God amid the Israelites or his representation there by an agent, his "angel." We are informed in this digression that up till this moment in time, YHWH's Presence has not abided among them. He has appeared on the mountain peak and, at its base or elsewhere, only at the entrance of the Tent of Meeting, which was pitched, not among the Israelites, not even at their geographical center, but "outside the encampment—at a good distance."[18]

Episode I

> 12.Moses said to YHWH, "Look now, You say to me, 'Lead this people onward,' but you have not let me know just who it is you are sending with me. You who have said, 'I have singled you out by name, and you have, of a surety, gained my favor!' 13.Now then, if I have, indeed, gained your favor, let me in on your way that I may know you—that I may know that I continue in your favor—and [I plead,] acknowledge that 'that' nation is your people.'
>
> 14.He responded, "Suppose I go myself in the lead—will I satisfy you?" 15.He said to him, "If you do not yourself go in the lead, do not make us stir from here. 16.For how else, say, make it known that I have gained your favor—yes, I and your people—if not by your accompanying us, revealing how distinguished we are, both I and your people, from every other people of the earth?"

We have seen that the concluding verse of Episode G refers to a point in time far ahead of the "present" time of the narrative's action. We saw further that the time frame of Episode H is earlier than that of Episode G's concluding verse; that is, the latter has Israel standing under judgment but not yet punished, while the former jumped ahead to report that at a certain time there was punishment and expiation. Similarly now, we must realize that Episode I is unconcerned with the question of Israel's punishment. Its concern is with the other end of the spectrum, the generous extent of God's grace in his dealings with Israel, symbolized in the issue of his Presence in person among his people. It is Moses who now raises this issue, but in a circumspect way, not head on. In citing YHWH's command to lead the people forward, he harks back to Episode G (32:34) and Episode H (33:1). He then raises the question as to the identity of the "angel" whom God has indicated he will appoint to show the way to Moses (32:34), and to represent him in the people's midst (33:2–3).[19] He implies, further, that YHWH's failure to reveal to him the identity of the angel is somehow not in consonance with YHWH's declaration that he, Moses, has gained YHWH's favor.

While we could wish for more information on the ancient Israelite's view of angelology, three passages do give us part of that picture. In Genesis 32, Judges 6, and Judges 13 we have tales of mortals confronted by a mysterious stranger in human form who only at the tale's conclusion is clearly recognized by the mortals as a numen, a supernatural being, an agent or angel of God. In all three cases, the moment of recognition is attended by an expression of awe, of fear that one may yet die as a result of such contact with divinity or of wonder that one had indeed managed to survive such contact. In Genesis 32:30 and Judges 6:22 this expression of awe features exactly the same words: "I have seen divinity [*a god* or *God,* in Genesis; *an angel of* YHWH, in Judges] face to face." In Judges 13, where *face to face* does not appear, the idea is implicit in the words, "for we have seen divinity." The accounts in Genesis 32 and Judges 13 have another feature in common: the mortal bluntly asks the numen to reveal his name; and the identical response—equivalent to "Don't ask!"— implies that in asking the question the human is overstepping his bounds: "How is it that you venture to ask for my name?" (In Judges 13, the angel adds, "since it is Mysterious.")

These accounts supply us with something of a context for Moses' complaint (is there a note of petulance in it?) that YHWH has withheld the angel's name and for the play on words in citing YHWH's singling out Moses for favor; the Hebrew idiom is, literally, "I have known you by name." These words also supply a bridge between the notice in Digression A that YHWH spoke to Moses "face to face" and Moses' interest in the name of the angel YHWH proposes to send along with him, in Episode I.

But the narrative pattern is different in this episode than in the other passages treating of angels. Moses does not ask for the name of the angel.[20] He does not pause to give YHWH a chance to respond to the complaint. He indicates that his complaint may be laid to rest in another way: you may demonstrate your favor, he says, "that I may know you" (note, *you,* not *your angel*) if you will "show me your way."[21] What is meant by this "way"? The answer becomes clear if we render this word by a synonym, *route,* and read it in context. Moses is asking for a road map! He does not want the name of the angel, because he does not want the angel. The whole purpose of the angel is to serve as guide to Moses. Let us dispense with him altogether, says Moses. Give me a map of your intentions, and I shall be guide enough. The general failure to discern this meaning of Moses' words is due to two factors. The first is the failure to credit the narrator with full capacity for artistic subtlety. YHWH is the ultimate monarch. And Moses, for all his stature, is his subject. It ill behooves the subject to reject outright a proposal by his sovereign that is put forth to show him favor. He therefore expresses his wish in an oblique way or, to put it more honestly, slyly.

The second factor is the failure to grasp YHWH's response, a response in which YHWH amusedly reveals that he, for his part, can dispense with Moses' coy hinting. The response of YHWH is almost universally rendered as a declarative statement, along the lines of (to pick the best of translations) "I will go in the lead and give you rest" or "and lighten your burden."[22] But the issue at hand is not rest nor easing of a burden. It is the putting to rest of Moses' complaint. And it is regrettable that the absence of punctuation marks in the Hebrew have obscured this deliciously playful thrust of YHWH's.

Moses is not abashed. Having been caught out, he might as well hang for a sheep as for a lamb! "You ask if that is what I will settle for? Yes, that and only that. No angel. You yourself—or nothing!" And in explicating his ultimatum—an ultimatum that resonates with his ultimatum in Episode G—he again underlines that what is involved is not merely the question of YHWH's favor to him personally. When he was still being coy in his approach, asking for the favor of a road map, he asked it as a favor to himself but added, ". . . and acknowledge that that nation is your people." Now, again, he emphasizes that there is no separating him from his people: "Your accompanying us in person can alone distinguish us—'me and your people'—from all the rest of mankind."

Episode J

> 17. YHWH said to Moses, "With this thing also which you propose will I comply, [to show] that you have, indeed, gained my favor, that I have singled you out by name." 18. He said, "Make visible to me your Presence!" 19. He said, "I will parade my benevolence before you, herald for you the name YHWH, to wit, who it is to whom I show favor, who it is to whom I show love." 20. [The meaning of what] he said: "You may not see my face, for no human can gaze upon me and survive." 21. YHWH said, "Here, close by me, is a spot—where you are to position yourself—on the crag. 22. Lo, the moment that my Presence goes on parade I shall place you in a cleft of the crag, with my hand block your vision, until I have passed by. 23. Then I shall remove my hand, so that you may see my back. My face, however, may not be seen." (Exod. 33:17–23)

Verse 17 is both the conclusion of Episode I and the beginning of this one, so tightly are the two joined. YHWH's unqualified acceptance of Moses' ultimatum leaves us totally unprepared for Moses' response. At a point where we might have expected "Finis" or, at the most, "Thank You," we get instead a peremptory imperative from the mouth of the mortal. And the jarring tone of Moses' command is matched by the audaciousness of its substance. The Hebrew word *kābōd,* correctly translated Presence here, most often appears in the sense of "honor, glory." In a few cases it has the sense of "body" or "person"; and with the pronominal suffixes of the first, second, and third person, it stands for I/me/myself, you/yourself, he/him/himself, respectively. Moses' abrupt response to God's gracious yielding to his own cautiously developed petition is, "Show yourself to me, now!" And YHWH's response is an unruffled, albeit qualified, compliance.

To appreciate the audacity of Moses and the complaisant response of God one must recall that throughout Scripture YHWH is pictured in terms of fearful fire and incandescent light. Mortals who trespass against him or his sancta are incinerated or electrocuted. Frequently they are smitten by *sanwērīm,* a light so intense that it burns out the optic nerve when it does not cook the brain. Even the heavenly courtiers must be shielded from this power. For example, in the vision of Isaiah (chapter 6), the seraphim (a word connoting blazing creatures) who minister to YHWH in his celestial throne-room have six wings apiece: one pair for flying, one pair to cover their legs (a euphemism for pudenda), and one pair to cover their eyes.

Such is the dazzling Presence that Moses asks to see! What would the narrator have us understand by this request of Moses that Deity should, so to speak, step out of the closet? The possibilities are as unlimited as the power of imagination. One particular possibility cannot, in this context, be excluded. This episode is an expansion of the preceding one. Moses has insisted, and YHWH has agreed to accompany Israel in his own person. Moses now wants to be able to recognize that person. To this desire, too, YHWH—to the extent that his nature and human limitation allow—is amenable. The narrator exploits the imagery of YHWH's deadly brilliance to tell us that Moses was granted a view of God, only a partial view (his nonradiant back) but at that, a fuller view than ever was vouchsafed to mortal being.[23] That part or aspect of Deity is also characterized as his goodness or benevolence, from which we may infer that the aspect denied to Moses was his awesomely destructive or malevolent power. But Moses will not only see, he will also hear. God will proclaim the essence contained in the name YHWH. In the episode that narrates the carrying out of this promise of self-revelation, we shall see that both benevolence and malevolence are constitutive of that essence. But in this episode the proclamation to come is characterized only by the attributes of benevolence: the self-willed display of grace and love.[24]

Episode K

1. YHWH said to Moses, "Carve yourself two stone tablets like the first, and I will inscribe upon the tablets the words that were on the first set, which you shattered. 2. Be ready at dawn. At dawn, ascend Mount Sinai and present yourself to me there, on the mountain peak. 3. No one is to come up with you—indeed, no one is to appear anywhere on the mountain. Even flock and herd are not to graze opposite that mountain face. 4. He carved out two stone tablets like the first. Promptly at dawn, Moses climbed up Mount Sinai in obedience to YHWH's command, two stone tablets in hand.

5. YHWH descended in the cloud, stood by him there and made proclamation of the name YHWH. 6. As YHWH paraded before him, He proclaimed: "YHWH—YHWH is Deity, loving and gracious, slow to anger, abounding in graciousness and truth, 7. extending grace to the thousandth generation, forbearing in regard to iniquity, transgression, and sin. Yet by no means does he wipe the slate clean: for the iniquity of parents he calls to account children and grandchildren down to the third generation or the fourth."

8. Moses quickly touched head to the ground and did obeisance. 9. He said, "If I have truly gained your favor, my Lord, let my Lord, I pray, ever continue in our company—however stiff-necked a people it is—forgiving our iniquities and sins, taking us for your own. (Exod. 34:1–9)

This episode gives the context of YHWH's self-revelation to Moses. It is to take place on Sinai's peak—confirming that the dialogues between God and Moses narrated in the preceding episode took place at the Tent of Meeting described in Digression A. The occasion is YHWH's act of grace, his reaffirmation or reinstitution of his covenant with Israel, symbolized by his inscribing a duplicate set of tablets. (Unlike the first set, which were carved as well as inscribed by YHWH himself, these tablets are carved by Moses—perhaps to suggest how important a role had been played by him in winning YHWH over to a renewal of the covenant.) The idea, incorporated in the proclamation of YHWH's attributes, that he credits ancestral merit endlessly to later generations and calls only great-grandchildren, at the worst, to account for ancestral sins, is, of course, not new here. It appears first in the Decalogue after the first two commandments, in expansion of the prohibition of worshiping other gods along with YHWH or the manufacture and use of icons to represent Deity. In the Decalogue, however, this expansion is essentially negative, for all that its substance is essentially positive. The infinite extension of reward and the limited extension of punishment is unquestionably an affirmation of his grace. In the Decalogue, nonetheless, this affirmation follows a prohibition, the threat precedes the promise, and the single explicit attribute of YHWH is that he is "an exacting Deity" (that is, in punishment). In this episode, by contrast, the promise precedes the threat, the latter included altogether only in order not to distort the full picture of YHWH's action in human affairs. Here, the attribute "an exacting Deity" is notably absent. For it has been replaced by the attributes of compassion or love, graciousness, patient tolerance, abundant and enduring favor. And that favor is characterized not just by his carrying *iniquity* on the books, but *iniquity, transgression, and sin,* that is, offenses of every kind and degree. The formulation here is, then, altogether in consonance with the promise uttered in Episode J, the parading of YHWH's benevolence, a benevolence particularly manifested to the people he now reclaims as his own, the benevolence spelled out in the words, ". . . who it is to whom I show favor, who it is to whom I extend love."

Moses understands the full measure of grace that YHWH has affirmed; his posture of obeisance is in expression of grateful thanks. He has now witnessed the gracious Presence of YHWH, which will now accompany Israel as it resumes its journey to the promised land. But even at this point the narrator gives the prophet the last word, once again appealing for an ever-greater measure of that grace. He has no illusions about the temptations his people will face before and after they enter that land nor about their resoluteness in the face of those temptations. This is what he has in mind when he now characterizes his people as incorrigible—stiff-necked—and asks nonetheless for YHWH's awful Presence among them. Now for the first time he introduces a new word in his entreaty. He asks not for forbearance but for *forgiveness,* which God— like a father—will be able to manage if, as he says to him, "You take us for your very own."

Episode L

10.He [YHWH] said, "Lo, I commit myself in this covenant: with all your people as witness I shall work wonders never before wrought on earth, among any of its nations. And all this people in whose midst you are will witness YHWH's doings, how awesome is the manner of my treating with you. (Exod. 34:10)

This single verse, which we have marked off as a separate episode, is followed by sixteen verses that constitute a second digression, Digression B. To present the detailed comment it deserves would be inadvisable here. We might find it almost impossible to resume the thread of the narrative. Instead, we shall present its salient points.

Whereas YHWH's address in verse 10, in the second person singular, is to Moses, that same second-person-singular address in the digression is to the people of Israel in its collectivity.

Digression B

1. God will see to the defeat of the inhabitants of the promised land. Israel is forbidden to enter into any treaty of accommodation with them, lest they lure them into the snare of their pagan practices.
2. Israel is not to make castings representing Deity.
3. The firstborn male of any animal belongs to YHWH. If of edible animals, the male is to be gifted to the Temple; if it is of an animal that may not be eaten, it must be either redeemed with one that is edible or destroyed; human firstborn males must be redeemed.
4. Only YHWH is to be worshiped. Homage is to be paid him by the observance of the weekly day of cessation from labor and by attendance at his earthly court, sacrificial gifts in hand, on the thrice-yearly pilgrimage festivals.
5. Bread offered together with animal slaughterings is not to be leavened; the meat of the paschal sacrificial meal is to be consumed by the celebrants or by fire before dawn; the choicest firstfruits of the harvests are to be brought to YHWH's court; the flesh of a kid is not to be prepared in milk taken from its own mother. (Exod. 34:11–26; see pp. 119–21 for complete text)

Every one of these details (with two exceptions) appears, often in identical phrasing, in Exodus 23:12–33.[25] The duplication of these catalogues of prescriptions for the proper worship of YHWH—and YHWH alone—is due neither to accident nor to the whimsy of an incompetent editor. One exception is item 2, prohibition of idols, absent in chapter 23 and present in this catalogue, which appears in the context of the golden calf narrative! The second exception is the appearance in this digression of item 3, the dedication of the firstborn males to YHWH. Its appearance here points to the ordination of the Aaronide priests, whose dedication to YHWH's service releases the firstborn of other Israelites from this obligation—upon the payment of a redemption fee. A third link connecting this digression with the narrative context is the expression—in connection with the three yearly pilgrimage obligations—that literally reads, "to see the face of YHWH." No one, to be sure, not even Moses—as Episode J take pains to underline—can literally gaze upon the face of God. But even those words—a metaphor for the privilege of audience with a monarch—were regarded as too blatantly anthropomorphic by the rabbis who transmitted the Hebrew text. To avoid the boldness of an image that, to their minds, bordered on the blasphemous, they vocalized the verb as a passive, yielding the sense "to be seen (by)" or "to appear (before)" YHWH. Little is lost to the reader in most instances where this substitution occurs. In this digression, however, it does serve to conceal the link between the privilege of the pilgrim visit to YHWH's shrine and Moses' beholding the Presence of YHWH. And this last, YHWH's Presence in Israel's midst, in the Tabernacle constructed for his residence, is the fourth link to the golden calf narrative; for, as pointed out earlier, it is between the instructions for its building and the carrying out of those instructions that our entire narrative is set.

This recognition, that chapters 32–34 constitute an episodic narrative within a larger narrative framework, helps us to solve another problem in Episode L. Episode L precedes Digression B. YHWH, addressing Moses, refers to Israel as "this people in whose midst you are" (so in my translation, as in the older standard, more literal translations). More recent translators seem to have felt that the narrator could not have intended such an obviously superfluous phrase on the part of YHWH. For example, the new Jewish Publication Society translation reads, "and all the people—who are with you," with a note, "Lit. 'in whose midst you are.'" NEB fails to translate these words altogether and provides a version of God's words that is without any basis in the Hebrew original.[26]

The clue as to the intent of the literal Hebrew, "this people in whose midst you are," is provided in Exodus 23:20–24, which follow the catalogue of prescriptions that are duplicated in our digression. These verses report YHWH's words:

> 20.Lo, I am sending an agent [or angel] before you to guard you on the way and to bring you to the place I have appointed. 21.Take care in his presence, obey him; do not play the rebel with him, for he may not forbear in regard to any transgression on your part—since my name is in him! 22.If you will only be obedient to him, doing all that I dictate, then shall I be enemy to your enemies and foe to your foes. 23.With my agent [angel] going before you—when he will have brought you to the Amorites, the Hittites, the Perizzites, the Canaanites, the Hivvites, and the Jebusites—then shall I wreak depredation among them. 24.You are not to worship their gods, nor do them service, nor follow their practices. (Exod. 23:20–24)

This passage contains the first reference to YHWH's plans to send an "angel" to lead Israel to the promised land. It is to this reference that YHWH therefore reverts in Episode H when he first explains why he cannot accompany Israel in his own Person. It is this angel whose identity Moses, in Episode I, declares that YHWH has not revealed to him and whose company he respectfully declines. In Episodes J and K Moses is granted a peek at YHWH's Presence that will accompany Moses, obviating the necessity for an angel. Now in Episode L we learn at last who the "angel" was whom YHWH had in mind back in chapter 23, the agent who would lead Israel to the boundary of the promised land but would not himself enter it. That "angel" that YHWH promised Israel, that "angel" in whom YHWH would be present ("since my name is in him") is no angel at all. It is the man, Moses! The wonders that YHWH will work "with all your people as witness"; the "wonders never before wrought on earth"; the wonder of the "awesome . . . manner of my treating with you," Moses, to be witnessed by "all this people in whose midst you are," Moses—is that I, YHWH will be present in you and will thereby be present in the midst of the people "in whose midst you are."

Will we ever again underestimate the craft of an author who can bury the prelude to the narrative of the golden calf in Chapter 23 and then provide a clue to its existence by a seemingly pointless commonalty of context? All to achieve what? A surprise ending? Yes, that too, but primarily (as we shall see) to cloak a breathtaking metaphor, a metaphor expressing God's Presence in man.[27]

Episode M

27. YHWH said to Moses, "Write down these words, for these words are the conditions for the covenant I make with you and with Israel."

28. There he abided with YHWH forty days and forty nights. He ate not a morsel of bread, took not a sip of water. While he [YHWH] inscribed upon the tablets the terms of the covenant—the Ten Commandments.

29. When Moses came down from Mount Sinai—the two tablets of the pact in Moses' hand and Moses unaware that the skin of his face had become radiant in the course of his [YHWH's] speaking with him—30. then, when Aaron and all the Israelites beheld Moses and lo, the skin of his face was radiating light, they shrank from coming near him. 31. But Moses called out to them, so that Aaron and the elect of the assembly came back to him, and Moses addressed them. 32. After this all the rest of the Israelites drew near, and he charged them with all that YHWH had addressed to him on Mount Sinai. 33. When Moses finished his discourse to them, he placed a veil over his face.

34. [And so it was that] whenever Moses entered YHWH's Presence to speak with him, he would put the veil aside—even until he had left the audience. He would then go out and address to Israel what he had been charged with. 35. And the Israelites would behold the face of Moses, how the skin of Moses' face did radiate with light. Only then would Moses put the veil back over his face—until his next entrance to audience with him. (Exod. 34:27–35)

In this, the concluding episode, we are told that the second audience of Moses with YHWH on the mountain equaled the first in duration: forty days and nights (a round number for an indeterminate, but lengthy, stay), during which he consumed neither food nor drink, sustained rather by the grace of God. The catalogue of prescriptions for the proper ways to do homage to YHWH, the contents of Digression B, are the words that Moses is bidden to write down for himself. In all probability, this catalogue itself, characterized as the conditions imposed on Israel for the covenant with God, are only intended as an abstract of the fuller instructions which follow in the rest of Exodus as well as in parts of Leviticus and Numbers—hence the lengthy stay on the mountain. The Decalogue itself, as consistently described elsewhere, was inscribed by YHWH himself.

The concluding notice bears out our interpretation of Episode L. The radiance emanating from the face of Moses is a visible token of the presence of YHWH, the Radiant Presence, in the person of the prophet. The symbolism of the radiance is sharpened by the detail of the veil. The prophet, for all the immanence of Deity within him, continues to be a human and mortal, flesh and blood, living out his own personal concerns and aspirations. In the privacy of his own person and preoccupations, the immanence is yet there, dimmed or veiled. In each audience with his God, the veil is removed and the radiance is renewed or recharged; and the veil remains removed until the people and its leaders can witness the radiance of God's truth even as they hear its words from the mouth of the prophet who represents them to him and him to them.

Recapitulation

The piecemeal presentation of this three-chapter narrative and analysis of its formal structure and story line has been necessitated by a number of factors: that it was composed in an ancient tongue reflecting the conventions of an ancient civilization; that it therefore required a new translation, one often at odds with the many standard and often mistaken, or even obtuse, renderings; and, lastly, that it constitutes but one unit in a much larger narrative in whose context alone it can be comprehended. The reader who would arrive at a full appreciation of this achievement of narrative rhetoric or who would question my analysis of it in whole or in part would be well advised to reread the narrative without the interrupting commentary. He would then see, I believe, that the narrative contains not a superfluous word. Not a word or phrase is repeated except to advance the author's argument. There is not, in these ninety-three verses, a single contradiction or inconsistency, except for those occasioned by the narrator's conscious shift of voice or variation of metaphor. We may review the narrative in outline form.

Episode A. A golden artifact is mysteriously produced and is hailed as a god responsible for a chapter of history that took place before its production. Its projected role is to replace a human leader who never returned from a solitary audience with an invisible Deity on the cloud-enshrouded peak of a mount whose name is Sere or Horeb (Barren).

Episode B. Up there amid the clouds the lost leader, Moses, is informed by his God, YHWH, after forty days and nights of discourse, that he must return to his people camped at the base of the mountain. The reason: they have committed the sin of making an idol, a golden bull, in direct contravention of one of YHWH's commands so recently addressed to them.

Episode C. This bottom line—permission to rejoin his people and resume his leadership role—was not easily arrived at. Moses had first to implore his God not to punish the people with a well-merited annihilation.

Episode D. Moses descends from the peak, picking up his faithful attendant Joshua halfway down the slope. Moses cannot make out the nature of the sounds emanating from the camp. This, and his shock at beholding the idol, suggests that the narrator, in putting mention of the specific nature of the sin in the mouth of YHWH was employing the device of free direct discourse; that is, God told Moses of a heinous sin but did not give the specifics. The reason for the specific is to emphasize for the reader the heinous nature of this particular transgression—"idolatry"—in the eyes of God. Moses in anger and despair shatters the covenant tablets, reduces the idol to gold dust and has the entire people drink water upon which the dust has been sprinkled.

Episode E. Before Moses so treated the idol, he charged his brother Aaron with the responsibility for leading the people into the sinful act. Aaron denied responsibility and claimed that the production of the bull was a miraculous event. This claim of Aaron's, readily confirmable by the many witnesses, goes unquestioned.

Episode F. Moses calls for volunteers to rally to God's cause against the instigators of the idolatry. The Levites answer the call and at the command of Moses, spoken in the name of YHWH, traverse the camp, executing the criminals, numbering some three thousand. Their ability to single out the instigators is implicit; it goes without comment. The narrator permits the thought to dawn upon us that that ability is related to the gold dust potion administered by Moses to all Israel. Before the execution (we are told in conclusion) Moses had the Levites dedicate their hands to YHWH. This precaution we are told is to ensure YHWH's blessing on the Levites, a word pointing explicitly to the favored role this tribe will play in the Temple cult and, explicitly, to the immunity from blood vengeance for their slaying of their kinsmen.

Episode G. Moses returns to YHWH (at a spot undisclosed) to propitiate him and win a stay of sentence for the Israelites who tolerated the idolatry. Harking back to Episode C, YHWH's proposal to destroy Israel and replace it with Moses' descendants, Moses achieves his goal. Moses is given permission to resume the march of Israel to the promised land. He is told that YHWH will supply him with an angelic guide. Stay of sentence, YHWH emphasises, is only a suspended sentence. At some late date, we are informed, Israel will be punished for its role as accessory to idolatry thus expiating this offense.

Episode H. YHWH's reason for sending an angel, rather than himself accompanying Israel on its trek, is given. The reason, a judgment on Israel's incorrigibility, eventuates in Israel's assuming a stance of mourning. This stance is further explicated. Israel is to remain under judgment from Mount Horeb on—until the occasion of expiation anticipated in the concluding notice of Episode G.

Digression A. Inasmuch as we are not here given information on where—at which camping site in the Sinai wilderness—the Tabernacle was constructed (or, for that matter, at which point in time), the last mention of Horeb provides a felicitous point at which the narrator may fill in a number of details whose absence would make the episodic narrative technique even more confusing than it has already proved to generations of interpreters. It tells us how and where Moses had audience with YHWH on other than the two occasions when he climbed Horeb's peak. More important, it contrasts the occasional descent of YHWH in the cloud pillar to speak to Moses at the entrance of his improvised Tent of Encounter with his taking up residence, so to speak, in the Tabernacle Tent of Meeting that is yet to be constructed. And it thereby prepares us further for the issue at stake in the episodes to come: Will YHWH be present in person or in the person of an intermediary angel?

Episode I. YHWH divines Moses' reluctance to settle for an angel in place of his own Presence.

Episode J. YHWH accedes to Moses' request. Moses boldly asks for a view of the Presence that will accompany Israel and himself. YHWH agrees to show him as much of it as a mortal may glimpse. The revelation will take place on the mountain peak, the walls of a rock cleft serving as blinders on either side and YHWH's outstretched palm blocking a frontal view until he has passed by. The back of YHWH, which Moses is to see, symbolizes—if nothing else—a parallel expression of what YHWH will disclose: his attributes of benevolence and love.

Episode K. Moses carves out two stone tablets and ascends the peak, where YHWH will duplicate the words he had inscribed on the first set. He parades, as he has promised, before Moses, proclaiming his benevolence, a benevolence that does not, however, preclude punishment for past sins. Moses gratefully acknowledges this reassurance. He asks, however, that YHWH, accompanying Israel in his own person, show his love not only by granting stays of sentence but by full forgiveness of the people he is reclaiming as his own.

Episode L. YHWH responds with a promise to Moses personally. He will treat Moses in a way unheard of since time began.

Digression B. Then follows a catalogue of prescriptions for the proper worship of YHWH. This digression, identical in large measure with a catalogue in chapter 23 of Exodus, points to the significance of an earlier episode; for this first catalogue is immediately followed by a promise of YHWH to Israel that he will be present among them in the person of an agent, or angel. The reader now can grasp the intent of that first promise, now clarified, in Episode L in the promise to Moses. The agent in whom YHWH's Presence will be immanent is no angel but the prophet Moses.

Episode M. The tablets are inscribed by YHWH. Moses' second stay on the peak equals in duration his first audience on the mount. In the course of forty days he receives additional details of YHWH's covenant stipulations with Israel. YHWH's Presence in Moses is disclosed in a divine radiance illuminating his face, a radiance whose intensity will be renewed in his every subsequent audience with YHWH, a radiance displayed to Moses' Israelite audience every time he relates to them the instruction of YHWH.

The Kerygma

With our analysis of the text completed, we are in a position to address ourselves to the synthetic question, What is the essential message of the narrative? Or (since there may be several messages, all essential) what is its central, or core, meaning? In addressing this question we must raise a number of considerations deriving from a broader acquaintance with the literature of the Ancient Near East and with the biblical books as a whole.

The first of these considerations is the nature of the phenomenon that we label *idolatry*. This term is variously used in senses ranging from the narrow and literal sense to the broadly figurative or metaphoric. Literally construed, idolatry means the worship of man-made images as gods, which is to say that the sculpted or cast form is not for the worshiper merely a representation of superhuman or supernatural powers but rather the very embodiment of those powers. The ascription of such powers to an object or collection of objects is covered, however, more precisely by another, newer term: *fetishism*. If we then consider that fetishism derives from the realm of magic (in which powers are manipulated or coerced), whereas idolatry derives from the domain of religion (where powers are petitioned, or addressed in worship), we shall have to conclude that idolatry in the literal sense (that is, worship of powers that are manipulated) is essentially a self-contradiction. Either an idol is merely a representation of powers that exist beyond it and serves to help the worshiper focus on those powers, or it is a fetish—in which case it is not, properly speaking, worshiped at all.

Voltaire, more than two centuries ago, perceived that no pagan in antiquity, however benighted, would have recognized himself as an idolater in the literal sense. And everything we know about the religions of the Ancient Near East (or anywhere else, for that matter) confirms Voltaire in his perception. Gods—cosmic, regional, tutelary, or whatever—were represented in various forms or shapes, often mounted on pedestals in the shape of one animal or another so that the pedestal image by itself might represent the majesty of the god in question. And though these images were venerated, they were not worshiped. No more than a flag is worshiped, for all that its desecration will be deemed an insult to the national sovereignty it symbolizes. Idolatry, then, is a term rarely (if ever) meaningfully used in the literal sense. And in its metaphoric usage it is not so much a descriptive as it is a value term. Idolatry is a judgment, the attribution to another of the worship of false gods or the pursuit of false values. As today, so then; as then, so today. And the figurative, polemical deployment of this concept is never more clear than when it appears in the hyperbole of lampoon. As for example, when the prophet in Isaiah 44 describes the idolater who fells a tree, uses part of the wood for a fire to warm himself, part to cook his meal, and carves the rest into a shape before which he prostrates himself in worship with the prayer, "Save me, for you are my god!"

Another consideration is the predilection for the use of images in worship (iconoplasm) in some religions and its proscription (iconoclasm) in others. Monotheism need not be iconoclastic, although biblical monotheism was; and polytheism need not be iconoplastic, although pagan religions have been so. It would seem, indeed, that religious iconoclasm is a phenomenon attested only for biblical religion in antiquity and derivatives of that tradition in our own day. And the question that we must ask is *why*. Why is Scripture so adamant in its iconoclasm? What is wrong with rendering the intangible into the concreteness of three-dimensional imagery?

It is these very questions, I would submit, that lie implicitly behind the narrative of the golden calf and its aftermath and that are answered in the strands of its mythopoeic development. The crux of the matter lies in the first two commandments of the Decalogue, the two so closely connected as to be almost indivisible.

The introduction to the Decalogue in Exodus is 20:2, "I, YHWH, am your God, who liberated you from the land of Egypt, the house of bondage." The first command is verse 3, "You shall have no other gods alongside of me." This prohibition does not, in itself, deny the existence of other numina, divine agencies of major or minor order. In contrast with pagan practice, which often portrays a chief god's hospitable relations with kindred deities, often providing them with quarters in his own palace or niches in his own temple, YHWH forbids the association with him of any such divinity. The second command is verse 4, "You shall not make for yourself a sculpted image, or any likeness that exists in the skies above, or on the earth below, or in the waters under the earth." Verse 5, underlining the seriousness of what precedes, applies to both preceding commands, "You shall pay them no homage or service." A point that often fails of appreciation is that the proscription of images is not in reference to representations of other deities; for worship of these and, a fortiori, of representations of them, have already been precluded in the first command. The second command forbids the representation of YHWH by any image. In Deuteronomy 4, when Moses expands on this prohibition—listing the likenesses of male or female, beast, bird, lizard, fish, celestial bodies such as sun, moon, stars—he makes this explicit in verse 15, "Take care for your very lives! For you saw no likeness (of any sort) at the time when YHWH spoke to you at Horeb, from within the flame." In other words, you cannot represent YHWH in any natural likeness—fauna or celestial luminary—for you saw no image when he revealed himself to you.

Now let us review the elements featured in our narrative.

1. Moses as the intermediary between YHWH and Israel;
2. a representation in three-dimensions—its substance gold, its normal manufacture by casting in a mold—which turns out to be a bull image to replace the missing human intermediary, Moses, and the characterization of this image as "a god," clearly representing the Power that brought Israel up from Egypt;
3. the question of YHWH's grace in forbearing, after the destruction of the bull image and the execution of those who instigated its production, to punish the rest of Israel for their complicity—repeatedly formulated in terms of the expansion that in the Decalogue follows the first two commandments;
4. the theme of the Presence of YHWH in the midst of Israel, first in the context of the ordaining of the Tabernacle in which YHWH will be present in his person—and in connection with that Tabernacle, the election to YHWH's ministry of the priestly Aaronides and their Levitical tribesmen;
5. the theme of the Presence of YHWH in the midst of Israel for the duration of the trek to the promised land and his Presence in person in contrast to this Presence in an angelic intermediary;
6. the resolution of the question in favor of YHWH's direct Presence—despite the risks entailed by the awesomeness of this powerful Presence in the midst of a people with a propensity for sinfulness—on the strength of YHWH's grace and love;
7. the inseparability of Moses and Israel, leader and led;
8. the full measure of YHWH's grace revealed in the unprecedented Presence of YHWH not in an angel nor in the Tabernacle but rather in the person of the mortal prophet Moses—the immanence of Deity in man symbolized by the divine light radiating from his face; and
9. the dawning awareness of the reader that the catalogue prescribing proper worship of YHWH, inserted in the narrative as a digression, is a repetition of the catalogue already given in chapter 23 (but this time with the addition of one element and one element alone from the Decalogue, the prohibition of manufacture of mold-casted "gods") and the realization that the angel/agent promised by YHWH to Israel in connection with the catalogue in chapter 23, the angel/agent in whom YHWH will be present ("My name is in him"), had as its original intention no angel at all but, as now emerges from the denouement, the human prophetic agent, Moses.

Biblical iconoclasm is not in derogation of man's aesthetic creativity. Painting and sculpture are no more proscribed than is poetic imagery—imagery so bold as to picture a mortal given a glimpse of God's person. Only the representation of YHWH in sculpted or cast form is prohibited. And this prohibition is limited to forms copied from nature. Fantasy images, not copied from nature, such as the cherubim that constitute the throne of the Invisible God atop the pedestal of the Ark of the Covenant, are not only not proscribed, they may be prescribed. The immanence of God among men is not to be sought in form of fauna or celestial bodies. It may be manifest on occasion in numinous "angels"; but it is ideally best manifested in God's most special creation, the species that alone was "created in the very image and likeness of God" (Gen. 1:26). This species alone can know and fulfill the will of God. This potential is realizable in the prototypical human-at-his-best and most faithful to God, the prophetic exemplar. Is it any wonder that when the prophet Hosea chides the patriarch Jacob/Israel for having traffic with angels instead of invoking God directly by his name YHWH, he reminds the Israel of his own day: "But it was by a prophet that YHWH brought Israel up from Egypt, and it was by a prophet that it (Israel) was preserved!" (Hos. 12:14).[28]

Poetical Review

Repetition (words). The Hebrew term for "face" (*pānīm*) appears more than twenty times in these three chapters, often independently, often in a bound prepositional form. Not a single instance of this repeated occurrence is factitious. The author employs the term in a number of different idiomatic expressions that, in their separate contextual meanings and taken together, all point to the central theme, the philosophical or theological problem of how God reveals himself, formulated in terms of where, and when, and in what or whom the divine Presence manifests itself and may therein or thereby be represented as present. The very word *pānīm* (literally "face, aspect, mien") may appear in a context where the metaphoric sense of "Presence" is the one and only way of translating it. Thus, in 33:11 *face to face* means "directly, without intermediary, unambiguously." In 33:18 Moses asks to see God's Presence, expressed (as elsewhere) in the term *kābōd;* and YHWH's reply is that Moses cannot see his Presence (*pānīm*) in one sense but that he can see it in an antithetical sense—not the forepresence but the hindpresence, perhaps not the full dazzling glory of that Power in its future manifestations but in the traces of its activity in the past (33:20, 33). But even the graciousness of God's self-revelation to his prophet is made pointedly clear by God's proclaiming his name to Moses "for your benefit" in 33:19; passing himself in review "before him [Moses]" in 33:19 (*'al pānāyw*); and sweeping away the enemy *mippānēykā* "from before you [Israel]" in 34:24, as in 32:2 God will dispatch his angel *lefānēykā* "before you/ahead of you [Israel]." In contrast is the idiom (with the angel as subject) *lek lifnēy* "to go before/to lead you [Moses]" in 32:34 and in the same sense of the god to be manufactured in 32:1, 23. Again in subtle contrast to this last idiom is *pānīm hōlᵉkīm* (literally "face going") in 33:14–15, which denotes "going in person." In contrast to the almost literal metaphor of not being able to see the forepresence of God and survive the experience is the idiomatic *lir'ōt pnēy,* "to see the face of [Deity]" in the sense of "make pilgrimage to his earthly seat" in 34:23–24—pointing up the Tent of Encounter (in Digression A) and its successor Tabernacle, the projection of which and the execution of which surround Exodus 32–34 as the embracing metaphor for God's presence. Finally, there is the beaming radiance of Moses' *face* in 34:29–30, 35 and the covering of it by veil and its disclosure in 34:33–35 so that "when the Israelites saw the *face* of Moses, that is, "that the *face* of Moses beamed radiance," they recognized the radiance or revelation in him of YHWH.

Repetition (incident). At least six instances of **synoptic/resumptive technique:** (1) the synoptic/resumptive in Episodes B and C; and (2) Episodes G, H, and I as resumptive to the bottom line of C; (3) the synoptic Episode D followed by Episode E, which takes place in time before the bottom lines of Episode D; (4) in regard to **treatment of time,** within Episode F, the synoptic verses 32:25–28 and resumptive verse 29 and the time treatment in 32:35 in contrast to 32:30–34 (Episode G); (5) Episode H with 32:1–4 as synoptic and 5–6 as resumptive; (6) Exodus 23:12–24 as synoptic resumed in Exodus 34:10–26 and, in regard to hypotactic parenthetic flashback, Digressions A and B.

Dialogue. In the Book of Jonah, I noted, the ratio of dialogue to narrative is fifty–fifty. The ratios in these three chapters are far more lopsided. In chapter 32, the ratio of dialogue to narrative is three to two; in chapter 33, it is three to two if we count the digression and six to one if we do not: in chapter 34, it is two to one if we include the digression (dialogue) and Episode M (which is almost completely narrative). Such extensive use of dialogue makes it possible for our author to "speak volumes" by **showing** what happens from different points of view. For example, Aaron's matter-of-fact description of how the golden calf came to be (in Episode E) is supported in advance by the narrator's own description and report of Aaron's words in Episode A. Moses never accuses Aaron of actually making the bull, and God never charges Aaron with its manufacture; yet both Moses and God repeatedly charge the people with its making. The narrator himself does not describe the making of the bull but suggests in 32:4 that Aaron somehow produced it, while in the concluding verse 35 of chapter 32 he manages to charge the people with the making of the bull that Aaron made. Another subtle use of dialogue is the reintroduction of Joshua, so that in the exchange between him and Moses on the noise in the camp we can see that Moses was not really told on the mountain top just what was going on in the camp at its foot. Note the **free direct discourse** in Episode C.

Figure of speech. Note the **oblique expression** in Moses' taunt of Aaron in Episode E and, in that same episode, the *opus alienum.*

Literary and metaliterary conventions. A review of the paragraph on dialogue, above, will reveal how the author crafts the narrator's stance and the dialogue of characters (and plot elements, here the presence of witnesses who may be questioned) to convey the narrative truth of Aaron's version of the production of the bull. As compared with these purely literary conventions of **ancient literature**—it makes no difference to the story whether in real life bulls spring out of liquid metal blobs—metaliterary considerations are involved in our analysis of such terms as *idolatry* and *fetishism* and their deployment in various contexts to establish whether (or rather to what degree) literal or metaphoric **expressions** for worship are present.

Genre. This unit in itself—the philosophical fable, virtually without peer or precedent in other literatures—points to the need to develop a unique classification of genres or subgenres for the narratives of Scripture.

Characters. As in the case of Jonah and in consideration of the matter of genre just discussed, characters in a parable or fable, for all the roundness that may be achieved for them in a given narrative, serve essentially as types. As such, for example, it would seem pointless to try to integrate the character of a Moses or Aaron (examplars, respectively of prophet and priest) as it appears in various narratives.

Poetic integrity and genetic provenience. The narrative strategies of gapping and bridging (as in the making of the calf and the identification of the idolaters), as also the synoptic/resumptive technique, do away with the problems of inconsistency and contradiction that led to the promulgation of documentary hypotheses. Similarly, the distribution of the *opus alienum,* concepts of angelology, patriarchal traditions and variations on them in prophetic texts throughout the Bible support my argument for the poetic integrity of Scripture. This is particularly borne out by how prescriptive and cultic material (Digressions A and B) is inserted into a narrative to advance the purport of characters and plot; this in a narrative (Exodus 32–34) itself set into a cultic context, the prescriptions for, and the execution of the prescriptions for, the cult center, which is itself set into the narrative frame of the journey from Egypt to Canaan, from slavery to freedom.

Prescriptions for Proper Worship

The two texts are reproduced in parallel columns to facilitate a study of their remarkable similarities. To be noted is a difference in juxtaposition that reveals how craftily the ordinances are fitted into their respective contexts. In Exodus 34, the catalogue is preceded by God's promise *to Moses* and seems to be a non sequitur to that promise. In Exodus 23, this cultic catalogue follows an extensive list of moral prescriptions (the so-called Covenant Code) regulating relations between man and man and is followed by the promise *to the people* of an agent who must be obeyed, for his being the vessel of "my name" will preclude him from exercising the kind of forbearance that Moses will in the golden calf narrative plead God to exercise.

Exodus 34	*Exodus 23*
10.He said, "Lo I commit myself in this covenant: with all your people as witness I shall work wonders never before wrought on earth, among any of its nations. And all this people in whose company you are will witness YHWH's doings, how awesome is the manner of my treating with you.	20.Lo, I am sending an agent [or angel] before you to guard you on the way and to bring you to the place I have appointed.
	21.Take care in his presence, obey him; do not play the rebel with him, for he may not forbear in regard to any transgression on your part, since my name is in him! . . .
11.Take heed of what I at this moment charge upon you—I am about to drive out before you the Amorites, the Canaanites, the Hittites, the Perizzites, the Hivvites and the Jebusites.	21.With my agent [angel] going before you and bringing you to the Amorites, the Hittites, the Perizzites, the Canaanites, the Hivvites and the Jebusites—then shall I wreak depredation among them.

12. Take care not to make any covenant with the inhabitants of the land upon which you are marching, lest they be a lure among you.

13. Rather, their altars you are to demolish, their pillars you are to smash, their sacred posts you are to fell.

14. Lo, you may worship no other god (for YHWH, exacting in his name, an exacting God is he)

15. lest (you making a covenant with the inhabitants of the land) in their whorish rites of their gods when they make sacrifice to their gods, they invite you and you eat of such sacrifice;

16. and (you taking wives for your sons from among their daughters) those daughters of theirs, a-whoring after their gods, lead your sons as well into whoring after their gods.

17. Gods, cast in molds, you shall not make for yourselves.

18. The Feast of Unleavened Bread you are to observe, eating unleavened bread for seven days as I have charged you, at the set time in the month of Abib—the month of Abib when you went free from Egypt.

19. Every first issue of any womb belongs to me, that is, the male first issue of cattle, large and small;

20. while first issue of ass you may redeem with a sheep—or if you are not minded to redeem it, you must break its neck; and every firstborn of your sons you must redeem. None shall appear in audience before me [literally, "see my face"] empty-handed.

32. Make no covenant with them and their gods.

33. They are not to remain in your land, lest they make you sin against me; being a lure to you, that you serve their gods.

14. Three times a year you shall celebrate a festival for me.

15. The Feast of Unleavened Bread you are to observe, eating unleavened bread for seven days as I have charged you, at the set time in the month of Abib—for in it you went free from Egypt.

15b. Nor shall any appear in audience before me [literally, "see my face"] empty-handed.

21. Six days you may work, but on the seventh day you must desist; even at plowing time or harvest time you must desist.
22. The Feast of Weeks also you are to enact, with the first yield of the wheat harvest; and the Harvest Feast, as well, at the turn of the year.
23. [These] three times a year, every one of your males is to come for audience before the Sovereign, YHWH, God of Israel.
24. Inasmuch as I dispossess nations from before you and extend your border's stretch, there will be none to make design on your land when you go up for audience with YHWH your God [these] three times a year.
25. In slaughtering, offer not the sacrificial blood that is my due together with leavened bread. Of the paschal sacrifice nothing is to remain by morning.
26. The choice first yields of your soil you shall bring to the house of your God YHWH. You are not to boil a kid in its mother's milk.

12. Six days you may perform your labors, but on the seventh day you must desist.
16. [You are also to observe] the Reaping Festival, with the first yield of the produce that you sow, and the Harvest Feast at the end of the year when you harvest your labor.
17. [These] three times a year, every one of your males is to come for audience before the Sovereign, YHWH.
18. In slaughtering, offer not the sacrificial blood that is my due together with leavened bread. Of the paschal sacrifice nothing is to remain by morning.
19. The choice first yields of your soil you shall bring to the house of your God YHWH. You are not to boil a kid in its mother's milk.

5

Tales of Elijah

Uncanny is the word for Elijah. And uncanny is the image of this prophetic figure that the author of six or eight chapters in the Book of Kings manages to project so as to have him dominate this book as Moses does the Pentateuch; and as a purely narrative feat the effect achieved in Elijah's case is incomparably more marvelous. Moses, after all, is the paramount creator (under God, to be sure) of Israelite history and institutions: liberator, legislator, wilderness guide and judge, paramount chief, author of the priesthood, and paradigmatic prophet. Even the prophet Samuel, whose figure seems to shrink when juxtaposed with Elijah's, was both priest and prophet, chief magistrate and anointer of Israel's first two kings. Yet in terms of civic office, political involvement, and specificity of social and political values Elijah's roles and program are almost negligible.

His story is told in the compress of eight chapters, two of which feature great military crises and critical prophetic activity in which he does not appear at all. Of the six chapters in which he does figure, the first deals with a few personal and seemingly trivial experiences and the last with his death (or at least disappearance, as in the first chapter, from his people's ken). In a third chapter, other human characters are absent or marginal, the action being limited to a transaction between Elijah and his God, in which the former tenders his resignation from service and the latter accepts it. In terms of public appearances, he is limited to three—four at most; indeed, his prophecy is limited to the pronouncement of three oracles. One oracle takes up half a verse at his career's beginning, forecasting a drought; a second oracle, a pronounce-ment of doom on Ahab and Jezebel, he utters at the tail end of a story of royal corruption, in which story he plays no part; the third oracle is, again, a prediction of death for the son who has succeeded Ahab on the throne. In this last story Elijah is characterized as a vague figure from a distant past, unknown to the royal court; and his identity is recovered by the speculation of the king to whose parents this prophet played nemesis. Cast as he is as a central figure in a historical account of a great religiopolitical crisis in the Northern Kingdom of Israel, Elijah's activity (or lack of it) takes place in the twenty-four years spanning the reign of Ahab and the two-year reign of his son Ahaziah. Of the birth, rearing, education and vocational call of this paramount prophet we are told nothing. We know neither his age at his first prophetic experience nor why his critical prophetic activity—presumably in the prime of life— was cut short. Indeed, we are left to guess how Elijah occupied himself during more than 99 percent of the twenty-two years of Ahab's reign. In retrospect of his three public appearances, one scene looms large—Elijah's duel with the Baal prophets on Mount Carmel. Savage as is the role he plays in this historic confrontation, there is in the telling of it some of the humor—sometimes grim, sometimes Puckish—that characterizes his appearance to the YHWH-fearing Obadiah, to the not-so-YHWH-fearing captains of Ahaziah, and the incongruity of his victorious marathon race with Ahab's horse and chariot. How marvelous the secrets of the storyteller's craft responsible for the emergence from history's action-filled past of this historic figure who appears so rarely, says so little, does so little, accomplishes less, dies a self-confessed despairing failure! Elijah lives on in memory (his career vastly extended in rabbinic tradition) larger-than-life and transcending history to reappear as messianic forerunner of the Promised One (perhaps, as in the speculation on Jesus' identity, the Messiah himself) and, in Malachi's concluding prophetic promise (or is it threat?), "Lo I am dispatching to you the prophet Elijah, in advance of that surely coming great and awesome day; he will turn the mind of parents to [their] children and the mind of children to their parents—else will I [myself] come and strike the land a blow of utter annihilation!"[1] (Mal. 3:23–24)

Tale 1: Elijah in Hiding

Introduction

1.Elijah the Tishbite, of the settlers of Gilead, announced to Ahab, "By the life of YHWH, God of Israel, in whose Presence I have stood, I swear: Neither rain nor dew shall there be in the years ahead unless I so do say."[2] (1 Kings 17:1)

This single verse, in which, without ceremony, the most famous of all Israel's prophets (save for Moses) is introduced for the first time, together with his mission, is instructive of biblical narrative style on several counts. The prophet is identified but not, properly speaking, introduced. He appears in midcareer, at an indeterminate age, with no biographical details preceding or to follow. No details are given as to his family, his class, his occupation, or his dress or, for that matter, the manner of his appearance on the stage of Israel's history. It is as if the narrator is assuming that the historic role of this prophet is too well known to require introduction and is telling us that all we really need to know about Elijah is what he announces about himself. He has stood in the throne room of Israel's God and received a commission from him: to announce the coming of a drought (or several droughts), which will not fail to come unless Elijah first announces the relief. The announcement is made to Ahab, who needs no introduction, for all the tales of Elijah appear as events in the larger setting of the reign of this king of the Northern Kingdom of Israel. Whether the prophet sent word to this king or appeared before him in person (and if the latter, how he gained audience and how he managed to escape the royal presence) are matters, we are left to understand, of idle or even frivolous curiosity.

Episode A

2.The command of YHWH came to him, 3."Leave this place. Face about to the east. Go into hiding in the Wadi Kerith, which lies east of the Jordan. 4.You will drink of the wadi['s water]; the ravens I have laid under charge to provide you with sustenance.'' 5.He proceeded in obedience to YHWH's command: he went and settled in the Wadi Kerith, which is east of the Jordan, 6.the ravens bringing him bread and meat in the morning, bread and meat in the evening, while [water] he would drink from the wadi. 7.After some time the wadi dried up, there being no rain in the land. (1 Kings 17:2–7)

We are not told precisely the "place" which Elijah was instructed to leave; it was in the territory of the Northern Kingdom, west of the Jordan. As no reason is given for YHWH's command to leave, so no reason is given for his destination; nor is the prophet informed how long will be his stay there. At this point we are informed that for some indeterminate period of the drought Elijah was to hide out in the barren no-man's-land of the wilderness east of the Jordan, in an unfrequented arroyo, whose gully will yet provide drinking water. We are left to understand that this wadi appears on no map: its very name in our story—from a root meaning "to cut or cut off"—might well be rendered in English as Wadi Boondocks. As for food, the all-powerful YHWH, who elsewhere (as we shall see) provides food by means of an angel, even a single meal which will sustain Elijah on a forty-day journey (one-way), here employs one of nature's scavengers as miraculous agent of his Providence. The concluding notice of this episode, that the wadi ran dry due to the absence of rain, must be understood as a sign that God was ready for Elijah to move on. The events that follow will disclose that this first episode does not appear here by accident. It is a necessary element in a single narrative, without which the point of the episodes that follow might remain irretrievable.

Episode B

8.The command of YHWH came to him, 9."Up! Go on to Zarephath—which is in Sidon's domain—and settle there. Lo, I have charged a widow woman there to provide you with sustenance.'' 10.He proceeded forthwith to Zarephath. When he arrived at the city's entrance, lo, there [was] a widow woman gathering sticks. He called out to her "Fetch me a bit of water in some container. I would drink.'' 11.As she went off to fetch it, he called out, "Bring me back, too, a morsel of bread in your hand.'' 12."By the life of YHWH, your God,'' she said, "I swear that I have not a single roll—nothing but a handful of flour in a jar, a little oil in a jug—and here I am, gathering a couple of sticks, so that I may go home, prepare what's left[3] for me and my son. After eating which, we shall wait for death.''[4] 13."Never fear,'' said Elijah to her, "go home, do as you have proposed. But first, make me from your store a tiny roll, and bring it out to me. After that you may prepare [the rest] for yourself and your son. 14.For thus has YHWH, God of Israel, promised: 'The flour jar will not give out nor the oil jug fail before that time that YHWH grants rain on earth's surface.'" 15.She went and did as Elijah had bidden. She ate, and he, and her family, for days and days. 16.The flour jar gave not out and the oil jug did not fail, in keeping with the promise YHWH had spoken through Elijah. (1 Kings 17:8–16)

With the signal of the wadi's drying up comes, in some fashion, the word of YHWH to Elijah. From the uninhabited wastes of Transjordan he is to proceed north and west to the Phoenician coastland, to the foreign town of Zarephath where a widow of an alien folk has been charged with the task previously fulfilled by ravens. Outside the town's walls, not far from its gateway, where trees grow and kindling twigs are to be found, the prophet comes upon a woman, recognizable perhaps by her garb as a widow, whose task he interrupts, bidding her to fetch him some water. She, who owes this stranger nothing, does not point the way to the town well. Without a word she turns to do his bidding, whereupon the stranger makes another demand of her. The puzzle of her prompt compliance with his first request is now resolved. She has, in some untold fashion, received advice from Heaven about the prophet who will come from the neighboring land of Israel. Her knowledge of both the stranger's native land, and in all likelihood, his prophetic status is disclosed in her identifying his deity as YHWH and his relationship to him by the attribute "your God," that is, "the Deity whom you serve." But she does more than reveal her knowledge. She acknowledges YHWH by taking her oath in his name. And the fact of the oath is itself significant. One does not swear to a beggar that one has nothing to give him. What an extraordinary expression of faith—not only in the majesty of a God one would not expect her to own but in his human agent! And, as if to put that faith to further test, the prophet asks her to share with him the last bit of food that stands between her and death, her death and (how naturally his existence is introduced) the death of her son. To be sure, the request is accompanied by the prophecy of a sustaining miracle, but she must show her faith in that prophecy before the miracle is activated: first she is to prepare the prophet's meal.

The concluding notice of this episode is that the miraculous replenishment of the flour and oil continued indefinitely, which is to say, until the end of the drought and the area's recovery from it. The episode that follows, therefore, took place before the drought was broken, when widow, son, and prophet were still surviving by grace of YHWH's miraculous bounty. Before going on to this episode, however, take note of a subtlety of Hebrew expression that will once again disclose the biblical narrator's economy of expression: not a single word is redundant. In verse 11, Elijah asks the widow to bring him a morsel of bread "in your hand." This expression, in (one's) hand or by (one's) hand, frequently appears in Hebrew in contexts where the English equivalent would indeed be redundant. In English one plays a lyre or lute. One does not add the obviously understood *by hand*. In Hebrew, by contrast, *by hand* is regularly added for the playing of a stringed instrument. One cannot therefore fault the standard translations that omit the expression altogether; yet in our narrative, the expression *in your hand* is purposeful and allusive. In verse 10, Elijah is described as asking the widow to fetch him some water "in a utensil." Inasmuch as she is unlikely to fetch the water in her cupped hands, the superfluousness of this word in the Hebrew must be accounted for. It suggests that the prophet carries no dipper of his own, no waterskin to be refilled; even in this he relies on Providence. The term *by hand of* in Hebrew is also the idiom for "by the agency of" or "through"—exactly what the Hebrew reads in our concluding verse 16: "the promise YHWH had spoken through (*by hand of*) Elijah. But agency in association with speech is normally rendered by *'al pī* (by the mouth of). Can we but conclude that the apparently innocent pleonasm, the "hand" from which Elijah asks bread is pregnant with meaning? In bringing food, as well as water, she will be the agent of Providence.

Episode C

> 17.Some time after this, the son of this woman, his hostess, fell sick, his illness
> growing worse until there was barely breath in him. 18.She said to Elijah, "What
> have you against me, O man of God? That you have come here to recall my sin and
> bring death to my son!" 19."Give me your son," he said, and took him from her
> embrace. He carried him up to his room on the upper story and laid him down on his
> bed. 20.He cried out to YHWH, "O YHWH my God, are you truly set on so evil a
> course against this widow whose guest I am, to let her son die?" 21.Three times he
> stretched himself out over the boy, invoking YHWH, "O YHWH my God, let the lad's
> life return to his body." 22.YHWH heeded Elijah's plea; the lad's life force was
> restored to his body, and he recovered. 23.Elijah then took up the boy, brought him
> down from the upper story, home to his mother. Handing him to her, Elijah said,
> "See, now, your son is well." 24.The woman said to Elijah, "This moment now I
> know that man of God you truly are, and truly in your mouth is YHWH's word."
> (1 Kings 17:17–24)

It would seem that the only puzzle in this episode is the sudden turnabout of Elijah's
hostess in her address to the prophet. When her son falls critically ill she charges him
with responsibility for her plight. The formulation of her charge is hyperbolic. The
prophet must harbor some ill will toward her; his very coming could have had but one
purpose: to encompass her son's death, and this as punishment for an unnamed sin.
The psychology behind the widow's words is clear to all who are intimate with the
biblical mind. Good fortune and misfortune are not haphazard occurrence. They are
dispensations of Providence—reward and punishment. No mortal is altogether
blameless; every human being has in his or her dossier an unexpiated offense, a
skeleton in the closet. And in the vicinity of a prophet, an agent of God—especially
one worthy of the supreme honorific "man of God"—closets have a way of springing
open to reveal their long-stored skeletons.[5]

The widow's anguished hyperbole has its desired effect. The prophet entreats
YHWH, who restores the child to health. To be noted is the detail that the child did not
die, contrary to such translations as "had no breath in him," "ceased to breathe."[6]
YHWH wrought a cure, not a resurrection. And Elijah's appeal to YHWH reveals that
the prophet does not for a moment believe that it is God's intention to kill the child.
What is the significance of these details? And what is the point of this story? Why
would Scripture reserve a chapter for a wonder tale of a prophet, especially when
strikingly parallel stories will soon be told about Elisha, the pupil and successor of
Elijah, stories in which the miraculous achievements of the master are made to pale
alongside those of his disciple—the cruse of oil that replenishes itself so that its gush
can be marketed to pay off debts and provide enough over to sustain a widow and her
son and the death of Elisha's hostess's son and his resurrection. The clues to the
answer must lie in the three episodes of this chapter. First, YHWH the dispenser and
withholder of rain, Author of plenty and famine, provides for the prophet in the most
inhospitable of environments (a desert wadi twice described as lying east of the
Jordan) by natural means: a brook that flows and omnivorous scavengers sharing with
him their bountiful pickings. Second, YHWH provides for the prophet, this time
beyond Israel's frontiers, by the agency of a non-Israelite woman of extraordinary

faith—faith in a YHWH she has never been taught to worship and in his servitor, an Israelite prophet who presents no credentials. And her faith is vindicated by a miracle, an event contrary to nature, evocative of the folklore fantasy of the goose that lays golden eggs. Third, a child falls critically ill, the prophet prays for his recovery, and the child is healed—not an everyday occurrence, perhaps, but far from extraordinary and hardly miraculous.

It is a lesson in human nature, in the vagaries of faith in the workings of Providence, and in the ironies in how human beings perceive or fail to perceive the true nature of what is miraculous in Providence.

The focus of this narrative is not Elijah, but his hostess. Her Phoenician nativity points to a moral about mortals in general, rather than Israelites in particular. Whatever the nature of her faith in her native pagan theology, she accepts life as it comes—rain and drought, life and death. Hence, the stoic equanimity or resignation, perhaps, as she gathers firewood to prepare the last meal for herself and her son. The moment in time when she received a revelation from YHWH, a message as to the coming of his messenger, is not made explicit. Her resigned wood gathering suggests that the revelation coincides with, or precedes by a few moments, the appearance of Elijah. In the revelatory moment, faith is—by definition—unqualified. She accepts YHWH; turns to do his agent's bidding; and, even when pressed to provide him with food, is concerned with convincing him that she is not niggardly or begrudging but has reached the last of her food. Death, after all, will not be long in coming. Perhaps most remarkable of all is the utter casualness (this at least is the author's design) with which she accepts the heralding of the miracle and its ongoing fulfillment.

Until her son falls ill. Even now she evinces no loss of faith. She does not rail against an unjust Heaven nor importune the prophet to intercede for her. She does not impeach his gratitude. She accepts the verdict, assumes an unexpiated guilt. She expresses a bitter wonderment that in a world in which no one is altogether innocent, it should fall to her to harbor a prophet whose presence springs open the door that conceals her guilt. And when Elijah delivers her son, so suddenly stricken and now so suddenly healed, her words conclude the story and drive home its moral with dramatic impact. Now, she—who had doubted neither her guest's standing as "a man of God," nor questioned the reality of "YHWH's word in [his] mouth"—confesses, as if for the first time, as if she had not until this moment believed, the legitimacy of the prophet, the presence in him of YHWH's all-powerful word.

The woman's last words, literally "YHWH's word in your mouth [in] truth" correspond to the last words of verse 1, where Elijah swore by YHWH that there would be no respite from the drought "except by my word"; the Hebrew preposition for "by" being literally "according to the mouth of." Thus, there can be no question that this opening narrative of the Elijah story cycle understandably stresses this prophet's legitimacy. But the arrangement of the three carefully crafted episodes makes for a deeper development of that phenomenon as it relates to every one of us: the true marvel of YHWH's exercise of sovereignty is his creation and sustenance of the world and its creatures. Yet this day in, day out phenomenon generally goes unremarked. We need the first episode, YHWH's providence for his servant, to remind us of this everyday marvel. Then comes the second episode, the "miracle" of Providence—flour and oil replenished in a pair of vessels; miraculous, unnatural, or supernatural yet ludicrously insignificant against the unremarked miracle of God's daily provision of our needs and properly unremarked-upon by the widow woman of Zarephath. When are we likely to acknowledge God's grace? Not in the macrocosm—in the daily and universal phenomena—but when our individual persons are touched in our concern for those we love most dearly, when the life of one child, our child, hangs in the balance. One such cure outweighs all his wonders—then we hail him, and his presence in the prophet and his healing word.

From another perspective, the central theme of this first Elijah tale is that of miracle. Miracle, it would seem, is ubiquitous in Scripture, whether in the loose sense of an event we would consider improbable but not impossible (such as the sudden and not-to-be-expected repentance of Nineveh's pagan sinners) or in the narrower sense of "an event or effect in the physical world deviating from the known laws of nature, or transcending our knowledge of these laws" (*Webster's Collegiate Dictionary*). Examples would be the emergence of a golden calf from a formless assortment of rings thrown into a fire, stick-to-snake-to-stick transformation, river water turned into blood, or a man's surviving three days in the belly of a fish. Biblical Hebrew has no single word (perhaps not even any combination of words) to express the narrower sense of our word *miracle*. The word in later Hebrew corresponding to "miracle" is *nes* (signal, standard), close in meaning to *'ōt* (mark, sign); while other terms for marvels are *mōphet* and *pele'*. And the marvels associated with these terms range from the loosest to the narrowest usages of *miracle*.

It is easy to see that whether the biblical authors believed in miracles in the supernatural or contrary-to-natural-law sense or not is a metaliterary consideration and that the nature of the answer to this question would have to play a significant role in the interpretation of any narrative featuring miracle. Years ago I suggested that the logic of the narratives in the Book of Exodus, whether featuring miracles deployed against Pharoah or to the benefit of Israel, was alike in that these prodigies had little consequence for the faith of either Egyptian or Israelite, victim or beneficiary.[7] It is my conviction that Scripture's stance on miracles must be derived from how the miraculous functions in biblical narrative; and it is my suspicion that Scripture is not so much interested in affirming or denying the reality of miracle as it is in exploring the extent to which faith is or is not affected by human perception of the miraculous.

In the case of this tale of Elijah, it introduces a number of stories about him in which the marvelous and the eerie, the supernatural and the uncanny, the incredible and the miraculous are a dominant feature. It is, therefore, perhaps not just a matter of thematic appositeness but of central design that the first tale set the stage with a commentary on just how our perception of the wondrous finds reflection and expression in the existential situations where faith or lack of it is a significant element of our experience. And the way in which our author constructs his philosophical story shows him a master of narrative strategy.

The chapter's first verse introduces not only the first tale, but subsequent tales as well; it helps to enrich the background for the first tale but is not absolutely essential for it. The first episode is meaningless as history or for history; that is to say, we may not, to the question, "Why is this told?" reply, "Because it happened." Even in histories many things that happened are omitted (as also, dare I say, many things are included that never happened). Histories include only events that are significant for nonparticipating humans, whether these are contemporaries or later generations. Episode A has only one character, no witnesses, and not much plot to speak of and comes to us only with the authority of an omniscient and arguably **reliable narrator.** Episodes B and C, by contrast, belong together, sharing a cast of two or three, who can witness—firsthand or by hearsay—to all the events. Episode B features a marvelous event, as does Episode C. Of the two the event in B is truly miraculous, contrary to the laws of supply and consumption. The event in C is, if anything, prosaic. A child falls sick, hovers between life and death, and recovers. Had a physician been present, it is questionable whether he would have been of any avail.

The prophet does nothing wondrous. He prays, pleads with God that to let the child die would be less than fair recompense to his full-in-faith hostess, stretches his own body over the lad's as if to symbolize his readiness to suffer the death blow in his stead, and repeats his action until the child's normal breathing shows that God has responded favorably to his servant's plea. Yet the woman who accepted the long-protracted miracle of cornucopia prophesied by the prophet who put her faith to test and made possible by her quick passing of that test and who accepted that miracle without a word of comment is prompted by the everyday occurrence of remission of illness to confess, as if newly discovered, a faith in YHWH and his prophetic agent, which she has demonstrated all along. Had we only these two episodes, we might well wonder about the point of this bathetic reaction, which ends the chapter. But we, the readers, are—unlike the woman of Zarephath—privy to Episode A. We know of the miracle of God's providing his prophet with running water and twice daily rations of bread and meat. An event less marvelous than the one in Episode A (as it is effected by the natural phenomena of stream and birds) but more marvelous than the event in Episode C (wilderness wadis being the first to run dry in times of drought and ravens being unlikely purveyors to eremitic prophets). We are, then, in a position to know that the woman would a fortiori have made her comment in Episode C against the backdrop of the event in Episode A, had she known of it. And then we ask, why are we told of Episode A at all? Have we now two marvels to compare with one another as against the one in C that evoked the woman's confession of her newly awakened faith? Is it possible that we are wrong in the assumption of a fortiori as between the events in Episodes B and A? Is not the ongoing universal Providence of God for all nature (as in Episode A) the greatest of all marvels, the wonder of a beneficent Providence that transcends the occasional and freakish event that operates as often for evil as for good? And yet is it not a truth to which we must each own for ourselves that we will acknowledge God's beneficence as it operates in our narrow personal interests (like the woman of Zarephath) even as we ignore the miraculous windfalls from magical technology and fail to acknowledge the unending miracles of daily existence and the unfailing grace that assures their continuance?

Like the philosophical disquisition on Jonah and the quest to understand his deepest motivations, like the philosophical fable on the golden calf and the quest for the understanding of what constitutes idolatry and what is the nature of its constitutive spiritual error, so this tale, too, on and about and around the miraculous and the prophetic and the relation between the two presents the reader with a problem-solving challenge: a challenge that we shall utterly miss if we assume the author to be naive

literalist or inept compiler of snippets of tradition; a challenge whose depth or
intricacy we shall fail to acknowledge unless we grant the author a wisdom and
capacity for abstraction that we usually reserve for the best minds of our own
generation; a challenge to which we shall prove unequal unless we employ the
sharpest tools of rhetorical criticism.

Poetical Review

Repetition (words). In dialogue and narration, in verses 4 and 6, 14 and 16, to
emphasize the marvel or the strangeness of how YHWH chooses to provide for his
prophet. The use of *hand,* in verse 11 in apparent pleonasm and in verse 16 for
agency—awkward idiom for a speaker, as shown by the idiom for vocal agency,
mouth, in the narrative's opening and closing verses (1 and 24). Note the words in
verses 4 and 9 to highlight YHWH's strange choice of agents.

Repetition (incident). Three miracles occur in three episodes, all different, each
one seemingly pointless in itself or in combination with any other; only from the three
read together does the kerygma emerge.

Dialogue (as dramatic effect). Note the function of oath in Elijah's address to
Ahab (verse 1) and in the woman's address to Elijah (verse 12) and the lack of
necessity for so strong an assurance in Elijah's address to the woman (verses 13–14).
Note the **showing** of the entire plot through the dialogic elements, including the
dialogue "gap" after the cornucopia miracle and the cluminating perplexing words of
the woman in verse 24.

Characters, plot, and genre. See chapter 1, pp. 6–9, 19–25.

History and fiction, poetic integrity, and genetic provenience. See chapter 1,
pp. 19–25.

Metaliterary conventions. Consider the place of miracles in the belief systems in
ancient literature, particularly Scripture.

Tale 2: A Duel of Prophets
on Mount Carmel

1.Many days passed. Then YHWH's word came to Elijah in the third year, "Go now,
show yourself to Ahab, that I may grant rain upon the earth. 2.So Elijah set out to
show himself to Ahab; [there] in Samaria the famine was fierce.

3.[Meanwhile] Ahab summoned Obadiah, his majordomo—Obadiah, now, was
one who held YHWH in great awe. 4.When Jezebel was eliminating YHWH's prophets,
Obadiah took a hundred of these prophets and hid them away, fifty to a cave,
providing them with food and drink. 5.Ahab ordered Obadiah, "Scour the land, for
all the springs of water, for all the wadis. Perhaps there is yet forage to find, to keep
alive a few horses and mules, so that we are not utterly bereft of animals." 6.They
then divided the land for exploration: separating, Ahab went one way himself and
Obadiah himself went another way.

7.Obadiah was proceeding on his mission when of a sudden there facing him was Elijah. The moment he recognized him, he fell face to the ground. "Is it really you, my lord Elijah?" he adked. 8."Yes, it is I," he said, "Go! say to your lord, 'Elijah is at hand.' " 9.He replied, "What offense have I committed that you would hand me, your servant, over to Ahab for execution? 10.By the life of your God, YHWH, [I swear that] there is neither nation nor kingdom to which my lord has not sent missions in search of you. When they returned answer, 'He's not here,' he put each nation and kingdom to oath that you were not to be found. 11.And now you say, 'Go! Say to your lord, Elijah is at hand.' 12.Should I leave you—the while YHWH's wind spirits you off to I-know-not-where—and reach Ahab to make this announcement, if he cannot lay hold of you, he'll put me to death, me, your servant, who since boyhood has held YHWH in awe. 13.Has my lord not got word of what I did when Jezebel was killing off YHWH's prophets, how I hid away some hundred of YHWH's prophets, fifty to a cave, providing them with food and drink? 14.Now, then, do you really command, 'Go say to your lord, Elijah is at hand'—so that he may kill me?" 15.Elijah replied, "By the life of YHWH-Hosts, in whose Presence I have stood, I swear that this very day I shall show myself to him." (1 Kings 18:1–15)

This long introduction to the meeting between King Ahab and Elijah, replete with dialogue, word-for-word repetition, hyperbole, and metaphor, invites the question, Why is it here? Why could not the entire episode have been omitted? Given the pithiness which is generally a characteristic of biblical narrative, would we have missed anything at all if in place of these fifteen verses we had one sentence, such as "In the third year, Elijah appeared to Ahab at YHWH's command." Assuming the historicity of the majordomo, Obadiah, why introduce at all a character of whom we shall never hear again, who is given six verses of dialogue (part of which repeats verbatim what the narrator has told us), especially when we shall soon learn that his heroic effort to save a hundred YHWH prophets came to naught: every last one was hunted down or driven out save for Elijah.

The answer lies of course in the nature of the narrative. Whatever may have been the historic matrix of this story, the author takes great pains to distance himself from the stance of objective, disinterested historian. In the portrayal of the characters, in the actions ascribed to them, in the diction employed to describe those actions, his use of hyperbole and metaphor are designed to convey a concern for symbolism (not fact), idea (not incident), homily (not chronicle), moral, theological, and psychological truth (not existential reality, that is to say, reality as we normally and wrongly perceive it). And in this he sets the tone as well for the entire corpus of the Elijah stories.

The embodiment of these features is, of course, Elijah himself. In this episode (as in all the others in which he figures), Elijah embodies the prophetic role to the exclusion of traits that distinguish one human individual from another. He is more role than person. He hails from Gilead, an area east of the Jordan river settled by Israelite tribesmen but not properly part of Israel's sacred land. He has no patronymic, and the label Tishbite refers to neither gens or locale known to us by name. He seems to have no family and no private life. His few appearances are as sudden and eerie as his disappearance, and how he occupies his time off-scene is anybody's guess. In the first tale—a parable of miracle, faith, and prophetic legitimacy—Elijah appeared once (if, indeed, he appeared at all in person) to announce the drought that would continue until such time as he would herald its end. He disappeared from Israelite territory and the view of his compatriots. How the narrator came to know of his solitary stay in Wadi Boondocks or of his sojourn with the widow of Zarephath we are not told. Would the author be surprised to hear his readers speculate on how Elijah passed his days and nights at the widow's home or whether the neighbors were not scandalized by her harboring this foreign stranger during these many months of the drought? This second tale begins "in the third year," two years after his announcement of the drought. YHWH has decided to end the drought, and Elijah previously declared that he must be the herald of the rain. The narrator is precise and consistent in his use of verb and preposition for Elijah's reemergence. He never appears *before* Ahab; for that is an idiom expressing the image of a subject in audience with his king, and Elijah serves a greater Sovereign. He appears or shows himself to the earthly king, whose power he mocks, at the bidding of that sovereign. To drive this point home, the narrator contrives a setting for the meeting far removed from the panoply of court, in circumstances that bespeak the paltriness of human majesty.

The call of YHWH to Elijah to show himself to Ahab is timed so that the meeting will take place when the king of Israel is, in his own person, engaged on a foraging mission, in the desperate hope to find enough fodder to keep alive a few of the royal mounts. Verse 6 reads literally, "Ahab went one way alone, and Obadiah went one way alone." The repeated word *alone* is not to be taken literally. It appears in Hebrew as a regular and natural redundancy, comparable to the English, "They parted company, each going his separate way." The context assures this; for neither king nor royal officer travel without soldiery, especially when the enterprise is one of confiscatory taxation. But (as the author will have us see) for all their seemingly lopsided advantage over Elijah, they might as well have been alone.[8]

The separation of Ahab and Obadiah serves at least two purposes. For one, it will eventuate in Ahab's appearing before Elijah (the mountain's coming to Mahomet) and not the other way around. For another it will contrast the clear-sighted self-interest of the YHWH-fearing majordomo with the fatuous self-reliance of the sinful king of Israel. Elijah appears to Obadiah, an apparition, looming suddenly out of the thin air. The royal captain reacts as if it is the apparition, not he, who commands a powerful host. He does abject obeisance, his mouth against the dirt mumbling his fear and wonder before the preternatural figure. Elijah assures Obadiah that he has seen aright: one majestic word in the Hebrew, "[It's] me!" Then he commands "Go! Announce to your master, 'Elijah's here!'" The image evoked by this bidding—the bidding of a visiting dignitary to the butler of the house—would be comic were the threat not so real. Were Obadiah to obey Elijah literally, he would appear like the member of a hunting party who stumbles into the embrace of a grizzly bear and calls out, "I've got him!"

Obadiah's first response to Elijah's command is (like the widow of Zarephath's question to Elijah) hyperbolic and oblique: I must have committed a great offense to warrant your ordering me to put my head in the noose. Before he clarifies his use of this image (a clarification required by the reader, not by Elijah), he describes what ensued in Israel after Elijah's first announcement of the drought (17:1). Elijah disappeared, if ever he had appeared. The drought, as predicted, began. The king instituted search for the herald without whose spoken word no rain would fall. The search itself is, to be sure, indication of the king's fatuity—as if, were he to lay hands on the person of the prophet, he could force from his lips—which only YHWH can open—the open sesame for rain. Such fatuousness on the part of a king so powerful that having combed his own land for trace of the prophet, he can force his neighboring kingdoms to swear that he has not found haven in their domains! (Now, for the first time, it is made explicit to the reader what Elijah was doing in Transjordan and Phoenicia—"holing up.") Now, Obadiah continues (citing again Elijah's command to him), suppose you—awesome will-o'-the-wisp that you are—are again wafted away by YHWH's spirit or wind (the Hebrew word *rūᵃḥ* means both) while I without prisoner in hand announce to Ahab that I had you in reach, won't Ahab vent his frustration by killing me? His argument continues with an implicit assumption that Ahab would be justified in so dealing with a subordinate who had been derelict in duty and that Elijah would be unjust in forcing him into so false a position. He pleads his loyalty to YHWH, cites his own risk taking in shielding YHWH's prophets and once again cites, in now-vindicated disbelief, Elijah's order that it would be suicide for him to obey.

Why would a writer who can pack so much into so short a space indulge in repetitiousness in regard to one detail? Why should he, in verses 3–4, anticipate the climactic plea of Obadiah in the matter of his proven piety? The answer to these questions is, to be sure, testimony to the author's control of his craft, his manipulation of "point of view."⁹ Our storyteller knows that people lie, especially when it is in their interest to do so. He knows that his readers know this as well. Obadiah's claim to piety may be true or false. The storyteller therefore uses his own "voice" to assure the reader that Obadiah will be telling the truth. Obadiah himself, when he says that Elijah must have heard of his sheltering of YHWH's prophets, is expressing more hope than confidence; for Elijah has been "out of it" for some two years now. For Elijah, who is privy to the mind of God, there was no question of endangering Obadiah. This, however, Obadiah could not and cannot know of a certainty. And so it is that Elijah's reassurance to Obadiah is formulated in a way that Obadiah cannot question, by an oath invoking Elijah's God, YHWH-Hosts.

16. Obadiah went to find Ahab. He told him. Ahab then went to find Elijah. 17. Ahab, catching sight of Elijah, said, "Is that really you, O troubler of Israel?" 18. He answered, "It is not I who have brought trouble on Israel, but you and your father's house, by your forsaking YHWH's commands, running after the Baal-gods. 19. Now then, summon for me all Israel to Mount Carmel, and the Baal prophets, all 450; and the Ashera prophets, the 400 of them—who dine from Jezebel's table.

20. Ahab sent summons through all Israel, and convoked the prophets at Mount Carmel. 21. Elijah stepped forward before the assembled people. "How long," he said, "will you stand on both sides of the fence? If YHWH is God, follow him; or if Baal, follow him!" From the assembly, not a word in response. 22. Then Elijah declared to the assembly, "I alone am left, prophet of YHWH's, the Baal prophets number four hundred and fifty. 23. Let us have two bulls. Let them choose one of them, dismember it, place it on the kindling, but set no fire. I shall prepare the other bull, place it on kindling, but set no fire. 24. You [to the prophets?] shall invoke your god by name, I will invoke YHWH by name. And so be it, the deity who responds with fire—that one is God." The assembly reponded as one, "Agreed!"

25.Elijah addressed the Baal prophets, "Pick a bull, prepare it first—you are the majority—invoke your god by name, but set no fire." 26.They took the bull which he granted them,[10] prepared it, and invoked the Baal by name from morning till noon, "Respond to us,O Baal!" No sound, no response—as they remained fixed in place round the altar that he had set up. 27.At noon, Elijah made mock of them: "Shout louder," he said. "God though he be, he may be preoccupied, perhaps in privy withdrawn, or on a journey gone. Perhaps he is dozing and may yet awake." 28.They called more loudly, gashing themselves with knives and spearpoints, according to their practice, until blood poured over them. 29.Noon passed and still they persisted in prophetic frenzy until the time of the afternoon offering—still no sound, no response, no sign of attention. (1 Kings 18:16–29)

As was indicated earlier, it is Ahab who seeks out Elijah and not vice versa.[11] Like Obadiah before him, Ahab can hardly believe that Elijah has shown himself. But whereas Obadiah greeted Elijah as "my lord," Ahab's greeting is an accusation. As disrespectful as the address is, it also constitutes a tribute to the power of Elijah and, in its overlooking of the prophet as the agent of YHWH, an act of disrespect to the God of Israel. Ahab implies that Elijah is autonomous, has of his own volition brought on the drought to trouble Israel, and has for two years capriciously withheld the word that will signal the end of the harassment. Implicit in this is the notion that Israel is the passive victim and that its leadership in the person of its king, Ahab, is in no way responsible for its plight. Elijah denies this attribution to him of both power and responsibility. The power is YHWH's, the responsibility is in the royal dynasty that even before Ahab's ascent to his father's throne and long before Ahab's marriage to the Phoenician princess Jezebel had forsaken YHWH's command and led Israel after false gods: the Baal-deities. Elijah then proposes—and Ahab accepts without argument—a convocation of the representative assemblies of Israel at large to witness a confrontation between the prophet of YHWH and the prophets of these false gods. Noteworthy (and reserved for further comment) is the puzzling insertion of "the Asherah prophets, four hundred of them" of whom we will not hear another word. Equally noteworthy is the far-from-casual notice that these 850 prophets of false gods are sponsored not by Ahab but by his Phoenician wife Jezebel—even as she, not Ahab, has been charged with a campaign to exterminate the prophets of YHWH.

The justification for our translation, "stand on both sides of the fence," in verse 21 (and "they remained fixed in place," in verse 26) will be presented later. For the moment, let us observe that the context supports our rendering and not the standard "hopping" or "limping" between two opinions. The Israelites, Elijah says, are not actively shifting gods in their worship; they are sitting on the fence. They refuse Elijah's challenge to get off the fence but agree to accept the verdict of the empirical test that he proposes. Dramatic purpose requires that the Baal be invoked first. This the storyteller achieves by having Elijah defer, with delicious irony, to the prophets who outnumber him, 450 to 1. These prophets obtusely accept Elijah's courtesy and patiently invoke their god all morning long. At midday Elijah taunts their efforts, mocking the anthropomorphisms of paganism. Their god's inattention thus far must be due to such human considerations as preoccupation with other matters, responding to the call of nature, or even to his being on a trip out of town from which he may be recalled. The prophets now intensify their appeals, resorting to the absurd practice of

self-multilation (which is pictured as only normal for them) and continuing for several
more hours—in vain. Now it is Elijah's turn:

> 30.Elijah now addressed the assembly, "Come forward." The assembly came
> forward. He repaired the altar of YHWH which had lain in ruins. 31.Elijah took twelve
> stones—in the number of the tribes of the sons of Jacob, to whom YHWH's word had
> come, "[Not Jacob but] Israel shall your name be," 32.He built the stones into an
> altar in the name, YHWH. Then he made a trench around the altar, [its area one
> requiring] two measures of sowing seed. 33.He arranged the firewood, dismembered
> the bull, placed it on the wood. 34.Then he said, "Fill four jars of water and pour it
> over the holocaust offering and the wood." [This done,] he said, "Again!" and they
> repeated the act. "A third time!" he said, and they did it a third time. 35.[This
> continued until] the water swirled around the altar, the trench itself he filled to
> overflowing. 36.At the hour of the afternoon offering, the prophet Elijah came
> forward. "O YHWH, God of Abraham, Isaac, and Israel" he called, "this hour let it
> be known, that you are God in Israel, that I am your servant, that at your command
> have I all this done. 37.Respond, O YHWH, respond, so that this people acknowledge
> that you, YHWH, are God—and that it is you who have turned their hearts backward!"
> 38. Down fell YHWH's fire, consuming the holocaust victim, the wood, the stones,
> the very earth; even the water in the trench it licked up. 39.At sight of this, the people
> fell face to the ground, crying, "YHWH! He is God! YHWH! He is God!" (1 Kings
> 18:30–39)

A lightning bolt coming out of a cloudless sunny sky in midafternoon to kindle a
single heap of wood would be considered a miracle, without further embellishment.
The additional elements of wood and flesh thoroughly soaked with water that burns
like gasoline (so that the very stones and earth are incinerated) even to the last drop in
an enormous moat[12] must have constituted a strain on the imagination of an ancient
audience no less than that of our disbelieving generation. If we consider that the aim
of rhetoric is to convince, one would have to judge that the overstatement of the
miracle constitutes a lapse on the part of the narrator. But this holds true only if we
assume that the author's primary porpose is to convince his audience or readers that
YHWH, not Baal, is God. This, however, is not his purpose; for the author knows that
if his readers do not acknowledge YHWH, neither will they put any credence in his
story.[13] The author's purpose is to convey a sense of the obtuseness of Israel, whom
only witnessing so awesomely compounded a miracle will bring round to acknowledging the truth that they should hold by faith, namely, that YHWH and YHWH alone is
God.

The symbolism intended in the hyperbole of the miracle should alert us to other
symbolic elements in this episode. In addition to the three instances of figurative
language (see notes 10, 12), there is one in verse 30 where the author's choice of
vocabulary, geography, and implied history conspire in the interest of symbolism.
Thus, the verb for Elijah's "repair" of YHWH's altar is *rp'* (to heal). This, on Mount
Carmel (a promontory of the Levantine or Phoenician coast), which is *never*
numbered among the hilltops bearing shrines to YHWH! Elijah then takes twelve
stones at random (not from any preexistent altar) in explicit symbolism of the twelve
tribes; these twelve tribes are identified as the sons of Jacob, whose name (suggesting
derogation, "He supplants") YHWH changed to Israel (simply, "God is sovereign").

And when Elijah invokes the Deity whom he serves, it is as "YHWH, God of Abraham, Isaac and Israel," emphasizing a continuity of worship going back to earliest ancestral times and suggesting thereby that the Israel he is challenging is guilty of an interruption of that tradition of loyalty. The fire which he invokes of his God is, he declares, to signify three things: (1) that YHWH alone is God (Power) in Israel; (2) that he, Elijah, is YHWH's servant, he and he alone, implying that unnamed others are laying claim to that office; and (3) that he does nothing autonomously, everything he does being at YHWH's bidding.

> 40. Elijah then said to them, "Seize the prophets of the Baal, not a one of them is to escape!" They seized them. Elijah brought them down to the Wadi Kishon and slaughtered them there.
>
> 41. Then Elijah said to Ahab, "Go up, eat and drink, for [I sense] the sound of torrential rain." 42. Ahab went up to eat and drink, while Elijah climbed Carmel's summit. There he doubled over to the ground, face between his knees. 43. To his boy he said, "Climb up and look out seaward." He climbed up, looked out, and reported back, "There's nothing at all." Seven times he said, "Back again!" 44. The seventh time, he reported, "There's a small cloud, the size of a man's hand, rising on the western horizon." "Go up," he said, "say to Ahab, 'Hitch up [your chariot] and make your descent, if you would not be detained by the rain'." 45. And to be sure, it took not long for the skies to darken with wind-driven clouds, and there fell a mighty rain. Ahab mounted his chariot and rode off to Jezreel. 46. YHWH's hand took hold of Elijah. He tightened his wrap high on his hips and raced on ahead of Ahab all the way to Jezreel's entrance. (1 Kings 18:40–46)

The story began with God ordering Elijah to appear to Ahab to announce the end of the drought. It ends appropriately enough with the onset of rain. In between, we have been treated to some delightfully instructive dialogue and action. We have learned something about the proper way to address a prophet in the service of the all-powerful YHWH, the proper stance of worshipful awe before him, and the dreadful fate that awaits the ministrants of his impotent rivals. To be sure, the slaughter of 450 proponents of a competing cult strikes us, in this age of toleration, as somewhat grisly. The question we must raise is whether it would have made a different impression on its ancient audience. I think not. And so I would regard the slaughter (and let us note that Hebrew verb here is šḥṭ "slaughter," used regularly of animals for food or sacrifice) of the 450 Baal prophets, like the disappearance of the 400 Ashera prophets (18:19), as intended to bear out the general metaphoric or symbolic thrust of the entire narrative.[14]

Be that as it may, whether the climax of the tale is in the hailing of YHWH by a thoroughly cowed Israel or in the triumphal butchery of competing cultists, the concluding episode cannot but strike us as a resounding anticlimax. After the disposal of the adversary clergy, Elijah invites the king (who up to this moment is the villain responsible for Israel's spiritual mess) to make a hasty banquet (which, in the circumstances, can only be a sacrificial meal) before the rains come. The king accepts and indulges, while the prophet (who has not eaten for how long?) abstains. Instead he crouches with ear to the ground and engages his lackey in a rigmarole exchange whose design can only be to demonstrate how nicely the prophet can gauge the arrival of the rain and the time required for a chariot to make the marathon distance from

Carmel to Jezreel. And, to cap all the preceding wonders, the fasting prophet is endowed by YHWH with such strength that he can outrace the king's horses on foot, over hill and dale, until their destination at Jezreel.

It is always an easy out for the critic to conclude that Homer is nodding. And the easy out is an alluring snare: the critic salvages his own conceit at the expense of the artist's reputation. The steadily accumulating evidence that the biblical narratives have reached us as they came from the author's pen, without corruption in the course of long transmission, without tampering of editor's tendentious pencil, should give us pause before we leap to such self-flattering judgment in regard to the incongruities in this last episode. It may be that the narrator is, so to speak, shifting gears in this episode, preparing for a changing grade as he makes a transition to the tale that follows. That this is, indeed, the case will emerge, we believe, from our analysis of Tale 3. Before we go on to it however, we need explore some ramifications of the author's extensive use of symbolism and figurative speech, in his own voice and in dialogue, in Tale 2.

Religious wars and religious persecutions are phenomena notably absent in early antiquity. In extrabiblical contexts we find suggestions of intranational religious jealousies and rivalries between devotees of indigenous cults. The earliest example, in Egypt, would seem to be the campaign waged against Akh-en-Aton's name and memory, after the death of this heretic Pharoah, by the priests of Amon who had suffered neglect during his reign. From Assyria we have traditions of attacks on two powerful kings (here, too, after their deaths) by champions of the cult of Asshur; the indications are that Tukulti-Ninurta and Sargon were inordinately cozy with Marduk, tutelary god of Babylon, whose imperial city these monarchs had conquered. All these, however, were intramural spats, of short duration, and limited to the higher circles of nobility and hierocracy. Campaigns, violent or propagandistic, to impose the worship of one god or another upon a population within a single country or upon a conquered people are totally unattested. Even in the Bible there appears but one account—in the late Book of Daniel—of an attempt by a king to impose such worship on his subjects; and this takes place in Babylon, to the glorification of the monolatrous Judean hero who gave his name to the book.

Israel, too, is never asked by its God to export its worship of YHWH or, for that matter, to require such worship by non-Israelite residents ("the stranger in your midst"). The exclusive worship of YHWH demanded of Israel and Israel alone spills over into a condemnation of an Israelite king for allowing his pagan wives to worship their native deities on YHWH's sacred soil. Intolerance, however, is confined to fellow Israelites, conceived as bound by a historic covenant to the worship of YHWH and YHWH alone. The religious tolerance of diverse cults, so naturally characteristic and to-be-expected of polytheistic paganism, has its attitudinal counterpart in Israel in the Book of Deuteronomy. In chapter 4, Moses warns Israel against iconoplasm in the worship of YHWH. He cannot be represented by any image, for no image was visible when he revealed himself to Israel. He continues, in verse 19 with the prohibition, "And that you do not—raising your eyes skyward and seeing sun, moon, and stars, all the heavenly host—be sidetracked into paying obeisance or worship to them: to them whom YHWH your God has allotted to all the other peoples under the heavens."

In other words, pagans are more to be pitied than scorned; no mention of proselytism, no encouragement of persecution.

What credence, then, can we give to the statement here that Jezebel, foreign wife to an Israelite king and queen only by virtue of her marriage, carried out—and successfully—a war of extirpation against YHWH's prophets? All the evidence suggests that this princess of Sidon would have tolerated YHWH worshipers in her native Phoenicia. Even had she been provoked by the intolerance of the zealots of YHWH, there is little reason to believe that she would have taken on the religious establishment of her adopted land even if her husband were of a mind so to indulge her. And in regard to this last item, the biblical account in 1 Kings argues quite to the contrary. Ahab, as in this very tale, is again and again let off the hook—once by YHWH himself, who accepts the king's submission (22:28–39). We are given several accounts, moreover, of Ahab's recourse to oracles at critical junctures in his reign; and each time, the oracle is requested only through prophets of YHWH. This last would also constitute a direct contradiction of any slaughter of prophets during his reign.

Other details in this tale (such as had they been omitted, we should never have missed) appear like red flags inserted by the writer to warn us not to take the charge of apostasy literally. Consider, for example, the word *baal*. It is essentially a common noun for lord, master, husband. Before it became associated with non-Israelite divinities it often appeared, like the term *'el*, as a general term for deity in sentence-names of Israelites (for example, with the verb for showing favor, Hannan-el and its Phoenician counterpart, Hanni-baal.) As the proper name of a god, Baal, it appears in early Canaanite mythology; during the period of Israelite monarchy it appears in conjunction with another name to stand for differentiated gods such as Baal-Melkart, Baal-Hadad, Baal-Zephon. In our story, Elijah characterizes the offense of Israel's royal house as forsaking YHWH's commands and going after the *be'ālīm* (the regular plural of *ba'al*), yet the pagan prophets invoke the singular Baal (verse 18). And in the next verse, Elijah asks Ahab to convene at Carmel, in addition to the 450 Baal prophets, 400 prophets of the Ashera. Whether there ever existed in Israel or outside of it a separate cult to a goddess named Ashera is questionable; in the Bible *'ashera* is a term for a wooden post set up by an altar, a practice attributed to Canaanite pagans and therefore proscribed in Israel's shrines.[15] Although our author sees to it that these prophets never show up for the confrontation, these prophets swell to 850 the number of pagan ministers whom he characterizes as "dining at Jezebel's table." Now surely this is hyperbole, whether the intent is that the queen maintained such a host of ecclesiastics in her own palace or that she financed the activities of such a corps throughout Ahab's domain.

The recognition of hyperbole or the possibility of a given instance of it is only the beginning of the critic's task. The determination of whether a particular feature in a creative work is indeed hyperbolic and, if so, in what degree requires an explanation of how the hyperbole serves the author's purpose, why he resorted to this figure at all. In the case of the hyperbole directed against Jezebel, one might, for example, speculate that the author is a political propagandist; he harbors a hatred for all things Phoenician and seeks to blacken an entire people in the person of a scion of its royalty. Militating against such a speculation is the consideration that of all its neighbors, it was with Phoenicia that Israel had the longest record of amity, collaboration, even alliance. Closer to hand, one might point to our author's choice of a Phoenician woman in Tale 1 for a remarkably virtuous role. A more cogent rebuttal, however, is the author's character as a religious, rather than a political, pleader. In keeping with the consistent thrust of Scripture, he opposes the religious features, not the material culture or political pretensions, of Israel's pagan neighbors. It is as a champion of that dreaded spiritual influence that a Sidonian princess reigning in Israel is ideally suited to our author's purpose. The battle between paganism and biblical monotheism is pictured as a war to the death, no quarter given or asked on either side. When the battleground is YHWH's sacred soil, the religious liberal is, at best, a person who cannot discriminate between right and wrong; at worst, an abettor of sacrilege.

I am suggesting, as a hypothesis to be considered in other biblical contexts as well, that as idolatry ("making graven and cast images" in connection with worship of YHWH) is a metonym for false values, so apostasy from YHWH ("worship of other gods, of *baalīm, gīlūlīm = turdlings,* etc.") is a metonym for false values or proscribed practices employed in YHWH worship. It may well be that the more ignorant of Israel's peasantry—like scientific scholars today—construed these terms literally. But for any Israelite sophisticated enough to be aware of the existing conditions in the cult centers of Israel or Judah the figurative senses of these rubrics must have been obvious. For the skeptics among my readers, a melancholy fact of our own experience may bear some cogency. The most bitter religious conflicts are not between widely separated religious systems but between competing branches of a single confession: Catholic and Protestant, Orthodox and Reform Jew, Sunni and Shi'ite Muslim. The smaller the gap separating them, the more internecine— frequently—is the conflict.

We have saved comment on the most breathtaking hyperbole in our story, perhaps in all Scripture, for the last. In verse 37, Elijah repeats what YHWH's response is to demonstrate to Israel, "that you, YHWH, are God." He then adds, "and that it is you who have turned their hearts backward." The Hebrew could not be more plain; yet the new Jewish Publication Society translation notes, "Meaning of Hebrew uncertain," and the New English Bible offers an alternate translation in a note, "thou that dost bring them back to allegiance," a rendering which does not even attempt to ground itself in the Hebrew. The literal meaning is that YHWH is to accept ultimate responsibility for Israel's backsliding. How could the ancient author have intended such a hyperbole, especially when the hyperbole bespeaks a philosophic sophistication?

The hyperbole here is of the same category noted in YHWH's miraculous production of the golden calf. It is that **oblique expression,** the *opus alienum,* the *strange act* of God, wherein he is responsible for bringing about the conditions that alienate his creatures from obedience to him. The nonexistence or impotence of other deities is implicit in the failure of the Baal to respond. YHWH therefore is alone God, alone Creator of the world, alone the Creator of the human species, to which alone he has granted a consciousness capable of witnessing his self-revelation, the revelation of his will, and, yes, the freedom to conform to his will or to rebel against it.

A mind capable of such hyperbole is capable of another, one in which we all indulge: we are never so apostate as when we are treasonous to God's values in the very act of worship in his name—hence our conclusion about the context of biblical idolatry as concerns Israel. Israel's idolatry inheres in the false values and mistaken practices associated with a protestation of loyalty to YHWH and to his norms.

Poetical Review

Repetition (words and incident). See examples in verses 4 and 13 by the narrator and by a character in **dialogue,** exploiting the strategy of **point of view** and in verse 16, to express symbolism of rank. See examples for **dramatic effect** in the prayers of the Baal prophets to heighten the futility of their invocations; in the drenchings of the altar, ostensibly to heighten the miracle of the flame, expressive of YHWH as God of fire and water, drought and rain; and in the impatient descents of Elijah's servant to report nothing while Elijah waits in faith for the sign he knows must come.

Dialogue. This occupies about 55 percent of the text. Note the pointlessness of the whole first episode (the meeting of Elijah and Obadiah) except for what is achieved through the dialogue. Note the subtlety of the seeming shift in Elijah's address (in verse 24) as though he is suddenly speaking to the Baal prophets, to whom he only turns in verse 25; he thereby is accusing Israel of active recourse to Baal in preference to YHWH, whereas before he has them merely straddling the issue. See n. 10. For **free direct discourse** see verses 10–12, where Obadiah explains for the reader the meaning of his oblique question in verse 9.

Characters. Note the unrealistic portrayal of the king: first, in going out in person to forage, a king powerful enough to dominate the surrounding kingdoms; then in his volte-face to search out Elijah; in his foolish charge of Elijah with responsibility for the drought and his quiet acceptance of the countercharge; his meek obedience to send out the summons for the duel to come; his tolerance of the slaughter and his capacity to down a hearty meal immediately thereafter.

Oblique expression. Note, in addition to verse 9 treated above under **dialogue** the *opus alienum* in the mouth of Elijah (verse 36).

Symbolism and metaphor. This relates to **metaliterary** factors, especially the ancients' differing from us in respect to belief in miracles. The question of **history versus fiction,** is also relevant.

Poetic integrity. Is there a single inconsistency or discordancy in narrative or dialogue that is not owing to the rhetorical purpose of the author?

Tale 3: Elijah at Horeb

1. When Ahab told Jezebel all that Elijah had done, particularly the whole story of his putting all the prophets to the sword, 2.Jezebel sent a messenger to Elijah with this word: "May the gods do such-and-such to me and worse, if by this time tomorrow I do not level your life to the level of one of theirs." 3.In fear, he took flight for his life. He reached Beersheba in Judah; there he left his attendant. 4.He himself continued a day's journey into the wilderness. At that point he sat himself down under a broom-bush and wished for an end to his life. "Enough!" he thought, "Now, O YHWH, take

my life. I am [deserving of] no better than my forebears." 5.Lying down under that solitary broom-bush, he fell asleep. There of a sudden an angel was nudging him, "Up! Eat!" 6.He looked about, and there by his headrest—a roll, hot-stone-baked, and a jug of water! He ate, drank, and lay down again. 7.Once again the angel was back, nudging him and saying, "Up, eat, the journey [ahead] is too much for you [as you are]." 8.He sat up, ate and drank; and on the strength of that one meal he went on for forty days and forty nights to the Mount of God, Horeb. 9.There he entered a cave, to spend the night there, when lo, YHWH's word came to him: "What brings you here, Elijah?" 10.He replied, "I have acted with zeal for YHWH, God of hosts— how the Israelites have forsaken your covenant, overthrown your altars, slain your prophets with sword!—I alone am left, and now they seek to take my life too." 11.He said, "Go out, present yourself on the mountain slope to YHWH!"

Lo, YHWH in procession! At the behest of YHWH, a mightily fierce wind, splitting off crags and shattering boulders. Not in the wind, YHWH! After the wind—an earthquake. Not in the quake, YHWH! 12.After the quake—fire. Not in the fire, YHWH! After the fire—the sound of thinnest silence.

13.On hearing [this], Elijah wrapped his cloak about his face, went out and presented himself at the cave's opening. (1 Kings 19:1–13a)

Jezebel, it would appear, is not daunted or intimidated by the slaughter of her minions. She receives the news from her husband, Ahab, with a sangfroid that lulls us into forgetting that he, not she, is king of Israel. The awesome miracle that cowed Israel into submission before their ancestral God, that converted Israel's tribal assemblies into a lynch mob obedient to Elijah's every word, makes no impression upon her. She does not upbraid her husband for his passivity.[16] Without so much as a by-your-leave to the king, she sends a warning to Elijah (this will-o'-the-wisp prophet who can disappear at will into thin air), a warning embodied in an oath expressive of her icy self-confidence. May the gods damn her if she does not, within twenty-four hours, send him to keep company with her protégés in the netherworld! (Note, incidentally, the scope of free direct discourse, which allows for her use of *one of theirs,* the antecedent being *all the prophets,* in the indirect discourse of Ahab's report to her.)

The normally intrepid prophet, for his part, takes fright and flight, fails to consult his God, and makes his way to the kingdom of Judah's southernmost town, at the edge of the Negev wastelands. He leaves his body servant there (even this lone wolf requires an attendant to indicate the dignity of his prophetic rank?) and makes his way into the arid wilderness. For what purpose? To achieve the very end with which Jezebel had threatened him. As it is not for him to allow Jezebel to triumph over YHWH by killing his prophet, so it is not for him to commit suicide. He may, however, go into the desert and let nature take its course—subject to YHWH's decision to intervene. His wish for death, his prayer to YHWH to take his life, are an expression of his hope that YHWH will, indeed, let nature take its course. As in the case of the prophet Jonah in the book bearing his name, Elijah's wish for death and prayer that YHWH take his life (the Hebrew expressions are virtually identical in both stories) constitute his submitting his resignation from YHWH's service. YHWH's refusal to accept Elijah's resignation is expressed in the angel's first appearance with the miraculous freshly baked bread and water in a jug. There is no question as to Elijah's awareness that his visitor is no mortal fortuitously passing by. He obeys the command, peremptory as it is curt, "Up! Eat!" But his resolve is unaltered; he lies down again. The angel's second appearance, again with a prod (not a gentle touch) and a command to eat, reinforces the first message: Elijah's resignation has been rejected. But to what immediate end? The order to eat is now amplified: YHWH has an arduous mission in mind for Elijah, a mission that turns out to be a summons to headquarters!

YHWH's earthly headquarters, or court, is variously sited in the Bible. Depending on the legitimacy attributed to one or another shrine at various times dating back to the patriarchs, YHWH holds court at such places as Bethel, Gilgal, Shiloh, Gibeon, and, to be sure, Jerusalem's Mount Zion. The rubric for site legitimacy is, as in Exodus 20:24, "in every place that I mark with my name," that is to say, "where I reveal my Presence." Scripture offers no clue as to the location of "the mountain of God" named Sinai or Horeb, where YHWH's revelation to the liberated Israelites culminated in the covenant he there concluded with them. Almost certainly a purposeful omission: Sinai–Horeb is not a mountain–fetish, it is a metaphor; Scripture wants it fixed in the human heart, not located for myopic pilgrims.

And in this tale, it requires forty days and forty nights for Elijah to reach it. A round number meaning a long time but also the same round number as represented the duration of each of Moses' two stays on that summit. And when Elijah reaches a certain point on that sacred slope, he enters a cave with the idea of spending the night there. But this is not a place for sleep. In his mind's ear YHWH's incongruous question resounds, the question of a startled host at the appearance of an unexpected guest. What business have you here? Elijah either fails to catch the sarcasm or chooses to ignore it. The reader fairly aches to answer for him, with a presumption that the narrator cannot allow him: "What is he doing here? Why, you summoned him, gave him the superhuman strength to come here, sent an angel to guide him here!" Whatever the reason for the summons, YHWH's question addresses Elijah's purpose in seeking audience with him. Elijah has submitted his resignation. It was declined at a distance. He is here now to present it in person. Now as he quails before his Lord, he does not dare to repeat his decision. Instead, he voices the despair that brought him to it. He thought that he had dealt the opposition a fatal defeat only to learn the next day that Jezebel was spoiling for a rematch!

The phrasing of Elijah's answer is revealing on several counts. He is made to sound like an officer reporting to his commander on his execution of a mission, as if the commander had not been there himself disposing of the day: "I have acted most zealously in the cause of YHWH, God of Armies!" Well, that is understandable. Elijah is pleading his case. He has done the best he could. The opposition, alas, is obdurate and of overwhelming number. It is his identification of the opposition that is so startling. As if to confirm our earlier conjecture that Jezebel and the Baal prophets served but as a metaphor for pagan influence, it is Israel that has breached its covenant with God. Israel has "overthrown" YHWH's altars—the same word used to describe the altar of YHWH on Carmel, ruined by neglect, which Elijah somehow "restored to health" when he constructed a new one of twelve stones representing the tribes of Israel. Did Elijah repair a literal altar? Or is it a metaphor for the restoration to a sound basis of Israel's worship of YHWH? And it is Israel that has put YHWH's prophets to the sword! When one considers how cautiously Israelite kings and nobles elsewhere approach the question of punishment for a single, treason-charged prophet, this hyperbole looms like a mountain. But hyperbole has its uses; and the final one—Israel is out to murder the last true prophet of YHWH—must serve to explain why a prophet rendered invulnerable by an omnipotent God has decided to throw in his hand, withdraw from the game.

Elijah's reply elicits the command to leave the cave and present himself to YHWH, who is about to make an appearance. Before he can do so, however, there are manifestations of YHWH's approach—an avalanche-causing wind, earthquake, and fire—in none of which phenomena YHWH is present. Then there is, literally, "the sound of thinnest silence." The mixing of metaphors, especially in relation to sense experience, is common in our own speech, for example, "thick darkness" (in Exod. 10:21 the darkness in Egypt is such that it can be touched or felt); and "The Sounds of Silence" is the name of a Simon and Garfunkel song. The intent in this narrative is clear: we rarely, if ever, experience absolute silence: the very elimination of noise from the foreground brings background sounds to our ears—if only the beating of one's own heart or a ringing in the ears. But the eerie hush that must fall on all nature when nature's God appears is what Elijah "heard." Then and only then, taking precaution against the resplendent radiance of God, which no man can see and survive, did Elijah, with cloak as blindfold, go out to meet his God.

In Exodus 19 the theophany on the mount is attended by falling rocks and sounds like those emitted from a ram's horn (the wind keening and soughing?); and YHWH descends upon it in fire. Here in the theophany witnessed by Elijah, the very same phenomena are cited—but to deny the presence of God in any of them! The zealot Elijah had thought that YHWH's triumphal response with fire on Carmel and the violence wreaked on the Baal prophets were tokens of the decisive defeat of paganism, history's last apocalyptic chapter. The sweet taste of success turned the next day to ashes in his mouth. Disillusioned and dispirited, he wanted to retire from the struggle. Jezebel's threat, a reminder that the opposition does not so easily surrender, is too much for him—but not for YHWH. YHWH's progress in history may be marked by the cataclysmic noise of violence, but the ultimate manifestation of his transcendent majesty is in silence. The time for that absolute silence is not yet. Elijah understands the message. He goes out to meet his God. Will he persist in his first resolve? Or will he now withdraw his resignation?

13b.He went out and stood still at the cave's entrance. And lo, a voice to him, its gist "What brings you here, Elijah?" 14.He replied, "I have acted with zeal for YHWH, God of hosts—how the Israelites have forsaken your covenant, overthrown your altars, slain your prophets with sword! I alone am left, and now they seek to take my life too." 15.YHWH said to him, "Go! Retrace your tracks—to the steppe of Damascus. On arrival, you are to anoint Hazael as king over Aram. 16.And also, Jehu ben-Nimshi you are to anoint as king over Israel; and Elisha ben-Shaphat of Abel-meholah you are to anoint as prophet in your stead. 17.And it shall be that he who escapes the sword of Hazael, him will Jehu slay; that he who escapes the sword of Jehu, him will Elisha slay. 18.I will leave in Israel only seven thousand, only the knees that have not knelt to the Baal, the mouth that has not kissed him."

19.Proceeding thence, he came upon Elisha ben-Shaphat, he engaged in plowing, twelve yoke of oxen pulling and he himself with the twelfth. As Elijah passed by him, he flung his mantle over him. 20.He, then, abandoned his oxen, ran after Elijah, calling, "Let me kiss my father and my mother goodbye, and I will follow you." "Go back," he replied. "What sign have I made to you?" 21.When he did turn back, however, it was to take the yoke of oxen; he slaughtered them, boiled their flesh using the ox yokes [as fuel], and distributed it to his crew. When they had feasted, he rose to follow Elijah, whose retainer he became. (1 Kings 19:13b–21)

We saw in chapter 4 of the Book of Jonah, where the prophet tenders his resignation, how a narrative may eventuate in alternative endings, incompatible with one another if taken literally but complementary in developing the story's moral. This double ending technique is signaled by the exact repetition of a sentence in dialogue. The very same technique is employed here.[17] In the first ending, Elijah is asked the purpose of his seeking out YHWH; and his reply, for all its indirection, is in explanation of his submission of resignation from service. YHWH himself does not answer, but the narrator answers for him in the description of an event. The narrator's answer does not make it altogether clear whether the event constitutes acceptance or rejection of the resignation or perhaps some sort of qualified answer. Now, in the second ending, the query to Elijah and his reply are repeated word for word. This time YHWH himself gives his answer. The answer is a qualified acceptance of the resignation. Elijah will be excused from further service upon the completion of three missions, namely, arranging for the rise of three men who will continue YHWH's campaign against the Baal-worshiping House of Ahab.

The three men are Hazael, an Aramean general who usurps the throne and presses war against Ahab; Jehu, an Israelite general who brings Ahab's line to an end; and Elisha, whom Elijah is to search out and anoint as his apostolic successor. As the story of Elijah and Elisha continues to unfold, it will become clear that the anointing of the first two by Elijah never took place, that the narrator knew that it would not take place, and that his introduction of it here is to reinforce a metaphor that is yet to come. The chapter concludes, however, with the recruiting of Elisha, the one event that did take place. And the narrative of this literally intended event is comprehensible only in the light of its metaphoric significance.

Elisha is supervising a plowing crew driving two dozen oxen. In an age when many peasants pulled the plow without the aid of any draft animal and a single bull represented a small fortune, the purpose of this detail emerges: to convey to us that Elisha is a scion of a wealthy family. Elijah's gesture requires no explanation. In English, the word *mantle,* like the word *aegis,* signifies more than a garment. It is a robe of office and can stand, by metonymy, for the office itself. Elisha's response to this symbolic invitation to join the prophet in service to God is instantaneous. Elijah does not break his stride (he may still be wearing his cloak, having merely swept it over Elisha in passing), so Elisha, "abandoning his oxen," runs after him; his request for enough time to bid his parents farewell would in itself without his additional words bespeak his acceptance of the call. And for all this show of alacrity, of his readiness to leave all behind—at a moment's notice, no second thought required—his request elicits a rebuff from Elijah. Elijah dissembles. The sweep of his mantle was an accidental brush, it constituted no signal. And the moral is evident. Nothing takes precedence, not even filial love or duty, over the call of God. The elliptical economy of the narrative should not confuse us as to the storyteller's intent. Elisha does not return home. Elijah waits. And there on the spot Elisha makes a sacrificial feast of the two oxen (two oxen to feed a dozen men when one would suffice for a hundred!), and the hyperbole reinforces the symbolism of the fuel for the fire: not hastily gathered brushwood but the arduously crafted wooden yoke. Our own metaphor "burning his bridges behind him" comes close to, but does not quite hit, the symbolic mark here. Elisha celebrates his call in joy and thanksgiving, he breaks with his comfortable and mundane past, he burns the tools of his trade. The overseer of his father's plantation and heir to a noble estate resigns past and prospects to become apprentice to the man of God.

Poetical Review

Repetition (words and incident). Note Elijah's twice lying down to die and twice prodded awake by the angel, verbatim repetition of lengthy **dialogue** in verses 9–10, 13–14, and **synoptic/resumptive** episodes.

Dialogue. This constitutes about 45 percent of the text. Note the **free direct discourse** in verse 2, Jezebel's use of *their* in her address to Elijah in reference to the Baal prophets of Ahab's communication to her.

Character and metaphor. The queen is undaunted by the miraculous powers at the disposition of Elijah, and the intrepid Elijah, in fear for his life, escapes to the wilderness in the hope that God will let him die there.

Inconsistency and poetic integrity. YHWH charges Elijah with three missions before he leaves the scene. Elijah fulfills only the last of these, and not literally at that; for though he recruits Elisha, he does not anoint him. What of the other two charges? They represent gapping (the bridging is yet to come) and also the difference between the recruiting and anointing of Elisha.

Tale 4: The Sin of Ahab

1.Sometime after the foregoing events (Naboth the Jezreelite, who was in Jezreel, owned a vineyard adjoining the palace of Ahab, king of Samaria) 2.Ahab made a proposal to Naboth, "Let me have your vineyard—to make myself a pleasure park, it is so convenient to my palace—and I will give you in exchange a better vineyard. Or, if you prefer, I will pay its price in cash." 3.Naboth replied, "YHWH forbid that I make over to you the land-heritage of my ancestors!" 4.Ahab returned home in a seething temper over the retort of Naboth the Jezreelite, "I will not give up to you the land-heritage of my ancestors!" He lay down on his bed, turned face to the wall, and refused to eat. 5.His wife Jezebel came in to him, "Why this distemper, why this refusal to eat?" 6."It's like this," he told her, "I made a proposal to Naboth the Jezreelite. I said to him, 'Sell me your vineyard, on a cash basis. Or if you like, I will give you a vineyard in exchange.' And he said, 'I won't give up my vineyard to you'." 7.His wife Jezebel then said to him, "You, now, are going to exercise kingship over Israel! Get up, have something to eat, and cheer up. I will get you the vineyard of Naboth the Jezreelite." (1 Kings 21:1–7)

Scripture everywhere and consistently ties YHWH worship to morality, to the norms for ethical relations that YHWH wills between man and man. This story is therefore of crucial significance for our understanding of the moral consequences of apostasy from YHWH. For it is the only story about Ahab, and clearly meant as an exemplar, which concerns itself with an outrage against morality. Indeed, we shall see, as the story proceeds, that five of the Ten Commandments are broken.

But the story cannot be fully appreciated without some sense of what land—real property—meant for ancient Israel (as, indeed, it did for all the ancients). Land is sacred. It is a gift from God apportioned to the various tribes descended from Israel (which constitute an extended family) and to their subdivisions: clans, sibs, families. Land may not be alienated from the family (sold in perpetuity, that is), for the felicity of the ancestral spirits in the afterlife depends in large measure on their legitimate descendants' continuing to hold tenure of the sacred land. It is likely that in actual practice the securing of this imperative was more often honored in the breach. But it remained at the core of the Israelite ethos, and its every breach was regarded as an outrage.[18]

Let us note, now, how virtually every word (especially those that appear superfluous) serves to reinforce the gravity of what is involved and the chain of impieties that ensue.

The story line proper begins with Ahab's address to Naboth on a matter that to us would appear as a mundane commercial transaction. But this first sentence hardly begins before it is interrupted by an aside. The Jezreelite Naboth, who owns the vineyard, is not an absentee owner; he lives on his patrimonial estate. King Ahab, on the other hand, who rules over Israel, is identified as king of Samaria, the city that his father Omri, founder of this dynasty, purchased from its aboriginal possessors (1 Kings 16:24). Thus, Naboth is the native and Ahab is the interloper. The motive for the encroacher's offer is another example of free direct discourse. It is of no interest to Naboth why Ahab desires his vineyard; but the narrator informs his readers that it suits Ahab's fancy to convert another man's vineyard into a garden for himself![19] The generosity of Ahab's offer sets off the motive for Naboth's refusal. No matter how great a profit he might realize from the transaction, it is out of the question; for filial piety guides him. And his invocation of YHWH to keep him far from such a temptation is evocative of the Fifth Commandment of the Decalogue. Of all the commands, this one—*Show honor to your father and mother*—is the only one provided with the specific expansion ". . . that you [the people of Israel at large] may long endure on the soil that YHWH, your God, is granting you." In other words, if you would not forfeit your future on YHWH's sacred soil, do not sell or exchange your patrimonial holdings![20]

Twice more, in the next three verses, the retort of Naboth is given: once in full, once curtailed. Pointless redundancy? Hardly. First, the narrator, in his own voice, gives the reason for Ahab's distemper: not the mere fact of the frustration of his whim but the temerity of a subject who charges his king with such insensitivity to a basic element in Israel's YHWH-ordained ethos. Second, in Ahab's explanation of his moodiness to Jezebel, he stresses the generosity of his offer but leaves out the reason for Naboth's refusal. The implication, of course, is that it is the subject who is arbitrary and unreasonable.[21] For all of his disgruntlement, Ahab's passivity reflects an acknowledgment that Naboth is well within his rights. His Phoenician princess, however, does not share his Israelitish scruples, either because they are not part of her pagan heritage or because the statecraft she learned at home suggests a way of getting around such obstacles. Power is the exercise of power. She will now show how royal power can be exercised to achieve royalty's ends:

> 8.She drew up writs, in Ahab's name and impressed with his seal, and dispatched these writs to the elders and burgesses of his city, Naboth's fellow magistrates. 9.The gist of her writs was,[22] "Proclaim a fast, seat Naboth as president over the council. 10.Seat two men—scoundrels—in opposition to him; they are to charge him so: 'You have committed outrage against God and king!' Then take him out and stone him to death." 11.The dignitaries of his city, the elders and burgesses who were his city's magistrates, acted according to Jezebel's commission to them, faithful to the writs she had sent them. 12.They proclaimed a fast, seated Naboth as president over the council. 13.Two men came—scoundrels—took seats in opposition to him. These two scoundrels laid charge against him, Naboth, in the council's presence, to this effect: "Naboth has committed outrage against God and king." They took him outside the city, there stoned him to death. 14.Then they sent word to Jezebel: "Naboth is dead, by stoning." (1 Kings 21:8–14)

In this episode, again, we must fill in the gaps left by ellipsis, and recognize the technique of free direct discourse. In regard to the latter, it must be clear that Jezebel did not characterize the two suborned witnesses as scoundrels—knavish though she knew they had to be to play their assigned role in her knavish commission. In regard to gaps, we have to fill in the omitted details in connection with the fast, the council, the people to whom the various writs of Jezebel were sent, and the nature of the capital charge against Naboth.

Proclamation of a fast is, in antiquity, standard operating procedure in times of national disaster, such as crushing defeat in war, pestilence, and drought. The last, the most commonly experienced on the littoral of the eastern Mediterranean, is probably intended here. Disasters—visitations from heaven—are sure signs that a great crime has been committed. At times, inquiry is made of oracles as to the identity of the criminal or criminals, who, when "fingered," are persuaded or coerced to confess the criminal act. But oracles, notorious for the ambiguities of their response, are normally the court of last recourse. First recourse is normally had to ad hoc tribunals, courts of inquiry or commissions of inquisition (which, alas, have been known to degenerate into witch-hunts)—hence, in this narrative context, the conjunction of a fast proclaimed and a court convoked.

But it is far-fetched to assume that Naboth's town was suffering an isolated disaster and waited for instructions from the palace. This reinforces our suggestion that the context of the story is a nationwide drought. The royal writs (note the plural) were dispatched throughout the land. But only to Naboth's town went an additional set of instructions—not, to be sure, to all the citizenry but to enough of the influentials, the corrupt, or lackeys to the crown to ensure the railroading of Naboth. And what a fine touch, to have honest Naboth presiding over the inquiry, to find himself impeached for treason and worse, while the spectators in the court wag their heads knowingly, as if to say, "Even the gentry cannot be trusted"!

What was the charge brought against Naboth? The Hebrew reads literally,

"Naboth has blessed God and king." The verb *blessed,* here and in a few other instances where its object is God, is universally recognized as an antonymic euphemism. It is a substitution for its polar opposite, and everybody knows that the antonym of *blessed* is *cursed.* The Hebrew word *bless,* however, has a number of antonyms. And that antonym (which appears with God as its object) is a verb whose root meaning is "light, slight, without honor or respect." The verb, it has been demonstrated, does not mean "to curse," that is to say, to invoke disaster (upon someone). The basic meaning of the verb is "to show disrespect" at one end of its semantic spectrum and at the other, "to abuse or treat savagely." The formula *to commit outrage against God and king* is a rubric for any offense so serious that it constitutes a flouting of the standards of conduct upheld by God and man, heaven and earth.[23] The narrator, employing free direct discourse, in placing this formula in the mouths of the accusers, thereby discloses that he does not know or does not care what the capital charge was in specific. And what matter, when all know that the charge was false?

15. Upon Jezebel's hearing that Naboth had been stoned to death, she, Jezebel, said to Ahab. "Up now, take possession of Naboth the Jezreelite's vineyard, the one he refused to let you have for a price. Yes, Naboth prospers no longer; he is dead."
16. Upon Ahab's hearing that Naboth was dead, he, Ahab, set out for the vineyard of Naboth the Jezreelite, to take formal possession of it.

17. The word of YHWH came to Elijah the Tishbite. 18. "Up, go down to face Ahab, king of Israel, he of Samaria! There in Naboth's vineyard where he has gone down to take possession of it! 19. To him say, 'Thus says YHWH: Murdered have you, and taken possession as well!' To him say further, 'Thus says YHWH: In retribution for the dogs' having lapped up the blood of Naboth will the dogs lap your blood, yes, yours'."

20. Ahab said to Elijah, "So you have got to me, enemy mine!" He replied, "Yes, I have got to you: because you have given yourself over to doing what is evil in the eyes of YHWH, 21. I am bringing disaster upon you—and I will make a clean sweep after your death, cutting off every penis-owning descendant of Ahab's, near and far in Israel. 22. Your house will I make like the House of Jereboam ben-Nebat, like the House of Baasha ben-Ahijah, for the fury you have provoked, leading Israel into sin."

23. And of Jezebel, too, YHWH spoke: "The dogs will devour Jezebel in the plat of Jezreel. 24. The dead of Ahab's line—those [who die] in town the dogs will devour, those who die afield the birds will devour." 25. (Of a fact, never was there another the match of Ahab in so giving himself over to doing what is evil in the eyes of YHWH, he who was incited by his wife Jezebel. 26. Most abominably did he go off in the wake of the fetishes, just like the doings of the Amorites whom YHWH dispossessed in favor of the Israelites.)

27. Upon hearing these words. Ahab rent his clothes, put sackcloth on his skin, and went into fasting. When he lay down, it was in sackcloth; and softly did he tread. 28. Then came the word of YHWH to Elijah the Tishbite, 29. "Have you taken note how Ahab has humbled himself in submission to my decree? Because he has humbled himself in fear of me, I shall not bring that disaster in his lifetime; in his son's times will I bring down disaster upon his line." (1 Kings 21:15–29)

Every word in this episode is there by design. Verses 15 and 16 begin exactly the same way. The narrator's intent might better be served if we rendered the opening words by *As soon as* or *The moment that,* for the intimation is that neither of the royal couple lost time in following up the juridical murder. This intimation is conveyed by repetition in each verse of the name of the subject of the sentence. The news that the deed has been done is brought not to the king but to Jezebel, who instigated it. Ahab is made to appear like the disingenuous male picked up in a police raid on a bordello: he knows nothing about the establishment, he only plays the piano. Ahab is content that his own skirts are clean; the traces of blood on his wife's skirts must be food stains. Jezebel's words, conveying the news to Ahab, reveal her gloating over a stubborn adversary, mocking her dead victim. The way to the object of your desire, she tells Ahab, is now clear of obstacle. The vineyard for which the misguided and stubborn simpleton refused a generous offer is now yours, free for the taking. And the word *ḥay* (plural, *ḥayyîm*), meaning "life" or "live" (like its Akkadian equivalent *balṭu*) also has the meaning "health" or "the enjoyment of good health." The standard translation, "[Naboth is] not alive but dead" is a redundancy not uncommon in biblical Hebrew. But so rendered, it suggests the hyperbole of our own "deader than a doornail." That hyperbolic sense is out of place here. Jezebel is sneering at the victim, whose refusal to jump at a bargain is attributed to a comfortable smugness: "So much for Naboth, who thought himself so well off. Well, now he's well off, all right!"[24] And Ahab utters not a single word. He rushes off to embrace his new toy. Without a thought even to the fortuitousness of the death of the man who had balked him! Finally, lest the reader overlook for a moment the heinousness of the deed or the centrality of the victim or that in the eyes of God the rightful owner is still Naboth, dead though he be, the storyteller attacks our consciousness subliminally. In two verses, consisting of forty-one words (including prepositions and conjunctions), Naboth's name appears five times—twice with the addition of his gentilic "the Jezreelite" and in conjunction with the vineyard, the patrimonial heritance that had cost him his life.

In contrast to Naboth, whose upholding of the Fifth Commandment led to his death, the encompassing of his death involved his enemies in breaches of the Sixth, Eighth, Ninth, and Tenth Commandments, to wit, "You shall not murder," "You shall not steal," "You shall not bear false witness against your neighbor," and "You shall not covet your neighbor's house; that is, you shall not covet your neighbor's wife, his slave (male or female), his ox or ass, [in short,] anything at all belonging to your neighbor." yhwh's message to Ahab, by Elijah, sardonically conjoins the Sixth and the Eighth. Murder—and now theft? Since the first of these two is by far the greater offense, the order here would normally constitute bathos, a comic device altogether inappropriate. The theft, however, is expressed in "and taken possession," which suggests that to murder is one thing, to profit from a murder is to compound that crime. But there is in this expression a far more subtle thought. The condemnation and execution of Naboth by his fellow townsmen was in no way traceable to the throne; Jezebel's cat's-paws could not speak without condemning themselves. But murder is not followed by theft fortuitously: murder is committed in order to take what is not otherwise obtainable. Ahab's move to appropriate the spoils is what points to his connection with the murder. Fool that he is, this had not occurred to him until the point is made by Elijah—three words in the Hebrew: "Murder, now expropriation!" These three words shock him into blurting out two words, in the Hebrew, which constitute a confession: "Trapped me, my enemy?" In addition to admitting that he can no longer play the innocent, Ahab repeats a previous error. He talks as if it is the mortal, Elijah, and not the God he represents, who is his adversary. From the narrator's standpoint there is in this an additional irony, for YHWH is not enemy to anyone. His absolute supremacy puts him far beyond that. It is the mortal who opposes his will, who rebels against his norms, who is the enemy. Thus, in the reply to Ahab, the narrator does not say "Elijah replied." The subject of the verb *replied* is YHWH (whom Ahab never fooled), who continues in the first person with the sentence of doom for Ahab's dynasty; "the clean sweep" will be completed only after Elijah has left the stage of history.[25]

The formulation of the sentence of doom pronounced on Ahab and his dynasty requires further investigation. Before we go on to this, however, let us turn back to Ahab's taking possession of Naboth's vineyard. Theft can take place without murder; and murder does not normally eventuate in the murderer's succession to the victim's estate. How then did it work this way for Ahab, who clearly is not in the line of Naboth's heirs? There can be but one answer: Naboth's heirs must have been done away with also. Had we no further testimony in the text, this would be a good guess; for Scripture specifically outlaws a practice that was followed in the case of the most serious crimes (such as violation of a treaty oath or treason). Deuteronomy 24:16 reads, "Parents shall not be executed together with [or *for*] children nor shall children be executed together with [or *for*] parents; only for his own crime shall anyone be executed." Another practice followed in the case of certain serious crimes was denial of burial to the corpse. Such exposure would take place afield (where the flesh would be devoured by jackals, vultures, or ravens) or in the form of impaling on a city's walls. This practice is outlawed for a length of time in Deuteronomy 21:22–23: "Should anyone, guilty of a capital offense, be executed, and you impale [literally, hang] him on a stake, you must not leave his corpse on the stake overnight. For such impalation [exposure] is an affront to God." (Of incidental interest, the word here translated "affront" is the same rendered "*commit outrage against* God and king.") In the case of Naboth's heirs, we need not content ourselves with a guess. In 2 Kings 9:25–26, where Jehu ben-Nimshi has the corpse of Ahab's last son, Joram, thrown into the field of Naboth the Jezreelite, he cites YHWH's promise of retribution in these words: "I swear—having taken note yestereve of the spilled blood of Naboth and the spilled blood of his sons—that I will requite you in full in this very plot!" The capital charge against Naboth, it thus emerges, eventuated in the king's confiscation of Naboth's estate, a confiscation made possible by the execution not only of Naboth but of his heirs as well.

The three requirements for a person's felicity in his afterlife—the continuation of his line, this continuation on the ancestral property, and burial—all these were denied to Naboth. And all three were, in one way or another, denied by YHWH to Ahab. Both Jezebel and Joram meet violent ends and are cast to the scavengers in Naboth's field.[26] Ahab himself is wounded in battle and returns home to Samaria, where he receives proper burial (1 Kings 22:34–40). The blood from his wound seeps into the chariot; when the chariot is flushed out by Samaria's pool, the dogs that lap the water also lap his blood in fulfillment of YHWH's oracle in 21:19. This thus confirms our translation of a word that while it usually means "in the place where," can also mean "in lieu of" or "in retribution for." All the standard translations, missing this meaning, create an inconsistency; for Ahab is buried, not exposed; he dies in Samaria, not in Jezreel. Our narrator may contrive for poetic justice in all its fullness, but he will not be guilty of contradicting himself on an item of information which he himself reports as a matter of (literal) fact.

Even verses 23–26, a parenthetic aside, are from the pen of the story's narrator. He likens the end of Ahab's dynasty to that which overtook the two preceding dynasties of the Northern Kingdom. Bad as his predecessors were, he notes, Ahab was worse, having been instigated by his pagan queen, Jezebel. As I observed earlier, this tale of Naboth's vineyard is the only one in the Elijah cycle that features a specific example of the moral outrages to which the following of pagan religious practices lead. In the Mount Carmel story, the rubric for these practices was Jezebel's Baal worship. Here the rubric is worship of *gīlūlīm*, perhaps "turdlings" (rendered in my translation by *fetishes*); and the practices are not of the contemporary Phoenicians but of the long-gone Amorites, the aboriginal pagans, who, for their immorality (not their polytheism) were deprived by YHWH of land and immortality—hence, another evocation of the Fifth Commandment, whose implications we have already discussed.

One last observation. Our storyteller is interested in telling a story, not in writing history as such. Just *when* in the careers of Elijah and Ahab this incident took place he neither knows nor cares. The setting for it is almost certainly a drought, like the one with which the cycle of Elijah tales starts. But that two-year drought came to an end in the Mount Carmel tale. The vineyard of Naboth episode cannot precede the Carmel tale, for the condemnation of Ahab here is inconsistent with the denouement of that tale. Or is it? There is yet another story of Ahab to come, one in which he is on the best of terms with prophets of YHWH. Well, as the rabbis knew, "There is no earlier or later" in Scripture: chronological history is the least of Scripture's concerns. So there *was* another drought. And Ahab *was* condemned. How did he survive until the next chapter? The storyteller supplies the answer in our tale's conclusion: Ahab did penance, so that the worst of the sentence was put off until the lifetime of his son, which is when his dynasty ends. But YHWH had said that his sons would be carrion, and so they became, so in what way did YHWH ease the sentence he had pronounced? Ahab himself—as told in the following chapter—paid for his crime, dying of a wound suffered in battle. But it could have been worse: he was spared witnessing the fate of his sons.

Poetical Review

Repetition (incident, **dialogue**). Note the threefold repetition in verses 2–6; double in verses 8–9, 9–12 (on the fast proclamation); and threefold in verses 14–16. See note 23. Also note the appearance five times of the epithet Jezreelite for Naboth.

Dialogue. This constitutes about 50 percent of the text. In addition to the uses of repetitive dialogue in the preceding paragraph, note the **free direct discourse** in the appearance of the epithet Jezreelite for Naboth in dialogue, where it is pleonastic, and in verse 18, where God, speaking to Elijah, refers to Ahab as "king of Israel, who is in Samaria [= he of Samaria]." And the **dramatic effect** of the subtle telescoping of dialogue from the mouth of God to that of Elijah (see note 23).

Characters. Note the almost incidental appearance of Elijah—not in the presence of courtiers or other Israelites, merely as the mouthpiece of God, upon whom Ahab may focus as his nemesis instead of upon the God whose messenger he is. Note also the play of active or passive responsibility in the words and actions of Jezebel and Ahab, not so much to contrast one character with another as to contrast the (supposed) ethos of Sidon with that of Israel.

Poetic Integrity. Note the consistency of this narrative with that which follows— the fates of Ahab, Jezebel, and their progeny as predicted here and fulfilled in later chapters.

Tale 5: Death of Ahaziah

2. Ahaziah took a fall through a lattice on an upper story of his palace in Samaria and was badly injured. He dispatched messengers with this order: "Go make inquiry of Baal-zebub, god of Ekron, as to whether I shall recover from this injury." 3. YHWH's angel, now, addressed Elijah the Tishbite: "Up! Go meet the messengers of the king of Samaria and address them, 'Is it for lack of a God in Israel that you are going to make inquiry of Baal-zebub, god of Ekron? 4. Therefore, now, here is YHWH's word: From the bed to which you have taken you shall not rise; die you shall!'" Elijah went off.

5. The messengers turned back, to him [the king]. He asked them, "Why are you back?" 6. They replied, "A man loomed up in our path. He said to us, 'Go now! Back to the king who sent you! Say to him, Here is YHWH's word: Is it for lack of a God in Israel that you must send to make inquiry of Baal-zebub, god of Ekron? Therefore, now, from the bed to which you have taken you shall not rise; die you shall!'" 7. "What did he look like," he asked them, "this man who loomed up in your path, who spoke such words to you?" 8. They answered, "A man in a hairy pelt, a leather belt around his waist." He then said, "Elijah the Tishbite, that's who it is!"

9. He then dispatched to him a captain with a full company. He moved up against him, and there he was—perched atop an outcrop peak. He addressed him, "Ho there, you man of God, the king commands, 'Come down!'" 10. Elijah spoke up, in answer to the captain of the company: "If it is 'man of God' that I am, then let fire come down from heaven and consume you and that company of yours!" Fire came down from heaven and consumed him and his company.

11. Again he [the king] dispatched to him another captain with a full company. He, now, took this tone in his address to him: "You there, man of God, the king commands, 'Come down, on the double!'" 12. Elijah spoke up in response, "If it is 'man of God' that I am, let fire come down from heaven and consume you and that company of yours!" God's fire came down from heaven and consumed him and his company.

13. Once again, a third time, he dispatched a captain with a full company. This third captain moved up, but when he arrived, he dropped to his knees before Elijah

and in supplication spoke thus: ''O man of God, let my life be of some value—and the lives of your subjects, these soldiers—in your eyes! 14.Lo, fire has come down from heaven, consuming the first two captains, and their companies—now, then, O let my life be of some value in your eyes!''

15.The angel of YHWH spoke to Elijah, ''Go down to him [the captain]—be in no fear of him [the king].'' So he went down to him forthwith, to the king. 16.He said to him, ''Thus says YHWH: 'Because you sent messengers to inquire of Baal-zebub, god of Ekron (Was it for lack of a God in Israel of whom to seek an oracle?)—therefore now, from the bed to which you have taken you shall not rise; no, die you shall!'' 17.And he did die, in keeping with the word of YHWH spoken by Elijah. Jehoram [his brother] succeeded him as king—this in the second year of the reign of Jehoram ben Jehoshaphat, king of Judah—because he [Ahaziah, his brother] had no son. (2 Kings 1:2–17)

Of all the tales we have so far examined, this one would seem to be from the point of view of fictional art the weakest and from the point of view of historical credibility the least grounded in conditions bearing any verisimilitude to a real-life setting.

First consider the point of view of fictional art. A king, recently come to the throne (his entire reign is less than two years) falls from a second-storey window and sustains serious internal injuries (nothing improbable about that, given royalty's penchant for tippling and carousing). But the next step is improbable to the point of absurdity: sending a delegation to a foreign state for the purpose of consulting an oracle never before heard of and never heard of again. And the question to be put to the oracle: Will I get better or not? Oracles are never consulted out of idle curiosity. For one thing, their charges are on the high side; for another, their responses are notoriously ambiguous; and for another, even though the oracular response is often ignored in the outcome, the purpose of the inquiry is to provide some basis for future action. (A not-far-fetched analogy to this mission would be an American president sending for a long-distance diagnosis [*prognosis,* rather] from a specialist in Vienna when his own physicians are stumped, or are divided in their opinions.)

The embassy is but a few hours departed from the capital when it is accosted by a single person, whom they neither can, nor seek to, identify; who, as we shall quickly learn, is dressed like the most primitive of savages; who bears no credentials; and who bids them to abort their mission and to return to their master and inform him that the God of the national established church is miffed at not having been consulted and volunteers an answer anyway, namely, the one least likely to be acceptable: Don't bother to ask, you are finished! And this empowered delegation returns, as if in a hypnotic trance, to their royal master's bedside, there to stand mute until he inquires as to their swift return. They then recite almost word for word the oracle delivered by the mysterious stranger. The king expresses no shock at these words, nor does he seem puzzled by his courtiers' obedience to the command of the stranger counter-manding his own. He asks for details about that unknown prophet's appearance and, on hearing the answer, jumps to a conclusion. It must be that long-unheared-from prophet whom neither he nor his court have ever known but about whom he has heard some tales.

Does the king accept the oracle? Does he assume that the prophet is legitimate, represents YHWH indeed, and has correctly understood YHWH's message? An indica-tion that he does is that he does not send again to Baal-zebub in Ekron; an indication that he does not may be read into his action of dispatching a considerable military force (the unit is more comparable to a regiment of our own times than to a company) to scour the area where the prophet was last seen, with one purpose—to arrest him. Lo, the prophet has not gone into hiding! He awaits their coming, placidly seated out of reach on a boulder. The captain's address to the prophet appears contradictory. On the one hand, he addresses him by the most honorific of titles, "man of God." On the other, he barks out an order to surrender, in the name of an earthly king. The prophet seizes on this implicit contradiction. If the sign of respect, he says, is well earned, then the sign of disrespect must be punished. And punished it is, by the equivalent of a tactical nuclear bomb. But the narrative does not report whether the explosion was witnessed by other than the prophet. We are informed only that the king dispatched a second force of equal strength. Again the captain addresses the prophet with the title bespeaking the highest respect, but the command in the name of the king is even more peremptory: "Come down, at once!" And again the prophet picks up the contradiction, another salvo is called in, and the second force goes up in smoke. Once again we do not know whether it was witnessed by other than Elijah. Nor do we know how much time passed before the king sent yet a third force. But send it he did. And this time the captain puts us in mind of Obadiah, King Ahab's majordomo in Tale 2. He falls to his knees before the prophet. He does not even beg him to come down. (Why, then, is he there at all?) He asks for mercy. On what grounds? That, like Obadiah, he has been on Elijah's side all along? No. On no grounds whatsoever—other than pity. He knows, so he tells Elijah, what happened to the two detachments that preceded his. Does he, now? Why, then, did not the captain of the second detachment know what had happened to his predecessor? And if he did know how the latter came to predecease him why did he not have enough sense to learn the lesson that his successor did?

Whtever the answers to such questions (if, indeed, we assume that the storyteller, being fully competent, must have known both questions and answers), the narrator chooses to tell us that at this point Elijah required new instructions from up above. Go along with the captain, says God's angel, and then adds (as if this intrepid superman is capable of qualms before secular power), "be in no fear of [the king]." Elijah goes to the king, and a third time we have the formula of YHWH's oracle to Ahaziah—with one slight change. There the story ends. How Elijah was dismissed, or escaped, from the king's presence we are not told. The concluding note is that the oracle was, to be sure, fulfilled and that Ahaziah, having no male heir, was succeeded on the throne by his brother Jehoram (this in the second year of the reign of the king of Judah, who, by coincidence, also bore the name Jehoram).

Second, consider the point of view of historical credibility. The proper names of Ahab's two sons, of the God of Israel and god of Ekron, suggest problems. First, the god of Ekron. The god of this city, as of its four Philistine sister city-states (Ashdod, Ashkelon, Gath and Gaza), is elsewhere identified as Dagon (or Dagan), a well-known deity in the area from the Mediterranean in the west to the Tigris River in the east. Here the god's name contains the Baal-element that we have already encountered in a Phoenician context. The second element is *zebub* (fly). Most students of the Ancient Near East are agreed that this term never appeared in the name of any god. "Fly Master" is a deliberate biblical invention, in derogation of a pagan deity (who, wrongly denominated "Lord of the Flies," owes his notoriety to Matthew 12:24, where, as Beelzebub, he is prince of demons, and to John Milton, who promoted many a pagan deity from biblical context to fallen archangels, coconspirators of Satan's in rebellion against God—notable among them, this Beelzebub). All are agreed that the element *zbb* replaces an original root *zbl* "prince, princely, royal." This complimentary attribute is featured in Canaanite myths in connection with mortals and gods, Baal among the latter; in the Bible we have it, in addition to other contexts, in the Phoenician Jezebel and the Israelite tribe Zebulon. However, Baal-*zbl* never appears either elsewhere in the Bible or in extrabiblical sources in connection with Philistia. (Nor, for that matter, is Ekron known for the oracle of any god, not to speak of a god of healing.)

The proper name of the God of Israel is YHWH, its vocalization unknown to us. Shortened forms of this name appear, given the various transliterations of Hebrew *yod*, particularly in compounds, as *Yah/Jah/Iah, Yahu/Jahu/Iahu, Yo/Jo/Io*, and *Yeho/Jeho*. Thus, to find YHWH's loyalists named Eli-Jah (YHWH is my God) or Obad-Iah (Worshiper of YHWH) is altogether to be expected. But how came it that scions of the Baal-worshiping royal family of the Northern Kingdom were given names that equally proclaim loyalty to the one and only God of Israel: Ahaz-Iah ("YHWH has taken [by the hand?]," an idiom for taking a mortal as protégé) and Jeho-ram (YHWH [has raised?] on high).

These considerations of historical credibility would seem to corroborate the suggestions I made in Tale 2 and again in Tale 4, that Baal worship—responsibility for which is variously assigned to Jezebel, Ahab, and Israel at large—is a metonym for improper practice (cultic or moral) in association with the worship of YHWH. The champion of YHWH, that is to say, of his proper worship, is the ideal prophet, epitomized—and satirized—in the portrait of Elijah. He is as elusive as a spirit, but he is present when the chips are down. He is under the special care of YHWH's Providence, which renders him proof against material scarcity and invulnerable to human dispositions of power—which, therefore, may account for his courage and boldness. But he is, like all prophets, human and vulnerable. He may despair of the war that never ends, no matter how decisive his triumphs in one or another battle. And (as in this tale) there are times when he quails at the prospect of putting himself within the king's reach and requires reassurance from heaven. He need not be a loner, but more often than not he is. And more often than not he is the one against the many. But as he speaks for the One God, so—however outnumbered by the opposition—he constitutes, in Thoreau's phrase, a majority of one. And it is to him alone, that is to say, to the God whom he faithfully serves, that Israel must have recourse when it seeks guidance in critical moments, whether crises in one's personal life or in the life of the nation. And one of the cardinal sins inveighed against by Scripture is a recourse to oracles that YHWH has declared illegitimate.

This, then, is the rubric for the faithlessness of Ahaziah, son of Ahab. The Baal worship of his Phoenician mother, Jezebel, was dealt a deathblow by Elijah on Mount Carmel; hence, another rubric for faithlessness to YHWH is required. And the continuity of this faithless streak in his family is subtly suggested by his paying tribute to the power of a "Baal"-deity, supposedly worshiped in one of Philistia's cities.

The garment of sheepskin or goatskin that Elijah is wearing is symbolic of someone who lives far removed from civilization's centers, where wool or flax are processed and woven into cloth. This clue alone is enough for Ahaziah—but not his courtiers—to guess the identity of YHWH's agent. This supports our earlier observation that the narrator is consistent in portraying Elijah as an 'outsider' in respect to the prophetic guilds, for the most part unknown in his own time to Israel, his role almost exclusively that of nemesis to the royal family. His role on Mount Carmel, played in sight of all Israel's tribal representatives, does not contradict this if we keep in mind that the narrator intends us to understand the confrontation there as metaphoric. And so here, too, the metaphor is extended. Ahaziah, like his father before him, tries to arrest Elijah, as if by laying hands on him he might contrive to make him sing another tune. This compliment to the autonomy of YHWH's prophet is in itself an insult to YHWH and thus compounds the offense of having ignored him (Was it because he was afraid in his heart of hearts to face his own God with the question?). Literally construed, Ahaziah's behavior is unquestionably nonsensical. But figuratively, it is as eloquent here as it was in the case of his father.

Now to the dispatch of three captains. The first one is necessary to the plot line. The obtuse officer who places his king's command above the mandate of God suffers his fate and carries his obedient sheep along with him. Dead men tell no tales, and witnesses there were none. The lapse of time before the second company is dispatched is not indicated. It may have been hours, a day, or days. (The very omission of the time factor is a feature of the author's metaphoric technique.) When a report from the field was long overdue, the king dispatched a second company. The second company commander may assume that his predecessor is incompetent. Afraid to report back on a failed mission, he is still beating the bush for the quarry or has lost himself in the rugged hill country of Ephraim. When this second captain comes so easily across Elijah, he congratulates himself on his skill, is emboldened to snap his command, and goes the way of his predecessor. Once again, no witnesses. But the third captain has reason to bethink himself. Two detachments have disappeared on this mission—into thin air, so to speak—on a mission to hunt and arrest a single man. When, then, he so quickly comes upon the object of his search, sitting so unconcern- edly out in the open, he takes a different tack. He does not know what happened to the others, but he can guess that whatever did happen was not totally unrelated to this composed prophet sunning himself on an outcropping rock. And the metaphor he employs for the failure of his colleagues to reappear—fire from heaven, a bolt from the blue—is yet another indication that the author intends us to take the destruction of the first two detachments as metaphor.[27] (This last suggestion may not be persuasive to those who have been inculcated with the theological and calumniating fiction of "the wrathful God of the Old Testament.") Our author is not interested in picturing YHWH as annihilating two companies of innocent soldiers for the discourtesy of their commanding officers. This should be apparent from the plea of the third captain. He could save himself and his men by simply leaving. But then, like Obadiah, he would risk death at his king's hands. His plea for mercy, then, is not just that he be spared the fire from heaven but that the prophet deliver him from his dilemma by putting himself into the power of the king. Little wonder then that Elijah requires explicit instructions from YHWH's angel that he is to go along with the captain's preposterous proposal.

Elijah appears at the king's bedside and delivers the oracle of doom but (as was earlier pointed out) with a slight change from the first two formulations. The Hebrew word translated "therefore now" in all three formulations of the oracle, appears regularly to introduce sentences of doom. It does not, however, specify cause. Elijah, this time, introduces an opening clause, "Because you sent messengers to inquire of Baal-zebub, God of Ekron." So only now do we learn that the injuries suffered in his fall by Ahaziah were not necessarily fatal. It was his lack of faith or his fear to put his faith in YHWH that determined the outcome he sought to avoid![28]

When we consider that this act of faithlessness is the only offense attributed to Ahaziah, it may appear that the sentence of death is somewhat excessive. What we must remember, however, is that Ahaziah can blame only himself for his fall. YHWH might have healed him had he asked. He did not ask, and YHWH did not heal. But we must not overlook another element implicit in this tale. The line of Ahab is under a sentence of doom, pronounced by Elijah in great detail in the preceding story. It would appear, then, that Ahaziah is doomed in any case; and if this be so, this entire story would seem essentially pointless. But a sentence of doom in the Bible differs from a similar phenomenon in Greek literature, such as, for example, the curse on the House of Atreus. As is well known, a controlling construct in the Greek mythic and tragic tradition is the ineluctable operation of Fate. In Israel, too, the causality explicit in the teaching that God punishes children in the third and fourth generations for their ancestors' sins is interpreted as constituting a fatal causality, a causality that would seem to contradict freedom of will and allow sufferers of God's wrath to deny reponsibility for their own fate—hence the saying in Israel, which is rejected by the prophets, that "the fathers have eaten sour grapes, but it is their children whose teeth are set on edge" (Jer. 31:28–29; Ezek. 18).

God does add in the weight of ancestral sin in determining punishment for a given generation, but only when that generation itself has sinned so grievously as to provoke punishment; for an innocent generation will not suffer at all for parental crime. So it is in the case of Ahaziah. The doom pronounced on Ahab's line is not determined for a specific generation. Ahaziah might well have enjoyed a long and prosperous reign. As it was (unlike his brother who succeeded him on the throne), he died in bed, prematurely and in the second year of his reign. This fact of history is interpreted by our story's narrator: Ahaziah did not die for the sins of his father Ahab (who himself won a reprieve in his own lifetime by doing penance): his life and career were cut short because of his own faithlessness to YHWH.

Poetical Review

Repetition, dialogue, point of view. Examples are the three communications of the one oracle from God to Ahaziah, with particular attention to n. 27, the three companies dispatched to apprehend Elijah, and the increasing impudence of the second company commander (as over the first) and the humility of the third in contrast to the first two. Also in verse 15 note the **free direct discourse** of the angel's bidding Elijah not to fear.

Character. Character is revealed in the picturing of Elijah as long unheard from, only dimly remembered, and recognizable, if at all, by his outlandish garb, a hairy animal skin cinched around the waist by a leather rope.

Literary and metaliterary conventions. Note the *Yh/Jh* theophoric element in the names of the Omride dynasty, pictured as apostate from YHWH and the seeking of oracles from beyond one's national borders. See also, in particular, n. 26.

Poetic integrity. Consider the function of this narrative in the developing theme of God's punishment of Ahab. Jezebel, and their progeny.

Tale 6: Elisha Succeeds Elijah

Introduction

1. The circumstances of YHWH's carrying off of Elijah, aloft in a whirlwind, were [as follows]: Elijah left (and Elisha) from Gilgal [in this manner:]

Episode A

2. Elijah said to Elisha, "You, stay on here. YHWH has dispatched me—as far as Bethel." Elisha said, "By YHWH's life and by your own life, I swear I won't leave you." So they made their way down to Bethel. 3. The guild prophets, there at Bethel, sallied out to Elisha. They said to him, "Are you aware that at this very time YHWH is about to take your master from command over you?" He said, "I also am well aware. Hold your peace!"

Episode B

4. Elijah there said to him, "Elisha, you, stay on here. YHWH has dispatched me toward Jericho." Said he, "By YHWH's life and by your own life, I swear I won't leave you." So they came to Jericho. 5. The guild prophets there in Jericho, presented themselves to Elisha. They said to him, "Are you aware that at this very time YHWH is about to take your master from command over you?" He said, "I also am well aware. Hold your peace!"

Episode C

6. Elijah there said to him, "You, stay on here. YHWH has dispatched me toward the Jordan." Said he, "By YHWH's life and by your own life, I swear I won't leave you." So the two of them continued on. 7. Now, some fifty members from among the guild prophets had come along and stood by at a distance as the two of them came to a halt by the Jordan. 8. Elijah took his mantle, rolled it up tight, and struck the water, which parted to one side and the other. The two of them went across on dry ground.

Episode D

9. The crossing made, Elijah said to Elisha, "Make a request. What may I do for you before I am taken from you?" Elisha answered, "May a double-share heritance of your spirit fall to me." 10. Said he, "No easy thing have you asked: if you behold me, as I am taken from you, just so, it is yours. If not, the answer is *no*."

11. There they were, going on and talking as they went, when lo! Fiery chariotry and fiery horses! These intervened between the two of them—then did Elijah go up

heavenward in the whirlwind, 12.Elisha, for his part, gazing on, he calling out the while, "Father mine, father mine, O horse-and-chariotry of Israel!" Then saw him no more.

Episode E

He took hold of his clothes and tore them into two sections. 13.Then he lifted high the mantle of Elijah, which had dropped from upon him. The he turned back and came to a halt at the Jordan's edge. 14.Taking the mantle of Elijah, which had dropped from upon him, he struck the water, declaring, "Where, now, is YHWH, God of Elijah, even he?" When he struck the water it parted to one side and the other. Elisha crossed over. 15.The guild prophets, those of Jericho, seeing him a way off, declared, "The spirit of Elijah has come to rest on Elisha!" Moving up to him then, they made obeisance to him, [face] to the ground.

Episode F

16.They then said to him, "Your servants number some fifty men, sturdy fellows. Let them go, pray, to seek for your master, against the event that YHWH's wind carrying him off, it has cast him exposed on one of the hills, into one of the gullies. He replied, "Dispatch not." 17.But as they pressed him without letup, he said, "Dispatch, then." So they dispatched fifty men; these searched for three days and came not upon him. 18.When they returned to him, he having stayed on in Jericho, he said to them, "Did I not tell you, "Don't go"? (2 Kings 2:1–18)

It is clear that the central theme of this sixth and last tale of Elijah is his succession by Elisha. That succession was ordained in Tale 3 by YHWH as a condition for acceptance of Elijah's resignation. It is noteworthy that unless the announcement of the drought at the beginning of Tale 1 was made in person at the royal court, the one and only public appearance of Elijah was his notice of the end of the drought and the consequent duel on Mount Carmel. That confrontation alone represents the entire career of Elijah before he offers his resignation. Subsequent to that scene at Horeb, Elijah's career, after his recruitment of the man destined to succeed him, consists of two brief appearances: one to announce (in private, for all we are told) his fate to Ahab, the other to announce his death to Ahab's son, Ahaziah. How much time passed between Elisha's recruitment and succession we are not told, and we can only guess what training Elisha underwent before the beginning of Tale 6. But this story is all about succession to Elijah's role, and there is nothing for us to investigate except the puzzling scene of the succession itself and the scenes that precede and follow it.

A comparison of my translation of the first two verses of this tale with such translations as the Jewish Publication Society's, the New English Bible, and the Revised Standard Version will compel agreement that mine alone is faithful to the tenses and syntax of the Hebrew. It also makes sense of the text, whereas the others do not.[29] Just as the first verse of chapter 17, beginning the Elijah cycle, identified him without introducing him, similarly, this opening sentence begins with the circumstances of his passing as if, the story of his being carried up to the skies by YHWH being a matter known to all, the narrator were now about to offer some additional information. The occasion having been specified, the author now breaks off the circumstantial clause and continues with the synoptic line, the bottom line he will lead up to again in the resumptive verse 2 of Eposide A. The bottom line is, "Elijah left from Gilgal; oh yes—so did Elisha."

Three times Elijah tries to go off somewhere without the company of Elisha, each time (so he says) at YHWH's bidding, on a mission that is neither specified nor fulfilled. Each time, Elisha swears that he won't be left behind. It is Elijah who is going off; but the oath Elisha takes is that *he* will not abandon Elijah, leave Elijah behind! Twice, at each destination, prophets come forward. They do not address the master at all. They speak to the disciple, telling him with a confidence that, one would think, could derive only from a special revelation, that his master's hours on earth are numbered. Each time, the disciple responds that he stands in no need of their information and brusquely tells them to hush up.

When master and disciple reach the Jordan, Elijah rolls his mantle into a cylinder (like Moses with his staff raised over the Red Sea) and causes the water to part—but here by striking the Jordan with the rolled-up cloak. On the other side of the river, Elijah casually mentions his imminent departure, asks Elisha to make a wish, and then—the wish having been made—tells Elisha that the wish will be granted only if Elisha sees him as he departs. The question is, How could Elisha fail to see his companion depart unless he were suddenly to be struck blind? This is followed by the sudden appearance of the fiery horse-and-chariotry (the Hebrew can be singular or collective), the puzzling utterance of Elisha, and the disappearance of Elijah. Elisha tears his own garments (why?), picks up Elijah's fallen mantle, and uses it to repeat the miracle of parting the water. Back on the western bank, he is recognized as Elijah's successor by the prophets, who insist on making a search for his master. The search is unsuccessful, and the story concludes with a seemingly bathetic I-told-you-so addressed by Elisha to the searchers.

In the detective novel genre, for all its not being rated high in the realm of fictional literature, it is assumed that a detail of decoration (say, a moosehead trophy on the wall of a den) will not be included in description unless it will somehow figure in the resolution of the mystery. Let us see now whether every detail in our narrative cannot be interpreted as plausible in itself and consistent in its function with the main theme of the story.

The reason for introducing the story with a synoptic notice of a departure from Gilgal by Elijah (and Elisha, too) is to draw our attention to something unusual about the circumstances leading up to that departure. If, since his recruitment, Elisha has been serving as an apprentice attendant (perhaps in some respect like Joshua to Moses), one would expect that his master would normally take him along on his missions (despite his not being mentioned as present when Elijah discharged his two missions, to Ahab and Ahaziah). Why, at the start of this mission, would Elijah suggest that his attendant stay behind? The answer, as we shall soon discover, lies in the meaning of Elijah's ascent heavenward. It was not an interlude in his career, it was that career's closing incident. Elijah is going to die, and he is not looking for company. We realize that YHWH's dispatch of Elijah from Gilgal is not in fulfillent of a mission but to meet the death he has asked for. Whatever role the apprentice might have fulfilled on a mission for God, there is no role for him in this projected journey.

In keeping with the nature of Elijah's journey and its destination is the preposition attached to the name of his destination: his orders from YHWH at this point will take him *'ad* (until, as far as) Bethel. But that is only a way stop. Once there, he will get further orders as to the next leg of his trip.

Why, though, does Elisha feel a need to take an oath that he will not be left behind? Apparently, he has guessed why Elijah is leaving and whither he is bound, why his company is declined and will continue to be declined—hence his notice to Elijah: neither command nor suasion will avail, he is determined to go along. And that he has guessed correctly the terminal nature of his master's journey is soon enough confirmed by the question put to him by the guild prophets at Bethel and again by the guild prophets of Jericho. The fatefulness of the journey for Elijah, the knowledge of it somehow revealed to the two prophetic communities, and Elisha's correct reading of its import are underlined by the three proposals that Elisha stay behind—at Gilgal, at Bethel, at Jericho—and his refusal to comply. But since nothing of note occurs at any of these places, why are they named at all?

The one place name that makes sense is Jericho, the last settlement before one reaches the west bank of the Jordan. What about the other two? Bethel's site is well known, centered in the Ephraimite mountains, miles removed from the Jordan as the crow flies, and a much longer journey uphill and down by foot. But where is Gilgal? Biblical atlases bear witness to two candidates for identification as ancient Gilgal, both located a mile or two from Jericho, a few miles west of the Jordan. But this Gilgal, where Israel under Joshua's command encamped before the assault on Jericho, cannot be the same as the one intended here. This would be "an identification patently absurd."[30] The only reason for the absurdity of such an identification is the literal understanding of this very story, for it would not make sense for Elijah to go west from Gilgal (What was he doing there to begin with?) to Bethel and then east from Bethel to Jericho, right back, almost, to his starting point. This is a fine example, and one of many we shall come across, of how scholars, gripped in the spell of literal-mindedness, will be thrown off the track by an absurdity in the biblical text; an absurdity placed into the text as a trail blaze. Everybody in biblical times knew that Gilgal and Jericho shared a location. And the whole point of these destinations is to tell us that in going westward to Bethel and then back eastward to Jericho, when the ultimate destination to begin with was the Transjordanian desert in the east, Elijah is acting out a metaphor. He is turning back on his tracks, a universal tactic on the part of a huntsman's quarry to throw the tracker off its scent! Elisha's prophetic instinct, so to speak, is being tested; and Elisha does not disappoint. Lest we fail to read the metaphor correctly, the message is reinforced by the question of the prophets at Bethel and Jericho to the yet-apprenticed Elisha. They know where Elijah is headed, being prophets; but does Elisha know? "Of course I do!" he answers. But why does he then go on with the apparently pointless rebuff "Be silent!"—often rendered "Say no more!" perhaps even with the force of "Button your lips!"?

The answer to this question must lie not in the knowledge of the coming event (which has now been established as common to both Elisha and his interlocutors) but in the manner in which they formulated their question. The Hebrew original does not have the prophets asking, "Do you know that YHWH is taking your master *from you* [or *away from you*]?" as rendered in the newer and freer translations. The Hebrew, as the King James Version attests, is "from your head." But even this is not altogether correct; literally, it reads "from over your head," that is, from having charge over you. And this is precisely what the narrator wants to convey, why he repeats the wording exactly from the lips of the two prophetic bands. They are not merely predicting the death of Elijah, they are interpreting the significance of the event for Elisha. Elisha might have reacted to his mentor's death in a number of ways. If he had become slavishly dependent on Elijah, it could have been with despairing grief. If his service to Elijah had become onerous, it could have been with a sense of relief, perhaps even with the prospect of abandoning a prophetic career altogether. If he had developed a sense of his own proficiency, he might have grieved over the loss even while he thrilled at the prospect of entering the prophetic office on his own. The context would indicate that the third is what the prophets think most probable. Even as they predict the event (no cause for rejoicing) there is, to be sure, a note of congratulation in their words to Elisha. Is it this congratulatory note that Elisha is repudiating? Or is it, perhaps, a nuance that we have not yet fully grasped? Once on the other side of the Jordan, Elijah acknowledges the reason for his coming there: "I am going to be taken from you. What can I do for you in my last moments?" Elisha's response is immediate: "I want to be your principal heir." This is why he insisted on accompanying Elijah in his last moments. And this is the point that the prophets had missed. It is not a question of Elisha's standing on his own, nor of Elisha making the transition from apprentice to journeyman prophet, He wants to be *the* successor, not *a* successor, to Elijah's prophetic preeminence. Elijah revealed his own standing as the successor of Moses in dividing the Jordan with his mantle. There were other prophets, clearly, in Elijah's time; but he was not just one of many. There will be other prophets in Elisha's time; but he, too, does not wish to be one of many. All this is expressed in his prayer that he be granted a double portion "in your spirit." A double portion is the share of a parental estate that falls by law to the firstborn. (If there are two sons, the elder gets double the younger's share, or two-thirds; if there are eleven sons, the eldest gets two-twelfths.) Elijah's estate is his "spirit." Elisha asks for twice the share in that spirit than will fall to any of the prophets who survive Elijah. And let us note that Elisha does not ask Elijah to grant this boon to him, he offers a prayer that the double portion be granted to him. To this Elijah's answer is that the decision is not for him to make anymore than a father was (according to God's word in Deuteronomy 21:16) given the prerogative to determine which of his sons was to receive the double portion. The sign by which Elisha may know whether God has granted his prayer will be whether or not he "sees" Elijah at the moment of his departure. What is the meaning of this "seeing"?

There is a common misconception that Elijah was "translated"—physically borne up to heaven on a fiery chariot drawn by fiery, flying horses. This is due, of course, to the graphic imagery here. In our text, however, it is by a whirlwind that Elijah is lifted up and the fiery chariot and horses never lift off the ground. They intervene, on the ground, a vertical barrier separating—and screening from one another's view—master and disciple. Elisha, on "seeing," cries, "My father, my father!" an epithet that here can refer only to Elijah. The word *father* in Hebrew can mean progenitor, sire, or master (as *brother* can mean sibling, peer, or equal and *son* can mean offspring, vassal, or subordinate). And Elisha is referring to his departing master. In apposition to "Master, master" are the words, "the horse-and-chariotry of Israel." Then Elisha saw no more.

Well, what was it that he did see? He saw—as if projected on a screen intervening between himself and his master—flaming horses yoked (the word *chariot* is a collective) to a squadron of flaming chariots. Not merely the visual faculty is intended by "seeing" here but perception, penetrating understanding. Only he who truly understands the critical role and function of the prophetic office is worthy of succeeding to it in its highest degree. Elisha perceives in the vision what it is that Elijah has been: in his one person, the armored divisions of Israel and in his representation of God's will to Israel, Israel's first and best line of defense! When the phantasm disappears, Elisha knows that he has seen Elijah as he was taken from him, that he has been chosen as his successor. His garments, probably symbolic of his old identity or status, he rips in two.[31] He does not merely pick up—he exultantly lifts on high—the mantle that could have disappeared along with its owner, the mantle twice described as "the one fallen from upon his [Elijah's] shoulders."

Elisha returns to the Jordan, with the mantle of Elijah, fallen from his shoulders, fallen now to Elisha; he strikes the water as did Elijah before him and exclaims in the form of a question, a question that rings as a challenge, "Where, now, is YHWH, God of Elijah, even he?"[32] The answer, to a degree, is in the effect on the waters. As YHWH was with Elijah, so is he now with Elisha. As Elisha crosses dryshod, the Jericho band of prophets witnesses the wonder and, not unnaturally, concludes that Elijah's spirit had settled upon his disciple. But the grandiloquence of Elisha's challenging question on the one hand and the banal conclusion that the prophets draw from the wonder, on the other hand, suggests that something of greater significance eludes us yet, as it eluded the Jericho prophets who now do obeisance to him. That this is so is supported by the fact that the story does not end here. The denouement is yet to come.

The prophets had remained on the west side of the Jordan, at some distance from the river bank. We are told in verse 11 that after crossing over, Elijah and Elisha went deeper into the wilderness—in all likelihood out of sight of the Jericho band. Surely Elisha was asked what had become of Elijah. And for answer he referred them to the windspout of the tornado, which would have been visible to them from afar. Well, prophets are reasonable and decent men, their heads often in the clouds, to be sure, but their feet firmly planted on earth. They know that what goes up must come down. The wind of YHWH, which sucked Elijah into its vortex, was seen headed in one direction. It must have cast (the same word used for *exposure* of bodies, in contrast with their burial) his body on some barren peak or into some forsaken gully. Is it proper so to neglect the honored prophet's remains? They volunteer for the arduous search. Elisha, apparently unfeelingly, declines their offer. He offers no explanation. They therefore insist. At last he gives in. For three days they scour the wilderness, fifty of them. Their search fruitless, they return exhausted to the successor who has waited for them in Jericho. Waited so that he could greet them with a sneer? "I told you not to go. I knew you would not find him!"

How did Elisha know that Elijah would not be found? Or would not a better question be, What is the meaning of their not finding the body, a meaning they did not grasp and he had known in advance? The clue to the answer is in the end of Moses, the

paramount prophet whose successor Elijah was. In Deuteronomy 34, Moses, at
YHWH's command (32:48–50), ascends the peak of Mount Nebo, east of the Jordan,
"east of Jericho," for a view of the land of his heart's desire, the land he may not
enter. There he dies and an unspecified "he" buries him there—the subject can only
be God—"no one ever having knowledge of his burial place until this very day." The
prophet is not in his body but in his spirit. As was the case with Moses, who
disappeared in the hill country of Moab east of Jericho, so now is the case with his
successor Elijah. The physical remains, the husk of the prophet, is tended to by
Providence. The prophet himself—his essential identity, his spirit—has taken up
residence in a new frame. The prophets of Jericho could not find Elijah because he
was waiting for them in Jericho, in the person of Elisha!

This, now, is the denouement not only of this tale but also of Tale 3. There we
were told of YHWH's charge to Elijah to anoint Hazael as king of Aram, Jehu ben-
Nimshi as king of Israel, and Elisha ben-Shaphat as "prophet in your place." The
succession of Hazael to the throne of Aram, as told in 2 Kings 8, is arranged by
Elisha. The succession of Jehu to the throne of Israel, as told in chapter 9, is arranged
by agency of a prophet who acts as Elisha's proxy. Only Elisha, of the three, was
literally named to succession by Elijah. But the metaphor is now complete. Elijah did
fulfill YHWH's commission—in the person of Elisha.

One metaphor is complete, but not another. Or to put it differently, one aspect of
the metaphor is complete, but not another. The Jericho prophets, on witnessing
Elisha's parting of the Jordan's waters hailed him as his master's successor with the
words, "The spirit of Elijah has settled on Elisha." The image, like that of the mantle
transferred from one set of shoulders to another, is external—the spirit alighting
upon. But Elisha dispensed with such imagery altogether when he struck the water
with Elijah's mantle, even with the attenuated imagery of "the spirit." His exclamation, "Where, now, is YHWH, God of Elijah, even he?" is a bold affirmation—even
as YHWH was hitherto in Elijah, he is now in Elisha!

Were Elisha telling this story about himself, it would bespeak a conceit and
arrogance bordering on egomania. But it is not Elisha who is telling the story. It is that
unnamed genius who exploits narrative art at its acme in the service of theology. He is
the author—or the identical twin of the author—who, in the story of the golden calf,
made the claim for Moses of YHWH's presence in his person that is made here for
Elijah and Elisha. Bold as the claim is, it is but a repetition of a metaphor that precedes
it, a metaphor in which the prophet is virtually identified with YHWH himself.

One of the common epithets for YHWH is God of Hosts. Another term for Israel's
God is YHWH–Hosts. It is commonly assumed that this latter term is an elliptical
contraction of the former; thus, YHWH–Hosts equals YHWH, (God of) Hosts. This
assumption is erroneous; for whereas in the one, YHWH is pictured as the God who is
in command of countless armies, in the other it is YHWH himself who constitutes those
armies. The confirmation of this is to be found in Numbers 10:35–36. The Ark of the
Covenant symbolizes the throne of Israel's invisible God. When the Israelites broke
camp, the ark would be carried in the lead "to scout out a secure campsite for them"
(verse 33). "When the ark was to set out, Moses would say, 'Arise, O YHWH, that
your enemies be scattered, that your foes flee from your Presence!'" (verse 35).
YHWH leaves his throne (so to speak) and clears the way for the Israelites who follow.

"When it came to a halt, he would say, 'Return, O YHWH, O Israel's myriads of thousands!' " (verse 36).[33]

YHWH constitutes "Israel's myriads of thousands," that is, Israel's defensive legions, in this passage from Numbers and everywhere in the name YHWH–Hosts. Elijah constitutes "Israel's horse-and-chariotry," that is, Israel's armored divisions, in the words of Elisha, hailing his departing mentor. Metaphor? To be sure. Anything else would constitute—blasphemy.

Poetical Review

Repetition (incident, **dialogue,** narrator's words). Examples are Elijah's bidding to Elisha and Elisha's reply (three times, verbatim); the question of the prophetic bands and Elisha's reply (twice, verbatim); the mantle of Elijah which had "fallen from upon him" (twice); and the striking of the waters with Elijah's (rolled up) cloak and its effect (twice). Note the **synoptic/resumptive technique** in verses 1, 2.

Dialogue. This constitutes about 45 percent. Note that verbatim repetition in the mouths of two widely separated bands of prophets suggests **free direct discourse** and metaphor—as does the pregnant expression *from over your head.* Also note free direct discourse in preposition *as far as* (verse 2).

Metaliterary conventions. Examples are the location of Gilgal, the itinerary, and the **metaphor** of a quarry doubling back on its track.

Poetic integrity. Within the Book of Kings there are the imagery of the attendant left behind in 19:3 compared to Elijah's attempts to treat Elisha similarly and Elijah hailed as Israel's horse-and-chariotry here as he departs the scene compared to Elisha on his deathbed hailed exactly the same way by the king of Israel (2 Kings 13:14). Across biblical books there are the imagery of YHWH and his prophets as Israel's first line of defense and the presence of YHWH in his prophets (Exod. 34; Num. 10:35–36; 2 Kings 2:12, 13:14).

Narrative genre. Is this story a fable, legend, parable, or what-have-you?

At this midway point in our treatment of prophetic narrative it might be of help to some readers if we attempted a review of the foregoing material with a view to synthesizing the various stories featuring Elijah, as well as synthesizing this block of narrative with the chapters on Jonah and Moses. On the other hand, there is the danger that our conclusion might strike some as a self-serving exercise justifying our arguments in Part I or as a rehash of kerygmas as made explicit in our discussions and of the referents already provided in the Poetical Reviews. There is another and greater risk—a descent into pedantry and (at that) a pedantry bespeaking a conceit and disrespect for the reader that we were at pains in chapter 2 to disavow. The nature of literary criticism and appreciation, as indicated in that chapter, is such that no single analysis should pretend to either exhaustiveness or airtight argumentation.

I am aware for example, that the reader will have made many connections between and among the units on Jonah, Moses, and Elijah long before I got to them. Similarly, I do not preen myself on modesty or humility when I confess my certainty that many readers will make connections within the material treated and with other material in Scripture that have not occurred to me. This certainty derives from my experience in classes and seminars where initial chagrin at being anticipated by students gave way to a sense of gratification that my approach was not utterly eccentric.

Many of these connections may persuade a skeptically inclined reader that the choice of narratives for presentation in this volume was not arbitrary. Thus, an examination of the Elisha stories, which might have appeared in chapter 7 but do not, will show that like stories about Jonah and Moses and Elijah (and Elisha), they fall into the category of fiction, rather than historiography. The arguments for this conclusion will continue to mount in chapter 6, where the historical nexus seems to be far more pronounced than it is in the Elijah stories. At this juncture I should like to expand upon a few points raised in chapter 1 that I believe have been illustrated in the narratives studies so far.

The works in the fiction **genre** are, all of them, of a parabolic nature. I have employed *didactic, fabulary,* and *theological* as descriptive of this so-very-serious fictional literature. Meir Sternberg characterizes it as ideational. None of these adjectives is wrong, none is quite adequate. The common denominator of all the stories treated so far is their kerygmatic constituency, the insights about the psychology of humans and their stances vis-à-vis one another and the One God whose attributes they so often forget, mistake, or ignore. These kerygmas (or preachments or morals), more often implicit than explicit, are the hallmarks of these parabolic fictions. And as this feature informs on the stuff of these narrative **plots** so does it dictate the nature of their **characters and characterizations.** Characters in histories are particulars—hence individual and essentially round. Characters in parables are essentially types, or flat. By some genius of creativity that may not be subject to analysis or mapping, the authors of Scripture manage in far briefer compass to achieve what a Dostoyevsky does on a large canvas. Like the four brothers Karamazov, their father, and Father Zosima, who loom so large as prototypes yet live in the flesh of sense and sensuality, Moses and Elijah appear as many-sided avatars yet seem to swell into three-dimensional personalities who invite us to an ever deeper sounding of their human depths. And somehow, in this process, the intangible God whom they represent (more or less faithfully at one time or another) grows in our consciousness as person and living reality.

6

The Prophets of Ahab's Wars

In chapters 3–5 I have analyzed and, to a degree, interpreted three units of biblical narrative. The first unit is explicitly defined as such by the Bible itself: a book of four chapters, featuring a single human protagonist, with a single plot line. The second unit is defined by its setting. It, too, features a human protagonist, a plot line that almost loses itself in the intricate design of its episodic development. The third unit consists of six stories, each with its own plot, held together by the one protagonist who figures in all six (and only these six) stories.

All three units have a common denominator. The protagonist is a prophet, a mortal in the service of a divine sovereign. And the plot lines of these various units, for all their variety, have in common that through them the role of the prophet is delineated—its weight and dignity, its burden and privilege, the ineluctability of the office, and the vagaries of its incumbents—not to speak of the mortals to whom the incumbents of the office must address themselves. The tensions in all these units are between a God concerned for humanity and a human constituency that fails to perceive him or his concern properly; between this God's disposition of reward and punishment in sanction of the norms he wills for human conduct and his sometimes perplexingly gracious toleration of mortals whose freedom to sin is increased in scope by his failure to punish as befits the crime; between the prophet's role as the prosecutor uttering God's threats and the pleader interceding with God for a stay of sentence; between the prophet's unquestioning commitment to the supreme service to which he has been called and his human fallibility, which renders him vulnerable to impatience, fear, even despair; and between the respect and reverence that the prophetic office should (and sometimes does) inspire in the prophet's audience and the cavalier treatment he often receives at the hands of the very people whose best line of defense he is.

We have far from exhausted the subtle plays on these themes in the several narratives treated. The richness and depths in them, which invite further exploration, are dimensions of the literary genre to which these ideological narratives belong. Whatever their matrix in real life, they are imaginative invention, fiction, fictional history, historical fiction, legend, myth, fable, allegory, what-have-you—almost anything but historiography, as this term is narrowly construed today. They are more fiction than fact yet lay claim to truth so real and lasting as to reduce "historical narrative," by contrast, to coincidental and incoherent chronicle.

Intervening in the tales of Elijah are two narratives in which this prophet does not figure. And in contrast to the Elijah narrations, which feature only one public appearance by that prophet and feature both prophet and royal foil in essentially personal and private preoccupations, these two narratives involve the king in the high business of state proper to his office: the waging of war—sometimes defensive, sometimes in a preemptive offensive, sometimes irredentist. In a measure, these two narratives fill out the "historical" picture of Ahab's reign. Who would credit the picture of a mighty king who does nothing for a score of years but forage for fodder in drought, indulge his foreign-born wife in murderous religious politics, abandon her cause and the cause of her church at the first setback, and engage in peevish hanky-panky with his summer estate neighbor? In a measure they serve to distract our attention from the scantiness of incident in the formidable Elijah's career. Both narratives, having as their setting war between Israel and Aram, are universally regarded in biblical scholarship as being of a primary historical nature.[1] Our examination of them will show that the opposite is true. Both belong to the same genre as do the Elijah stories. Both feature prophetic activity at the crux of their plots; and both are concerned with issues of war and peace as they are pursued by the state, abetted by prophets sometimes in true representation of divine will and at other times in sincere but dubious understanding of the God who is not always warlike.

Tale 1: The Prophets
in the Aramean Wars

Episode A: Aram Attacks Samaria

1.Ben Hadad, now, king of Aram had gathered all his forces, thirty-two kings accompanying him with horse and chariotry. He marched up and laid siege to Samaria in his attack upon it. 2.He sent envoys to the city, to Ahab, king of Israel, 3.with this address: "Thus declares Ben Hadad, 'Your silver and gold are mine, your choice wives and children are mine.'" 4.The king of Israel declared in response, "Just as you say, my lord king: yours am I and also all that I own." 5.The envoys came again with this word: "Thus declares Ben Hadad: 'In my dispatch to you, to wit, Your silver and gold, your wives and children you are to hand over to me, 6.I meant exactly this: tomorrow at this time I shall send my servants to you, they will scour your palace and the mansions of your liegemen; what you most prize they will seize and bear off.'"

7.Then did Israel's king summon all the land's elders. He said, "Take note and conclude: that fellow is only bent on mischief! When he sent demand to me for my wives and my children, my silver and my gold, I held nothing back. [But now—]" 8.The elders and the people all declared, "Obey not, consent not." 9.So he said to the envoys of Ben Hadad: Say now, "To my lord the king: All that you dispatched to your servant the first time I stand ready to do, but this last I cannot do." They brought their word back to him. 10.Then sent Ben Hadad as follows, "May the gods do such-and-such to me and worse if [when I am finished] there remains a handful of Samaria's earth for each of the host that follows me." 11.Then spoke the king of Israel in response, "Speak this: 'Let him who is buckling on the sword belt not brag as if he were already unbuckling it!'"

12.When he heard this reply (he now engaged in tippling, he and the kings under their bowers), he said to his men, "To the assault!" And they assailed the city.

13.Now a certain prophet came forward to Ahab, king of Israel. He said: "Thus has YHWH declared, 'Do you see that great horde? Lo I now deliver it into your power, that you may know that I, YHWH [am doing this].'" 14."By whose agency?" asked Ahab. Said he, "Thus has YHWH declared: 'By the guard-commanders of the shire chieftains.'" "And who shall signal the attack?" he asked. He replied, "Yourself!" 15.So he mustered the guard-commanders of the shire chiefs, they were 232 in number, and behind them he mustered the army, even the entire Israelite militia, some 7,000-strong. 16.They sallied forth at noon—the while Ben Hadad was carousing drunkenly under bower, he and the kings—thirty-two kinglets, his allied support. 17.With the first sally of the shire guard-commanders word was sent back to Ben Hadad; they told him, "Men have issued forth from Samaria!" 18.Said he, "If for truce they've come out, grab 'em alive and if for battle they've come out, grab 'em alive." 19.These, for their part now, had poured out of the city, guard-commanders of the shire chiefs with the army behind them. 20.Each struck down his man [and whoever rose in his place]. Aram took to flight, Israel in pursuit. Ben Hadad, king of Aram, however, made good his escape by horse and mounts. 21.[The circumstances of the defeat and the escape:] the king of Israel [himself] sallied forth, assailing the horse-and-chariotry and inflicting on Aram a grievous defeat.

22.That prophet came forward to Israel's king, he said to him, "Proceed, reinforce, consider well how you must act: a year from now the king of Aram comes up [again] against you."

23.Now the king of Aram's ministers, for their part, said to him, "Their gods are mountain-gods, so they proved stronger than we. But if we battle them on level terrain, [just see] if we do not prove stronger than they. 24.Yet this one thing is a must for you: remove the kings from their respective commands, replace them with [your] deputies. 25.And yourself, now, mobilize for your own force a battle unit for each unit you lost, even horse for horse and chariot for chariot. And doing battle with them on level terrain, [see] if we do not overpower them." Heeding their advice, he took such measures. (1 Kings 20:1–25)

Our author employs a number of strategies to show Ahab in a favorable light in this episode, as in this chapter generally. One way is in contrast to his adversary, Ben Hadad. Another way is in his stance vis-à-vis YHWH's prophets and theirs to him. A third way is in the outcome of the events, determined by God. A fourth signal is the name, title, or combination of title and name when explicit reference is made to either of the two kings. Thus, for example, Israel's monarch appears in this chapter thirteen times: twice as "Ahab, king of Israel," nine times as "Israel's king," and twice as "the king." Not once does he appear with that easy familiarity redolent of contempt as simply "Ahab."[2] Lest we miss the point (and it is perhaps the intent of the author to condition our predisposition subliminally), in the stories of Elijah's dealings with him (chapters 17–19, 21) the monarch is identified twenty-seven times as "Ahab." Only twice is he granted the epithet *king:* once in 21:1 with the sardonic label, "Ahab, king of Samaria," and once in 21:18, where YHWH speaks of him as "Ahab, *king of Israel, who is in Samaria,* yes there in Naboth's vineyard whither he has *descended* to take possession." Another contrast is the usage for Aram's monarch, ten times he is "Ben Hadad," once "king of Aram," and twice "Ben Hadad, king of Aram." This last usage is in 20:1, where the identification is demanded by the context, and in 20:20, where the epithet is a subtle sneer: "Ben Hadad, *king of Aram,* made good his getaway *by horse and steed.*"

In both the substance of the positions taken by Ben Hadad and by Ahab, and in the dialogue expressing these positions, our author is unabashedly intent on displaying the king of Israel at his reasonable best and the king of Aram at his unreasonable worst. First in regard to Ben Hadad's invasion of Ahab's kingdom. No motive whatever is given for the attack. It is not to secure his own borders against an attack from the south; it is not in the interest of taking or retaking border cities; it is not to force the king of Israel into an alliance against the Assyrian threat from the north and east; it is not a war of colonization, that most ferocious of wars of aggression in that it aims to exterminate the natives and to replace them with one's own surplus population. That leaves only one motive, the motive that impelled Assyria's imperial tide for half a millennium: to reduce its neighboring states to vassalage and to exact maximum tribute at minimum cost. For Ben Hadad, then, to escalate his demands to the point where his victim has no choice but to resist is not only brutal but self-defeating, even idiotically so. But let us look at the dialogue.

Twice, the weakness of Israel's king is expressed in Ben Hadad's ultimatums by a hyperbolic image of utter helplessness: all of Israel's king's possessions are as good as his! The first time, the king of Israel accedes, acknowledging Ben Hadad as "my lord the king." This signals his readiness to discuss the amount of tribute, to capitulate but not necessarily to surrender. But the Aramean, emboldened by this success, now demands what amounts to unconditional surrender. There will be no talk of a tribute to be paid, his soldiers will enter the capital and, if they choose, pick the city clean. This time he includes the property of Ahab's subjects as well as the royal estate. This now gives the king of Israel a way to put his case before his own senate, in a manner that can only stir them to resist, in self-interest and in loyalty to the king who has shown a selflessness matched by concern for his subjects. An interesting touch in the dialogue is that whereas Ben Hadad twice stated his demand for tribute in ascending order of value, first precious commodities, then the precious persons of the king's family (the latter perhaps, to be ransomed later or held as hostages to ensure future payments of tribute), Ahab inverts this order. He was ready to yield up his wives and children and, a fortiori, his wealth. The next remove from what he would naturally cherish would be his subjects' property, but that last demand he never mentions. As subtle a deployment of aposiopesis as ever would have delighted an Athenian rhetor! Having secured the support of his own subjects, the king of Israel nevertheless repeats his readiness to treat for terms, still addresses his tormentor as "my lord the king," only to provoke the Aramean into an oath expressive of a bully's rage or, better, a child's tantrum: for he threatens to cut off his nose to spite his face. He will so reduce Samaria that there won't even be enough dirt to distribute to his soldiers. Well, that should make everybody happy! This time, now, freed by this braggadocio to answer in kind, Ahab responds with manly firmness yet with regal restraint. Ben Hadad, however, is not only a braggart and a bully, he is also a bibulous general as inconsiderate of his own soldiery as he is impatient to bend an elbow. From under the shade of arbor or pavilion he orders another round of drinks for his craven client–kings while he sends his men to endure sun's rays and enemy arrows in an assault on the city's walls.

The point of view shifts from this picture of Ben Hadad, his carousing compan-ions, and his soldiers at Samaria's walls to a view from atop the city walls where a prophet holds converse with the king of Israel. This incident, which occupies verses 13–14, I shall reserve for later discussion. (Introduced by a nominal sentence, these verses are thus put into parentheses by the Hebrew itself; and the events that resume in verse 15 are in no way affected by our putting these verses aside.) Ahab musters his army, himself in direct command, the assault troops being the regular army units from the country's regional commands (rather than the king's own guard or the citizens' militia), all the forces together some seven-thousand-strong. As they sally forth from the city, the view shifts again to Ben Hadad now at high noon already in a state of advanced inebriation. There is something awry about the report of the sortie as it is brought to Ben Hadad, at least, if we take it literally: "Men have issued forth from Samaria." The **idiomatic** verb *yṣ'* can connote a delegation to hold a parley or surrender or a full-force sortie. But the context rules out ambiguity. The Arameans have been attacking the walls, which can only mean, at this point, with scaling ladders and, possibly rams at the gates. Suddenly the gates are opened and the attackers are pushed back as assault troops push through the gateway, clearing room for chariot charge and following infantry. Thus, the report to Ben Hadad is not ambiguous. The ambiguity exists only in his drink-muddled mind. He is so far gone that he cannot quite make sense of what it is that is being reported to him, but he does sense that he is being asked for orders. And the contingent orders he blurts out (or does he mumble them with thickened tongue?) are so nonsensical as to constitute high comedy: "Men are coming out, are they? Well, coming out for what purpose? To hold parley? Well, grab them and bring them to me alive." Alive? How else would one bring a delegation to sue for peace? And what need to seize them if they are coming out on a mission of truce? "Coming out? To make war? Take 'em anyway. Yah! Take 'em alive!" Take charging shock troops alive? Quite a trick, if you can do it!

The foregoing is another example of dialogue lines that are unplayable on stage, hence **free direct discourse.** In the interest of providing the reader with a chuckle or a guffaw? Yes, but only incidentally. Primarily, to complete the picture of the aggressor buffoon—bully, braggart, drunkard, derelict commander, and in the final scene coward—in ignominious flight, outstripping pursuit as he continues the gallop home "on horse and [re]mounts." In terms of poetic "interest," in whose interest is this character drawn? Not the God of Israel's, whose unlimited power exludes a straw man adversary and so, also, not his prophet. The only one left to benefit by contrast is his brother-king, Ahab of Israel, who has been benefiting by the contrast since the episode's beginning.

Verse 22 gives us a resumptive summation of the defeat of the Arameans, in keeping with the instruction of the prophet in verse 14 that the attack be loosed in person by the king of Israel. For all this commanding stature of Israel's king, however, and for all the buffoonery of the Aramean king, which led to his ignominious debacle, our author still has need of this menace from the north; thus, this epidode ends on a warning by the prophet that in a year's time the king of Aram will have sufficiently recouped so as to initiate a second campaign. The preparation for this second campaign is then given from the Aramean point of view. The defeat before Samaria is attributed by Ben Hadad's advisors to two factors: one divine, one human. Both factors may be corrected for. Since in enumerating both factors the advisors tactfully overlook any possibility of fault in the king himself, Ben Hadad readily accepts their analysis and their counsel. The divine factor is, to be sure, a nonsensical consideration. The potency of Israel's gods is limited to hilly terrain (where, perhaps, the Aramean edge in horse and chariots is cancelled out). In that case why did Ben Hadad choose that terrain to begin with? The human factor is equally ridiculous. The client kings who are being blamed for the rout did not mismanage their respective commands, they were not broken on the field of battle. On the contrary, they were at the time of the attack raising cups in toast to their tipsy overlord.

Episode B: Ben Hadad's Second Campaign

26. A year later Ben Hadad mustered Aram, advanced towards Aphek to do battle with Israel. 27. Now the Israelites, having been mustered and provisioned, went to face them. The Israelites encamped in opposition seemed like a couple of flocks of newly shorn goats, while Aram abounded everywhere.

28. Now a man of God came forward and made declaration to Israel's king. "Thus," he declared, "has YHWH declared: 'On account of the Aramean's having said, YHWH is a god of hills, not a god of vales, he—I shall deliver this great horde into your power, thus shall you know that I, YHWH, [do this]."

29. So did these war camps face one another for some seven days. On the seventh day the battle was joined. On that one day the Israelites inflicted on Aram casualties of one hundred thousand foot soldiers. 30. The remainder fled to Aphek, into the city. The city's wall fell, together with some twenty-seven thousand survivors; Ben Hadad the while retreated into the citadel ever deeper, room by room. 31. His attendants then said to him, "Here, now, we have heard say of the kings of the Israelite folk that they are magnanimous monarchs. Let us drape sackcloth from our hips, put ropes on our heads, and yield ourselves to Israel's king. Perhaps he will spare your life." 32. So they wrapped sackcloth round their hips, put ropes on their heads, and, reaching the king of Israel, declared, "Your servant Ben Hadad petitions, 'Spare my life.'" He replied, "Is he sound, my brother, he?" 33. These men now speculated—and quickly concluded—on whether he meant this sincerely, saying, "[Yes, yes,] your brother, Ben Hadad." "Go in, fetch him," he said. When Ben Hadad came out in surrender, he had him ascend into his chariot. 34. He [Ben Hadad] said to him, "The towns that my father took from your father I shall give back, and [duty-free] marts you may set up for yourself in Damascus as my father did in Samaria." "I, for my part," [said Ahab] "will release you for such a treaty." He concluded that treaty with him and set him free. (1 Kings 20:26–34)

If ever the political and military background of a campaign appeared murky to the historian, this one does. It is far from clear which side is the aggressor, which side makes bold to attack. If archaeologists have correctly identified the site of this city of Aphek, it lies about five miles east of the Sea of Gallilee. Apparently itself one of the contested border cities between Aram and Israel, it would seem from Ben Hadad's retreat into it after his defeat in the field to have been in the possession of the Arameans. This means that Aramean-held Aphek is merely the staging area for Ben Hadad's proposed incursion into Israelite territory. The king of Israel, having received prophetic warning about Aram's intentions and, doubtless, intelligence about Ben Hadad's having moved up to Aphek, moves his own army toward that border city to thwart invasion of his own territory. From then on, the logic of the narrative becomes more difficult to follow. The Hebrew word represented in our translation by "flocks" appears only here as a stative adjectival substantive. The root *ḥśf* has the sense of "bare, naked, exposed," hence "vulnerable." Therefore, the image is of two vulnerably small flocks, or flocks vulnerable to the elements by reason of recent shearing. But vulnerability is clearly the point of the description; Israel's army is hopelessly outnumbered by the swarms of Arameans. Why, then, do the two war camps face each other in inaction for seven days? Perhaps to suggest Israel's reluctance to charge so much larger a force and a Ben Hadad rendered pusillanimous by his previous experience with a tiny Israelite force? In any case, the battle is finally joined (by whose order we are not told), and the Arameans are signally requited for their sneer at the limited power of Israel's God. With one hundred thousand of Aram's soldiery dead, wounded, or fled, there yet remain more than a quarter of that number to take refuge behind the walls of Aphek. Israel's force, which in the previous year's campaign totaled seven thousand, now proceeds to lay siege to a city defended by four times that number. And since city walls do not come crumbling down of their own weight nor do beseiged soldiers crouch under them waiting for them to collapse, we must understand that the wall caved in at a point where it was undermined and that the wall did not "fall upon 27,000 defenders" (*pace* translations) but that *together with* (another meaning of the Hebrew proposition *'l,* "upon") the breach in the wall occurred the collapse of the defending army—except, that is, for Ben Hadad's royal guard, which resists fiercely as it is pushed ever deeper into the citadel's recesses.

It is hard to see what alternative to unconditional surrender was open to the cornered Ben Hadad. The narrative, which supplies the spoken, as well as interior, dialogue of Ben Hadad's courtiers suggests that nothing but a hereditary disposition to leniency on the part of Israel's king offered hope for Ben Hadad's survival. As they pleaded for their master's life, they were delighted by the solicitude of Israel's king for their master's well-being and concluded that his use of the term *brother* was not merely loose usage but betrayed the speaker's intent; in contrast to their own reference to Ben Hadad as "your servant" (that is, "your subject," by reason of defeat), his usage of a term bespeaking a continuing equality of status showed that Ahab had no desire to subjugate Ben Hadad. To this indication they give ready assent. But what if they had not given assent? What difference would it have made? They were all at Ahab's mercy! The answer can only be that once again, what we have here is **free direct discourse.** In verse 33, Ahab did not say directly, "He is my brother"; he asked, "Is my brother well?" So, also, when Ben Hadad's courtiers came out to speak to Ahab, they came not just to plead for a life that was forfeit, they came out to treat for terms. And they did have something to offer. What they had to offer is made explicit—again in free direct discourse—by Ben Hadad in Ahab's chariot. Both what Ben Hadad seems to be volunteering (as if for the first time and out of spontaneous good will) and the release offered by Ahab in reply are terms of "*the* treaty" (verse 34), which has already been negotiated.

Why then, if it was indeed to Israel's benefit for Ahab to negotiate a treaty that would bring peace, restoration of lost territories, trade concessions—all without further campaigning—does the author, who certainly knows all this and has managed to convey it to us, indulge in the rigamorale of all this scrambling of **time sequence** and free direct discourse? The answer is that he wants to focus our attention on those aspects of Ahab's motivation that can be characterized as generosity, mercy, forgiveness and peaceableness—judgment on which will be pronounced in the concluding episode.

Episode C: Aftermath of Battle

35.Now a certain member of the prophetic fraternity said to a compatriot, "By command of YHWH—deal me a blow!" The man refused to strike him. 36.He then said to him, "Because of your disobedience to YHWH's bidding, soon after you leave me a lion will strike you down!" (He left him, a lion came upon him and struck him down.) 37.He accosted another man and said, "Deal me a blow!" That man struck him, opening a wound. 38.The prophet then went to the road to await the king— disguised by a bandage wrapped above his eyes. 39.As the king came by, he appealed to the king in such wise: "Your servant was engaged in the thick of the battle when lo, an officer drew aside bringing an [enemy] officer to me. 'Guard this man,' he said. 'If he gets away, your life will be forfeit for his life, or you shall pay a talent of silver.' 40.Now your servant was engaged on this side and that, and—the man was gone." "That is the sentence upon you," said the king of Israel. "You have yourself made it final!" 41.Quickly he tore away the bandage from above his eyes. The king then recognized him—as one of the prophets. 42.He said to him, "Thus has YHWH spoken: 'Because you have released out of hand the man owed to me for disposal, your life is forfeit for his life, your people for his people!'" 43.Seething with distemper, the king of Israel made for home; thus he arrived in Samaria. (1 Kings 20:35–43)

The key to the moral of this story is the word *herem,* often rendered into English by *ban,* a synonym for *excommunication* in ecclesiastical context and, in broader usage, for *curse, proscription,* or *condemnation.* Here God speaks of Ben Hadad as "the man of my ban," which is perhaps less meaningful to the reader than "the man under my ban." More meaningful is "the man whom I doomed" (JPS); but for the implication that God wanted Ben Hadad put to death, it is unjustified. *Herem* is a term for the spoils of war that belong to God, not to the human victors. These spoils may, indeed, be destined for destruction, but they may also be devoted or consecrated to the exclusive use of God's service in his authorized cult. My rendering, "the man owed to me for disposal," avoids the possible but not necessary implication of divine bloodthirstiness in this particular case. The point—and the moral—is that the king of Israel was willing enough to accept God's oracle before the battle but did not see fit to consult his oracle in the matter of what disposition should be made of the enemy king. The two oracles of YHWH, preceding the two battles, were explicit in explaining why YHWH was granting victory to Ahab, namely, that he might acknowledge the power of YHWH. The victory was YHWH's, and Ben Hadad was YHWH's captive. By treating Ben Hadad as his own captive, Ahab was claiming the victory as his own, thus failing to acknowledge YHWH. Since there is no suggestion that these particular battles were in any way part of a "holy war," the moral must be read in its full metaphoric extension: all victories in battle are owing to YHWH; failure to acknowledge this is a sin, for which retribution will be exacted.

So far, so good—but in terms of literary analysis, nowhere near far enough, and far, far from good enough. Assuming that whatever the historicity of the events related in this chapter, we have before us a well-crafted composition designed not to entertain nor to convey information about people and events in a distant past but to edify, many questions remain to be raised, some purely literary and others of a **metaliterary** nature. In the category of the former, if the moral of the story as a whole is summed up in Episode 3, what need is there of Episode 1? Or if we need—for purpose of building up to the moral—two battle episodes and three prophetic interventions, why do we need the details of the dialogue between king and prophet? And what narrative purpose is served by the contrast between the sober and dignified king of Israel and the swinish buffoonery of Ben Hadad? On the borderline between literary and metaliterary is the question whether it was a normal occurrence (or thought to be normal) for deities to dispatch oracles when mortals had made no inquiry of them to begin with. In regard to the last of the three episodes in particular, was it normal for prophets to serve as infantrymen in warfare? Why would a prophet ask to be wounded? How many lions stalked the steppes of the Golan Heights or Israel's highlands at the time of this story's setting? And, moving back toward literary considerations, what is the point of the exaggerations in regard to the disparity of forces on the sides of Aram and Israel? In general, even if we are to proceed from the assumption that our author makes free and open use of fictional device for didactic purpose, is it possible for us—at our far remove from the Ancient Near Eastern setting—to separate out the fanciful elements from the real-life settings? Let us make the attempt, beginning with the conventions in Episode 3.

War is real. Armies were constituted of professional soldiery, expensively armed and armored. Such "uniformed" personnel would constitute, almost exclusively, a military force attacking an objective at a good distance from its home base. A defending force would also be augmented by a more crudely armed militia, recruited from the peasantry and townfolk of the invaded territory. The professionals fought for pay, rations, and a share of the spoils. Paladins would contest in individual combat, the arms and armor of the vanquished falling to the victor. A defeated champion might be taken alive, to be released after the battle on payment of a ransom, the richness of the ransom varying with the status of the captive. A talent of silver (about seventy-five pounds, the value of three thousand head of sheep) would represent a king's ransom, but it is in line with a standard penalty clause in cuneiform juridical decisions: if the defeated party in a suit contests the verdict, "he shall pay one mina (sixty shekels) of gold, one mina of silver." In the case of our own story, a king's ransom is particularly appropriate to the context. Life for life need not mean death—more likely it meant slavery. A common soldier entrusted with a noble captive would be obligated to withdraw from the fray and stand guard over the prisoner.

That a prophet would serve as a soldier in a battle is highly dubious. But a prophet might easily dress as an irregular combatant. We shall soon offer evidence that prophets bore identifying marks. To disguise his prophetic status the protagonist in our story would have to cover the identifying mark of the prophet, a cicatrix on his forehead, centered over his eyes and extending well past the hairline. Merely to cover the mark with a headband would invite a suspicion that the prophet cannot afford if he is, by his dramatic analogy, to trap the king into pronouncing sentence upon himself.

He requires an open wound, the blood seeping through the bandage, to disarm the king. But why not himself open a gash on his forehead?

The answer is, without question, that the author wants to tell the story not of the anonymous soldier who struck the blow but of the anonymous soldier who refused to do so and paid for his disobedience with his life. In the course of describing how an anonymous prophet came to pronounce YHWH's doom on the king of Israel, the author provides a prelude—after all, it requires only two verses—in which a parallel condemnation is pronounced (and fulfilled) on a disobedient soldier. But why should the author want to do so? How does the first condemnation relate to the second, in terms of dramatic development, of reinforcing the climax, of sharpening the moral? And what is the probability of the reader's assigning much credibility to this miraculous event? The charge of the prophet to the soldier is not witnessed by a third party (not explicitly, at least); and since the point of the wounding is to promote the prophet's subterfuge, it would hardly be consistent for the prophet to advertise it. Nor, similarly, is there any witness of the lion's attack, no indication of when or where it took place. Only the **omniscient narrator** knows, and he makes no effort to tell the reader how he came to know.

The problem is further compounded by another consideration. The context of the biblical ethos makes the moral itself highly unlikely. The person of the prophet is sacred, he is under the protection of an omnipotent Deity. Would the soldier's fear of laying a hand on such a person, a token of his awe before God, be impious? Or suppose the moral is that one must obey a prophet no matter how morally pointless the command. Would failure to grasp this point warrant a death sentence and the working of a miracle by God in order to execute the sentence? No, such a moral is not only unlikely, it is ludicrous.

Let us leave this question in abeyance and examine the moral of the king's condemnation. What was his sin? Unlike the poor soldier, who was addressed by a command in the name of YHWH, which he disobeyed, the king of Israel received so such command. The king of Aram surrendered to him and was at his mercy. He released his defeated enemy after concluding a treaty on terms favorable to Israel. An enemy is put to the ban or exterminated or the spoils dedicated to God only when the war is initiated at God's command and the ban is demanded by him at the outset (see, for example, 1 Samuel 15). Such is not the case in this war. Does a favorable oracle from YHWH before the battle constitute even a suggestion that the enemy is to be treated according to the rules of holy war? Not at all! The copious literature both in the Bible and preserved for us from Israel's neighbors reveals that consulting one's God or gods on whether and when to make war is the regular and ubiquitous practice; the kings of the Levant, Mesopotamia, and Transjordan in their victory inscriptions always give credit to their gods for the victories. But not a one of them nor any king of Israel is recorded as ever turning to his divine patron for guidance in the disposition of a defeated enemy. The word of YHWH (or, rather, the purported word of YHWH in the mouth of this prophet) can only be hyperbole for Israel's king's failure to acknowledge the support of YHWH. And even the charge that he had failed in this is gross hyperbole, for the battle has just ended and the king is on his way home, which is where thanksgiving to God is offered. Either God himself or the prophet who claims to speak in his name would seem to have it in for Ahab to condemn him on such flimsy grounds. And why the necessity for the condemnation or for basing it on such a transparently unconvincing basis? The stories before and after provide much sounder grounds—tolerating worship of the Baal, permitting Jezebel to slaughter YHWH's prophets. And the condemnation and doom is most convincingly pronounced in connection with the outrage against Naboth and his extirpated line! One is tempted to put the question, Who needs this story? But even to put this question would be to miss the point; for given the main currents of the biblical ethos, the entire story could be seen as a calumniation of the God of Israel: it renders him as nitpickingly redundant, risibly vain, and capriciously bloodthirsty—if, that is, the prophet is truly speaking for God.

Is it conceivable that a biblical author would present a prophet of YHWH's who does not properly represent God's judgment? The answer is that it is not only conceivable, it happens so over and over again. To cite but one example, of a most prestigious seer, Nathan—prophet par excellence to King David. In 1 Samuel 7 David in oblique fashion asks Nathan whether he should not build a temple for his God. Nathan's response is that the king should go ahead with whatever he is minded to do—YHWH is with the king. That very night, YHWH's word comes to Nathan: he must go to the king and inform him that permission to build the temple is denied. Nathan is not criticized for his earlier and opposite response to the king's question. Obviously, he had taken it for granted that YHWH would be favorably disposed to the building project. Quite as obviously, he was, this time, mistaken. God corrects the prophet's erroneous assumption but does not censure him for it, does not even chide him for the presumption of speaking for him on so weighty a matter without first consulting him.

So much for the possibility of a most reliable prophet's being mistaken. We need not labor the point—just recall the mocking of Jonah, the fun poked at the prophetic fraternities in connection with Elisha's succession to Elijah's authority. But the writing prophets are replete with condemnation of their rivals, whose oracles, contradicting their own, renders them "prophets of falsehood, who speak lies in my name." And in the next story with which we shall deal (1 Kings 22), we shall see how some four hundred of YHWH's prophets give one oracle in his name while one (and he alone vindicated) says the opposite. And in this we have another clue as to the subtle ways in which the biblical narrator, by arrangement of his story matter and his choice of prophetic protagonists, conveys his deepest intentions to the reader. Into the cycle of tales featuring the redoubtable Elijah, whose prophetic eminence almost equals that of Moses, he interpolates two narratives from which Elijah is absent, which feature anonymous prophets of YHWH's whose pronouncements may carry questionable authority or even none at all.

The biblical narrator has another way of indicating to the reader when a prophet is to be understood as speaking with unquestionable authority. In the Pentateuch, the reader may find himself becoming wearied with the repetition of the introductory sentence, "YHWH spoke to Moses as follows," which is often immediately followed by "Speak to the Israelites as follows" and this in turn succeeded (albeit rarely) by "This is the matter which YHWH has commanded." In the case of Elijah, who appears only a few times with an oracle from God, there are five occurrences of the formula, "The word of YHWH came to him,"[3] and twice the mission is relayed to him by "an angel of YHWH's.[4] In this chapter (as in chapter 22) no such phrase appears. While the presence of such a phrase signals the narrator's intent to convey that a prophet is speaking in YHWH's name and not his own,[5] its absence does not constitute a sign of the opposite—it merely leaves the question open. Now let us examine the prophetic appearance in each of the three episodes.

First, let us recognize that in terms of plot or action the prophetic appearance lies at the heart of, and is the raison d'être for, the third episode. The prophetic appearance in the first two episodes (as we saw, in our ability to skip them in our discussion) are incidental to the action there and could have been omitted without loss of continuity, congruence, or perhaps even moral. Their presence in these episodes, then, would have to be accounted for in terms of their indispensable and integrative function within the kerygmatic design of the narrative as a whole (thus pointing to the content of the third episode as the raison d'être for the first two episodes).[6] In the three episodes the mantic figure is a certain prophet, a man of God, and a certain one of the prophetic confraternity, respectively. There can be no question that in terms of the author's vouching for a prophet's credentials, the surest accolade is the rare honorific "man of God"[7] and the weakest certification is "one of the prophetic guild." In any case, there is no problem with the prophetic message in the first two episodes except perhaps its commonplaceness—even banality—in Scriptural context. (My teacher, Henry Slonimsky, would have called it "insipid with veracity," a poetical judgment expressing a double-edged truth, yet a truth nonetheless.) In neither of these two instances does the king ask for an oracle (God volunteers it). In neither instance does the oracle affect the decision to fight (the decision has already been made). Only the outcome is in question—very much in question, given the lopsided odds. In neither instance does the oracle feature merit on the side of the human protagonists—their own or ancestral carryover—as motivation for the divine intervention. YHWH is acting for reasons of his own—indeed, sending word in advance to make sure that the mortal will not take credit for himself. In the first episode, the purpose of the intervention (and of the announcement) is "that you [Ahab, king of Israel] may know that I, YHWH [am doing this]" (verse 13). In the second episode it is to requite the Arameans for the slight to Israel's God in the (designedly) clumsy metaphor of a deity whose power is contingent upon the nature of the battle terrain (verses 23, 25, 28). In both cases the king humbly accepts the prophetic word, going so far (in Episode 1) as to ask for (and abide by) instructions as to selection of assault troops and the deputizing of supreme command. Finally, in both cases the oracle is vindicated by the outcome of battle.

As against the foregoing, the guild prophet in Episode 3 shares one feature with the two preceding mantics: his oracle is also uninvited. But unlike the others, his prophecy comes after the event of battle, predicts a vague punishment that, in the final event, cannot be attributed to this antecedent act or prophecy. For Ahab's end will be traceable to his crime against Naboth, told in chapter 21; and as for the forfeit which Israel is to pay in place of Aram's, we do not know what the latter was supposed to be, so we have no way of knowing whether there was any fulfillment one way or another. All that we have is a story of a victorious king tricked by an anonymous guild prophet into pronouncing judgment on someone in a hypothetically analogous circumstance, a pronouncement that is neatly turned against himself—even though we do not for a moment know whether the king's judgment in the hypothetical case is to be considered fair or just. What we do know is that the analogous situation evoked by the prophet is—unlike Nathan's invention as parable for King David's guilt[8]—not at all analogous. For whereas the foot soldier in the parable is given fair warning as to his responsibilities and the penalties chargeable to him if he proves derelict to the charge, the king of Israel is given no mandate, and advised by the man of God of no charge to be placed against him for YHWH's intervention, an intervention not wrought for Ahab's benefit and not for Israel's deliverance but for Aram's discomfiture. There is thus neither need for, nor logic to, this condemnation of Ahab. One does not arraign a tyrant convicted of genocide (so Ahab in the atrocity against Naboth and his line) on a charge of stuffing ballot boxes or of concluding a pact on terms favorable to himself and his nations.

What, then, is the point of the entire narrative? The key to the puzzle must lie in the manner in which the guild prophet is introduced, in his roundabout way of achieving an excuse for a head bandage. (Incidentally, it is well known that even shallow head wounds bleed profusely.) As we have previously suggested, there is not much in our sense of morality or reason today (nor in what we know of Scripture's sense of morality or reason) to recommend the moral that seems to underlie the incident of the soldier who declines to raise sword against a prophet's sacred person. This poor Israelite soldier is placed by the prophet in a no-win situation. He is damned if he harms the prophet, damned if he disobeys him. So, at least, it must have appeared to him. Wisely enough, he decides that the best policy when in doubt is to do nothing[9]—for which wisdom, alas, he pays with his life.

For all the outrage that the fate of this innocent soldier may stir in us (and we shall return to this consideration), it would seem that the point of this questionable pronouncement of the guild prophet against an unwary trooper is to alert us to the equally questionable pronouncement of that same prophet against the equally defenseless king.[10] In regard to the trooper, to draw the moral that he must obey any prophet, or the prophet's say-so of God's say-so, however questionable the imperative, is either to create a bifurcation between Scripture's mind-set and our own or to draw an analogous moral to support the same claim for obedience to the licensed or ordained cleric today who pays yearly dues to one or another of a hundred ministerial associations. The minister in this case—the guild prophet—is the most hawkish of hawks. He is ready to turn a border conflict or a trade war into a nuclear Armageddon. The rectification of a border dispute, a reasonable accommodation arrived at after a decisive battle outcome, these mean nothing to him. Even clemency to a fallen foe (were that the issue) is, for him, heresy and blasphemy. And the tip-off to his character is there in his readiness to invoke a cruel death upon a simple soldier who, caught in a moral dilemma, has difficulty making up his mind and finally opts for nonviolence.

But let us return to the fate of the poor foot soldier. Is God not responsible for everything that happens? And is that not particularly the case when his intentions are communicated in advance by a prophet? And is it not proof of the truth of a prophetic message that what the prophet predicts comes to pass (as in Deut. 18:18–22) and all the more so if what the prophet predicts is of a miraculous nature, such as a lion appearing out of nowhere to strike down a seeming nobody?

There are, of course, many answers to these questions and few that are beyond debate. For we are touching upon questions that are among the most sensitive in any system of theology and central, beyond question, to biblical thought. For the present, proceeding in the reverse order of the questions raised, (1) on the principle of poetical methodology, the biblical attitude to the miraculous must be determined on the basis of sane analysis of biblical narrative, not the reverse procedure of assuming an approach to miracles in conflict with our own, an iron maiden mold into which we must force all biblical narrative—against which, I urge the testimony of Tales 1 and 2 in chapter 5 and the production of the golden calf in chapter 4; (2) in regard to a prophetic message's truth's being vindicated by its coming to pass, an examination of Deuteronomy 18:18–22 will disclose that this passage is concerned only with the matter *after the fact* (that is, in retrospect you will be able to tell which prophecies were true or not), not with miraculous signs to support credence in a prophecy when it is made (where this notion does appear, for example, Deuteronomy 13:2–6, the kerygma is that even a miraculous sign that comes to pass is no sign of truth or validation of credentials purporting to be from God, but quite the reverse!;[11] (3) God's ultimate responsibility for everything that happens (compare the *opus alienum* in Tale 2 in chapter 5) is not to be deployed in a mechanical one-to-one correspondence between felicity and reward, suffering and wickedness. This last message, the kerygma of the Book of Job, for example, is not a proven late development in a proven chronological arrangement of the books, or sections of the books, of Scripture. The "history of ideas" is by definition a diachronic venture. The poetical address to a body of literature, regarding whose chronological provenience we have only hints from within that literature itself, must begin with an assumption of synchronism.

In respect to the fate of the foot soldier, then, even the prediction of that fate—remarkable for its prescience and for the anomalous nature of that sudden stroke—proves nothing about the correctness of the guild prophet's judgment nor about the divine authority of the oracle against the king of Israel in that specific conflict; that is to say, the simple direct narrative style of the author here may be no more than an indication that this biblical author was already experimenting with the device thought to be so recent an invention, that of the unreliable narrator (or, better, the not-always-or-everywhere-or-in-all-respects-at-face-value reliable narrator).

We must, however, give due consideration to the facts of the history of interpretation. Generations of Bible readers, naive and sophisticated laymen, and shrewd as well as stuffy academicians have construed this narrative either literally, or as seriously intended history, or as a parable in glorification of the prophetic office (yes, even to the exclusion of concern with the worthiness of every incumbent of that office or the morality of his position). Out of respect for the many generations who so faithfully and painstakingly preserved and transmitted these texts to us, as well as respect—if not awe—for the artistic genius and intellectual probity of the biblical author, we must continue to entertain the likelihood that this narrative was composed to be read on several different levels—on some levels that perhaps we have not yet guessed at but at least on two levels: (1) to inculcate the ever-manifest power of God and to overawe the simple layman vis-à-vis the prophetic clergy and (2) on an intramural prophetic level, to caution that it behooves even (perhaps, especially) prophets not to leap to extreme positions on issues on which their competency is less than assured and the transmission from on high ambiguous or questionable.

Poetical Review

Repetition (incident). Note the two battles, three episodes—each featuring an oracle mediated by a prophet. Note both the common features of the prophetic incidents and the features that they do not share and how similarity and difference are manipulated to create a puzzle and provide the clues for its resolution. The parallelism of structure in chap. 20 and chap. 17 (see especially, n. 6.)

Dialogue. This constitutes about 50% of the whole. The point of the prophetic oracles as expressed in dialogue. Note the contrast with the narrator's **point of view** and the development of character as exhibited in dialogue.

Points of view. Consider this especially in regard to *narrator* of Episode 3.

Character. Ahab as stock sinner and idolater in the Elijah tales contrasts with the altogether different character of this Ahab, king of Israel. He trafficks only with prophets of YHWH and accepts their oracles, even the third and most dubious. Contrast the dignity, sobriety and statesmanship of Israel's king with the king of Aram— drunken braggart, coward, and fatuously susceptible to flattery.

Metaliterary conventions. These are conventions of warfare, miracle, and prophecy, especially as relates to **primitivity and sophistication in the interpretation of ancient literature.**

Tale 2: The Falling Out of Prophets and the Death of Ahab

The life and times of Ahab, King of Israel, are brought to a close in Kings 22. The death of Ahab is a consequence of a third war or battle against Arameans. As in the case of Tale 1, which uses a historical or putatively historical setting of military activity to explore various problems and quandaries attendant to the exercise of the prophetic office, so in this tale, too, we have a story of war and of prophecy. Presumably, the former is the concern of the historian, the latter of the theologian. A clue as to which of these two hats was the preferred headgear of our narrator may lie in a quantitative contrast. Of a narrative of 38 verses, the first 27 deal with prophetic preliminaries to the war itself; of the 11 verses dealing with the battle, 25 percent is dialogue, and the 5 verses that tell of the battle focus on how Ahab comes to sustain a critical wound. We shall present the second part of the tale first and discuss its narrative nature in terms of literary and metaliterary considerations.

The Death of Ahab

29. The king of Israel and Jehoshaphat, king of Judah, advanced on Ramoth–Gilead.

30. The king of Israel said to Jehoshaphat, "[I propose] to don disguise and go into battle; you, however, wear your [royal] garb." So the king of Israel disguised himself as he went into battle.

31. The king of Aram, now, had commanded the commanders of his chariotry, thirty-two in number, as follows, "Do not engage anyone, of low rank or high, other than the king of Israel, him alone." 32. So it was that when the chariot commanders

caught sight of Jehoshaphat—they thinking, "That can only be the king of Israel"—they maneuvered to close in upon him. Jehoshaphat shouted [to rally his forces]. 33.The moment the chariot commanders discerned that he was not the king of Israel, they drew back the attack. 34.One archer let fly with his bow—in ignorance of his target—and struck the king of Israel in a chink of his armor-joins. He ordered his chariot driver, "Pull about on the reins, get me out of the fray, I'm wounded." 35.Through the peak of battle that day, the king stayed propped up in his chariot facing Aram. With the fading of day he died, the blood dripping from his wound into the chariot's well. 36.As the sun went down, retreat was blown through the war camp: home—each to his native town!

37.So came the king to die. He arrived at Samaria and there was buried. 38.When the chariot was flushed out by Samaria's pool, the dogs thus lapped his blood and the whores bathed [in it]—in keeping with the decree that YHWH had pronounced. (1 Kings 22:29–38)

The one and only textual problem is the rendering of the two verbs *hithappes wābō'* in verse 30. The Jewish Publication Society renders them as imperatives addressed to Jehoshaphat ("Disguise yourself and go into battle"), which would constitute a contradiction of the immediately following clause ("you, however, wear your [royal] garb") as well as of the action actually taken. Most translations follow the readings of Septuagint and Targum, "I will disguise myself and go." This latter is certainly the ultimate intent of Ahab's proposal, but the Hebrew can only be made to yield this sense by emending the text. Our translation achieves the same result yet adheres to the grammatical forms of the Hebrew. The elliptical formulation is, in all likelihood, the author's way of underlining the dialogue as **free direct discourse.** It was not just a matter of one king's concealing his identity and the other's exposing his. Rather, Jehoshaphat was to remain with his chariot forces just where the army lines were drawn, while Ahab—known to his own warriors yet not flaunting his identity to the enemy—could inspire his men while minimizing the risk to his person. As in Tale 1, Ahab is always "King of Israel," not the stock figure of the king who cannot understand or cope with the will of YHWH in the Elijah tales. For all his sins (which we know from the other tales), he is a warrior both brave and capable. These capacities are further conveyed to us in the instructions of the king of Aram to all his chariot commanders. They are not to dissipate their strength or energy in indecisive charge or countercharge: their only hope is to bring down the royal commander who has in the previous two battles proven himself not only king but kingpin.

The battle itself is a set piece of field warfare. If the defending Aramean armor is decisively defeated on the field, the Aramean forces will presumably withdraw behind Ramoth–Gilead's walls to withstand a siege. If the forces of Israel and Judah can be checked by the lopping off of their head, the city can be spared a siege. And so it turns out. The war comes to an effective end with the critical wounding of Ahab. The retreat, it is clear, is orderly. Jehoshaphat does not suffer, and the King of Israel is brought home to an honorable interment. But while the battle rages, the Arameans seek for the hidden king in vain. Once they come close to Israel's defensive line and catch sight of a royal figure. Upon him they concentrate their attack. Once they recognize the royal person as merely the king of Judah, they disdain to press home the attack. Ahab's wound is caused by a chance arrow. The Aramean archer did not, as the translations would have it "draw his bow at random." He aimed at a target who was standing in a chariot. He did not know who that target was. The point of the wound is that just as none of the enemy knew that he was the commanding general of Israel's forces, so also no archer could have aimed at that tiny chink in his armor. And who can miss the moral of this entire story packed into five verses! Human wiles will not protect, nor human wiles succeed in bringing down, the man marked for YHWH's barb.

We need not dwell long on the denouement (which I have discussed before), the poetic justice of the dogs lapping the blood of Ahab in retribution for the exposure of the murdered Naboth—Naboth's corpse left literally to the dogs, Ahab's bloodstains flushed by the city's water source. The analogy of the dogs is a bit strained if we compare the actual fates of the two corpses, but what a fine metaphoric touch is achieved by the addition of two words (in the Hebrew): "and the whores bathed [in it]"! The water to flush the chariot would not have been deliberately drained into the town's water supply. Puddles would have formed by the chariot and from these the dogs may have lapped. But how do women bathing figure here, whores in particular? Modest women of modest means fetch water and bathe at home. Women of the wealthier class do not even appear at the water source: their baths are drawn and poured for them. But for the practitioners of the oldest profession modesty is counterproductive. The pool is an ideal place where the wantons may advertise their wares, exposing to the public gaze a fetching leg, perhaps even a flash of thigh.[12] Nor should we overlook our narrator's dramatic objectivity. The fate of Ahab speaks for itself and, in a measure, for his character. But he is human and a king. His tragedy is heightened and brought closer to us for self-identification in the last glimpse we have of him. No rout for him, no signal of despair to his soldiers. His lifeblood slowly seeping down inside his armor, he stands, propped in his chariot, in command to the end, face to the foe.

The Historicity of Ahab's Wars

Even if we could find evidence (from extrabiblical sources, for example) that would support the reliability of the facts or data regarding the three wars of Ahab with the Arameans, we would still have to conclude from the substance of the biblical narratives that these were not written as chapters in an historian's work, certainly not in the sense that we have in mind when we talk of history or historiography today. In this last narrative, for example, we are not supplied with a single significant detail of the kind that would concern a historian. Did the king of Aram have foreknowledge of Ahab's design on Ramoth–Gilead? Was the city actually invested? Did Ben Hadad come from Damascus with a relieving force? Or did he advance with armor and infantry to bar Ahab's assault at a good distance from the city's walls? What part was played by Jehoshaphat and his Judean army? How many casualties were suffered on either side? Who led the retreat (if retreat it was) to Samaria? Did Aram follow up the victory (if victory it was)? If not, why not?

The first two verses of chapter 22 even seem almost *designed* to confound our curiosity as to chronology. In verse 1 we are told that for a period of three years there was no war between Aram and Israel, while verse 2 informs us that in the third year (of that peace?) Jehoshaphat's visit to Samaria was the occasion for Ahab's proposal for a joint campaign for Ramoth–Gilead. Thus, if verse 2 means that there was peace for two years and a fraction of a third (in which third year Ahab sustained his fatal wound), the wars with Aram spanned the last five years of Ahab's life. If the third verse means that the conference with Jehoshaphat determined on a campaign for the following year, then these wars spanned the last six years of Ahab's lifetime. This difference of a year, in the absence of other and possibly problematic detail, does not loom large as a matter of poetical concern. It does, however, point up the narrator's lack of interest in historical precision.

But there are other problems. Why did Jehoshaphat, king of Judah, join Ahab's campaign for Ramoth–Gilead? Was he a vassal of Ahab's? Why, then, did he have to be asked? If he was not, for what reason did he risk his own life as well as the lives of his soldiers? But such omission of information by a serious historian is a trivial consideration when compared to other questions. The end of the second campaign at Aphek (chapter 20) was the treaty in which the decisively defeated Ben Hadad ceded to Ahab the border cities that his predecessor had seized from Ahab. Ramoth–Gilead perforce was one of these cities. Did Ben Hadad fail to keep his word? Why, then, did not Ahab immediately proceed to take what his all-triumphant forces had won for him? If, on the other hand, Ben Hadad had removed his garrison from that city and reoccupied it a year or two later, why are we not told? Why does Ahab make it sound (verse 3) as though Aramean occupation of that city is a long-tolerated injustice?

All the preceding questions are of a purely poetical nature; that is to say, assuming chapters 20 and 22 are essential historical narratives, why did the author write so much, tell us so little, and raise so many questions that would never have arisen but for the way he told us the little that he did. But aside from, and in addition to, these questions, we are in possession of extrabiblical documents that feature the kings of Israel and Aram in a context of warfare. And the testimony of these documents, while not unimpeachable in every respect, do provide us with **metaliterary grounds** for questioning the historical reliability of these biblical narratives centering on Ahab's warmaking. The chief of these documents for our purpose is an inscription of the Assyrian king, Shalmaneser III, featuring a reference to a battle fought in the sixth year of his reign in the neighborhood of the Syrian city of Qarqar against a coalition of some eleven or twelve kings. The chronology of the Assyrian monarchy in the first pre-Christian millennium is one of the more reliable of Ancient Near Eastern reconstructions and provides a canon for the reconstruction of the chronology of the monarchies of Israel and Judah. Inasmuch as Shalmaneser ruled from 858 to 824, the battle of Qarquar took place (so scholars are agreed) in the year 853. The inscription identifies the king of Aram as one Hadadezer of Damascus (to whom Qarquar belonged), the king of Israel as Ahab, and the third great ally as Irḥuleni of Hamath. But according to some calculations, 853 is the year of Ahab's death; and according to others, his death did not occur until the year 850.[13] If 853 is correct, then Ahab died of wounds received in battle against the very Aramean king with whom he was confederate in the very same year that they stood together against Shalmaneser III at Qarqar. If that seems somewhat improbable (and this improbability is largely responsible for the dating of Ahab's death to the year 850), then we are asked to believe that in 855 Ben Hadad, paramount king of the Aramean city–states, had nothing better to do than muster all his allies for a campaign to extract an impoverishing tribute from Ahab in Samaria (this despite the testimony of Shalmaneser III that in 858, the first year of his reign, he was already rampaging through northern Syria, along the west bank of the Euphrates, destroying towns belonging to Adini and Carchemish). The following year, 854, the king of Aram was preparing for another invasion—with an overwhelming force—of Israel, but was forestalled by a gallant preemptive strike by his uncooperative victim-to-be. Having saved his life by ceding to Israel's king all the contested border cities, he manages to recruit to his cause in the following year, 853, that very same king of Israel. We say *recruit* because the three main contingents facing Shalmaneser, as described in his inscription, rules out any possibility of Israel's being subservient to Aram: Irḫuleni of Hamath commanded 700 chariots, 700 cavalrymen, 10,000 foot soldiers; Hadadezer of Damascus commanded 1,200 chariots, 1,200 cavalrymen, 20,000 foot soldiers; Ahab of Israel's contingent consisted of 2,000 chariots and 10,000 foot soldiers. Thus, Ahab, who was farthest away from the Assyrian tide, brought to the aid of the Aramean buffer states a chariot force outnumbering the combined chariotry of his two major Aramean allies. And a mere two or three years later, in 851 or 850, after two or three years of peace with his erstwhile ally of Damascus, this king of Israel is prepared to recruit his Judean brother-king for a campaign to retake a city that must have been ceded to him only three years ago by treaty. We are supposedly asked to accept all this by a biblical author who, bent on preserving a historic account for future generations, seems oblivious to the threat of Assyrian encroachment and seems either never to have heard of the Battle of Qarqar or thought it too trivial for a historiographer's notice.

Thus, the purely literary considerations that point to the unreliability of chapters 20 and 22 as sources for the history of Ahab's Israel are complemented and supplemented by **metaliterary considerations** pointing to the same conclusion. But to say this is not to say that these narratives are falsifications of history. They simply are not intended as history at all. They are imaginative exploitations of certain historical or semihistorical settings to explore problems and quandaries attendant to the exercise of the prophetic office and the consequences of discriminating between those who, purporting to speak in yhwh's name, indeed do so always, sometimes, or perhaps never. Let us turn then to the first part of our tale, the part for which the end of Ahab is merely a confirmatory conclusion—the essential part of the story, whose central concern is the will of God and how humans strive to ascertain what that will is.

The Falling Out of Prophets

1.They were tranquil for three years—no war between Aram and Israel. 2.In the third year, Jehoshaphat, king of Judah, came down to [the court of] the king of Israel. 3.The king of Israel addressed his courtiers, "Do you realize that Ramoth–Gilead is rightfully ours, yet we hold our tongues, making no move to retake it from the king of Aram?" 4.Then he said to Jehoshaphat, "Will you come with me to battle at Ramoth–Gilead?" Jehoshaphat answered the king of Israel, "I am as one with you, my people as one with your people, my horses as one with yours." 5.Then Jehoshaphat added to the king of Israel, "Now, then, make inquiry as to the word of YHWH."

6.So the king of Israel convoked the prophets, some four hundred in number, and put to them the question, "Shall I proceed to war for Ramoth–Gilead, or shall I desist?" They answered "Go forward, YHWH will deliver it into the king's hands." 7.Jehoshaphat then said, "Is there not yet another prophet of YHWH here of whom we may make inquiry?" 8.The king of Israel said to Jehoshaphat, "Indeed there is yet another through whom we may inquire of YHWH, but I do loathe him—he never utters a prophecy in my interest, nothing but ill omen—Micaiah ben Imlah." "Let the king not hold to this tack," said Jehoshaphat. 9.So the king of Israel summoned a certain chamberlain and ordered, "Micaiah ben Imlah, but quick!" 10.Now the king of Israel and Jehoshaphat, king of Judah, were seated each on his throne, royally attired, on the threshing floor at the entrance of Samaria's gate, the prophets performing before them their prophetic exercises. 11.Zedekiah ben Chenaanah improvised horns of iron. He said, "Thus said YHWH, 'With these shall you gore Aram until you finish them off!'" 12.All the rest of the prophets uttered like prophecy: "Advance on Ramoth–Gilead and triumph! YHWH will deliver it into the hands of the king."

13.The messenger who had gone to summon Micaiah spoke to him in this vein: "Look, now, the replies of the prophets are unanimous, favorable to the king. Let your word conform to theirs all. Give a favorable answer." 14."By YHWH's life," exclaimed Micaiah, "what YHWH says to me—that I shall speak!"

15.He came before the king. The king said to him, "Micaiah, shall we proceed to war for Ramoth–Gilead, or shall we desist?" He replied, "[By all means,] go forward! And triumph, [to be sure!]—YHWH will hand [it] over to the king, [without question]!" 16.The king said to him, "How many times must I adjure you by the name YHWH that you are not to speak anything but the plain truth to me?" 17.Then he replied "I have received a vision: all Israel scattered over the hills like sheep without a shepherd. And YHWH declared 'They have no master, these—let them without hindrance each make for home'." 18.Thereupon the king of Israel exclaimed to Jehoshaphat, "Didn't I tell you? He never utters a prophecy in my interest, nothing but ill omen!" 19."Indeed?" said he. "Hear now YHWH's word. I beheld YHWH seated on his throne, all the host of heaven in attendance on his right and on his left. 20.YHWH asked, 'Who will undertake to entice Ahab, so that he moves against Ramoth–Gilead and falls there?' One said, 'I, in such a way,' and another said, 'I, in such a way.' 21.Then a certain spirit came forward, halted before YHWH and said, 'I will entice him.' 'How?' YHWH asked him. 22.He replied, 'I will go out, and be a lying spirit in the mouth of all his prophets.' He said, 'Yours it is to entice and succeed, go and do so!' 23.And lo now! YHWH has so put a lying spirit in the mouth of all these prophets of yours. It is YHWH who has decreed disaster upon you."

24.Thereupon Zedekiah ben Chenaanah stepped up and struck Micaiah upon the cheek, exclaiming, "How now, pray, did YHWH's spirit pass from me to speak through you?" 25."That you will learn," said Micaiah, "on a day very soon when you run from room to room to find yourself a place to hide."

26.The king of Israel gave an order, "Take Micaiah, turn him over to Amon, mayor of the city, and to Prince Joash. 27.Say [to them]: 'Thus has the king commanded: Put this fellow under lock and key, provide him with the barest rations of bread and water. This until my safe return.' '' 28.Micaiah spoke up, "If you do, indeed, return safe, YHWH spoke not through me!'' (1 Kings 22:1–28)

Biblical Prophecy:
Some Metaliterary Considerations

Inasmuch as all the information we have about the prophetic phenomenon in ancient Israel comes to us from the literary corpus we call the Bible, a proposal to differentiate between literary and metaliterary considerations for interpretation may seem, if not a foolhardy venture, one doomed to certain failure. Yet this is not altogether the case; for we have other data that may be brought to bear on the question, data from other societies that have come down to us in writing (both literary, as in narratives, and "scientific," as in omens and oneirocritical texts) and data that are ours by virtue of our own analogous experience. The failure of literary critics to discriminate consciously as to the nature of the considerations (whether literary or metaliterary) they are employing is a failure that makes for a muddled logic and decreases the likelihood of agreement on interpretation, since it keeps critics from recognizing wherein their disagreement lies. Such failure is not confined to biblical criticism. We may improve our perspective on the biblical problem by considering examples from other literary criticism.

Shakespeare's *Hamlet* will provide a few pertinent examples. The interpretive crux of this play is the character of Prince Hamlet; and that character in turn hinges upon the motivation for his acts or, more properly, his failures to act. At one critical juncture, shortly after Hamlet has, it would appear, ascertained that his uncle Claudius is his father's murderer and usurper of the throne, Hamlet has the opportunity to dispatch the villain as he kneels in prayer. In the prayer uttered by Claudius, we hear him confess his guilt. It is not altogether clear whether Hamlet has heard that confessional prayer or not, but there is perhaps an indication that he has in the reason he offers for not killing the king at this moment. To send him off to the next world when—just shriven of guilt by his penitence—his prospects for salvation are so high would be to do him a service rather than to inflict a punishment. The purely literary question is whether Hamlet himself believes what he is saying or whether he is actually cloaking a different—and from his point of view possibly indefensible—motive for not acting the avenger now. The answer to this question may in turn seem to depend on a metaliterary question or one or another formulation of that metaliterary question. Did church teaching in Hamlet's time indeed provide for the salvation of a murderer's soul by the simple act of confession and stance of penitence? (Was that church Lutheran or Roman?) Would Shakespeare have expected his Elizabethan audience to assume for Hamlet the same church teaching (whatever it was) that that audience itself shared? Would a hypothetical consideration of an enemy's possible immediate entrance to heaven stay the hand of an executioner long waiting for a judgment of guilt to free that hand to act? But again, whatever the correct metaliterary question, we may discard this entire question-and-answer process when we are reminded that in his prayer Claudius himself has expressed the notion that there is no penitence unless the sinner has proven himself ready to give up that which he has gained by dint of his offense and that he, for one, is not yet so disposed.

Closer to our concern with prophecy is the revelation to Hamlet by the ghost of his father that his brother Claudius had seduced his wife and administered to him a fatal potion. The purely literary question is whether Hamlet believed in ghosts, for if he did he stood under his father's mandate to avenge his murder. This question may be seen to hinge on the metaliterary question, Did Shakespeare believe in ghosts? Or (perhaps better) did Shakespeare assume that his audience would assume, believing in ghosts themselves, that Hamlet could not but believe in a revelation that he has himself just now experienced? Again, we might well put aside this metaliterary consideration if we remember the purely literary consideration that whether Hamlet the Dane—or Shakespeare or his audience, for that matter—would have believed in ghosts is neither here nor there; for even if they all did, the question of belief in ghosts is in no way determinative of Hamlet's motivation or lack of it; for as Hamlet's address to the ghost questioning whether his provenience is from the domains of God or of Satan shows, the question is not in the reality of the messenger but in the truth of the message he bears. Thus, the ghost may be as real to Hamlet as he is to himself and yet not be his father's spirit but a hell-sent impostor.

This last consideration ought to prompt us to reconsider our own existential stance in regard to prediction or foretelling, if not to divination or prophecy. We ourselves may wonder how shamans or witch doctors in "primitive" societies survive their high failure rate; how our neighbors impoverish themselves while supporting racetrack touts or horoscopists. Yet we do not marvel that we pay physicians and surgeons whether their patients recover or die and stay in the markets for stocks or futures despite disappointing tips from brokers who continue to live by our patronage even when we shift our custom from one house to another.

As then, so today—as today, so then. Humans have always had recourse to various avenues promising a glimpse into the future. The attempt to glimpse into the future and anticipate it remains the constant; the variables are the forms, the technologies, the presuppositions on which the latter two are based. Thus, divination in the biblical tradition shares much with divination in the pagan world which is both precursor and ambience of that biblical tradition. In the pagan world divinatory practice may be classified under two headings: scientific-objective-mechanical-fatalistic and religious-personalistic-voluntative. Under the first category we should have to list such practices as astrology, extispicy, and haruspicy, under the second category, dreams and messages sent through a prophetic agency. It is possible also that there is a middle ground between these two, such as the casting of dice or lots or arrows in connection with a prayer to deity to determine the cast according to a prearranged code. Biblical teaching excluded the first category altogether as well as phenomena in the second category that were associated with a deity other than YHWH, God of Israel. Only three avenues of divination were licit in Israel: the Urim and Thummim lots cast and interpreted by the Aaronide priests; oracular dreams of one sort or another; and prophecy.[14] But the last time we read of recourse to the Urim and Thummim is in the time of David; and specific instances of dream revelations are rare and appear mostly in narratives of early times. Thus it is that divination for biblical Israel becomes restricted, in effect, to prophecy; and prophecy is almost completely congruent with revelation, that is to say, revelation of God's will. The function of prophecy in general is to ascertain the will of God in terms of his norms for human behavior. In particular instances, the function of prophecy is to determine whether the God of Israel is favorably or unfavorably disposed to a project that is intrinsically neither heinous nor virtuous. Tale 2 is clearly such an instance.

The story is about the humans who seek prophetic oracles, the humans who pretend to be the channels for it, and one human's reception of the oracles that reach him. To discern the various intentions of our storyteller we shall have to examine the plot of the story and how it is developed—this in light of a search for the existential constructs that must have lain behind the belief of the ancients in this mode of reading the vectors of reality that determine the human quest. Among the most obvious lessons the story seeks to convey are that prophets may disagree, that a decision will be taken nevertheless, and that the overwhelming majority may turn out to have been wrong and a single prophet thus constitute a majority of one. But all of this could have been achieved without the inclusion of many details that the story presents. Let us then examine these item by them.

Jehoshaphat, king of Judah, comes on a visit of state to the court of Ahab, king of Judah. A hint that the former is in some degree under obligation to the latter may be present in Israel's king's raising of the agenda "to his courtiers." The chief issue is Ramoth–Gilead irredenta. His question to Jehoshaphat would indicate that the latter's obligation is limited. But Jehoshaphat's answer is an enthusiastic and unqualified *yes*, to which he appends a request that an oracle from YHWH be sought. Does this indicate a sudden change of heart? No, he agrees in principle (as did Ahab, in the previous story, to Ben Hadad's claim to tribute). Ahab convokes the collegium of YHWH prophets. He puts the question, To war or not to war? The answer is immediately forthcoming: To war and to triumph! Whereupon Jehoshaphat raises a question, incongruous if not impertinent. Four hundred prophets have given answer: Is there not someone who has not been asked?

King Ahab is not surprised by his colleague's question. He must understand, then, that the question is not unreasonable. We can only speculate on what may lie behind the question. Were Jehoshaphat's suspicions aroused by the very summoning of so huge a convocation? Were his suspicions reinforced by the very unanimity of so many? (So unanimity is not to be expected of prophets?) Does he suspect that the prophets are a rubber stamp assembly or too eager to tell their lord what they have divined he wants to hear? Or is it possible that prophets, too, may have divided by reputation along the lines of hawks and doves? If that were so, it is conceivable that Jehoshaphat noted the absence of a well-known dove and his question is a diplomatic way of inquiring as to his whereabouts. Whatever be the case, Ahab does not take umbrage. He admits to Micaiah's existence and indicates why Micaiah has not been invited to speak. This but confirms one of our guesses: Micaiah never gives the kind of oracle that Ahab wants to hear. Jehoshaphat's reply to the king is that Micaiah should nonetheless be consulted, and Ahab—apparently graciously—decides to humor his royal ally.

We shall never get behind any ancient story in which oracles figure prominently unless we recognize that all these accounts are elliptical, that behind the recourse to oracles there is a certain psychology whose axioms and postulates are so problematic that it is wiser not to subject them to too open a scrutiny. Oracles (as pointed out in connection with Ahaziah's mission to Baal-zebub of Ekron) are not sought out of idle curiosity. They are part of a decision-making process. Decision making involves its own peculiar algebra. Many factors in the equation are known, others surmised, yet others a complete mystery. In the stories that have come down to us, the formulation of the question put to the oracle is usually one that admits of a *yes* or *no* answer, as though there were only one simple equation to be solved. It is no wonder, then, that the response of the oracle is notoriously ambiguous. Ask a silly question, and you will get a silly answer. The oracle must protect itself from the inanities of the questioner. Thus, for example, the king who asks whether he should launch a war is really asking whether he will be victorious if he does. Hence, Ahab's prophets answer both the explicit question and the implicit one: yes, make war, you shall triumph. An oracle addicted to such categorical replies would not long stay in business. The king, having received such carte blanche assurance, might then proceed to dispatch a force too weak for the task, under an incompetent general, and blame the oracle for the disaster that his own imprudence determined.

It would seem, then, that a corps of prophets would be as necessary to the formulation of the questions to be put to the oracle as, say, a corps of theoreticians and technicians to make use of a computer. Behind the simplistic question of King Ahab—Shall I do battle for the recovery of Ramoth–Gilead?—there would be a host of other questions, associated or intertwined. Is the question confined to one battle? Will an assault on the city meet with initial success or repulse? Will the city surrender or will it require a siege to take it? If taken, will the king of Aram be content to let the matter lie, or will it be the beginning of a broader and longer conflict? If the latter, what is the prospect for a long war and for its ultimate outcome? Each of the prophets exercising his function before the two kings might well have addressed himself to a different aspect of the problem. The bits and pieces of their input and output fell together in an agreed pattern: the prospects for Ahab in such a war were favorable.

In like manner we ought to approach the notice of the iron horns made by Zedekiah ben Chenaanah. Most scholars are content to assume that a rite of sympathetic magic is to be understood in this act. Perhaps, but only if it is the narrator's intention to burlesque Zedekiah, as in the instance of the Ball prophets gashing themselves on Mount Carmel. The context here does not support such a suggestion. It is dubious whether generals in antiquity put any more store in magic than military commanders do today (a comparison that may be an injustice to the ancients). Certainly, a symbolic gesture improvised by a prophet at will would in no way have made for greater credence in his judgment. Micaiah says that "with these" you will gore the Arameans until you finish them off. The preposition *with*, meaning also "by, by agency of" here is the same as the preposition governing Ahab's question to YHWH's prophet in chapter 20: having been promised a victory over Aram, Ahab asks, "By whom?"—Who is to be the instrument of victory? And the answer is, "By the guardsmen of the provincial governors." (He then asks and receives answer as to who is to pick the moment for the attack.) In like manner, Zedekiah may be dramatizing the combined powers of the kings of Israel and Judah.

Instructive, too, is the chamberlain's advice to Micaiah. He probably intends no disrespect to this prophet in presuming to tell him what word he should deliver from YHWH. No more disrespect, that is, than may be inferred from it for the prophetic enterprise in general. The asking and giving of oracles is a highly complicated matter.

A prudent prophet will find a way of formulating the message he receives in such a way as to occasion the king minimal upset. The vote for war is unanimous. On this occasion a single holdout will count not at all. Why not, then, be prudent for once, Micaiah? And Micaiah's answer betrays no sense of insult. He merely swears that he will tell nothing but the plain truth, as he receives it from YHWH.

And what is the answer Micaiah gives to King Ahab? Without the bracketed words in our translation one would conclude that he is, after all, following the chamberlain's advice. The context assures what the written word cannot alone convey. Micaiah's repetition of the words of the 400 prophets, dripping with sarcasm, is Micaiah's way of letting all in hearing know that he, too, knows the answer the king wants, the answer he has so far got, the only answer he will accept, and the answer he will not get from Micaiah! The king's impatience with Micaiah's sarcasm, as translated by us, can be amply justified. The standard translations, keeping to the Hebrew word order, have Ahab not adjuring Micaiah in the name of YHWH to speak the truth plainly but adjuring him to *speak the truth in that name.* There is an almost unbearable irony (of which Ahab cannot be conscious) in this rendering; for it has Ahab accepting Micaiah's oracle as YHWH's word even as he is determined to deny it. Having learned that it is difficult to overestimate our narrator's skill, we would conclude that both renderings are right: the author intended the double entendre.

Micaiah's oracle is given in two parts. First, he presents the bottom line—Ahab will lose his life, and the leaderless army return home. The second part, the vision of YHWH's deliberation in the heavenly assembly (which adds nothing to the oracle itself) follows Ahab's I-told-you-so exclamation to Jehoshaphat. It begins with the word *therefore,* which we have previously noted as a prophetic rubric introducing a sentence of doom. Since that doom has already been pronounced, whatever Micaiah is about to say must relate to what Ahab has just said. Ahab has said, again, "He prophesies no good for me, only bad" (this is the literal rendering of the Hebrew). My own translation tries to render what Ahab had in mind rather than what he actually said. For what he said is probably not true in a precise sense; and what he intended is a hyperbole, which is by definition not literally true. We know of the cultic Urim and Thummim oracular device that it produced one of three answers: *yes, no,* and *no answer.*[15] And we know, further, that only the third answer is a sign of God's disfavor. Assuming that the questioner asks about a project on which he should like to embark, even the *no* answer is a sign of God's favor; for it warns the oracle seeker that a certain course is disastrous, and thereby—if he accepts it—keeps him from harm. God's anger against the oracle seeker is revealed by his refusal to say *yes* or *no.* He is denying audience to the supplicant ("hiding his face from him"), refusing him guidance. It is with this in mind that Micaiah responds to Ahab's complaint.

Ahab wants to go to war. A favorable answer—as far as he is concerned—is only when God acts as his rubber stamp. He is blind to the grace of God as represented in the guidance and discipline that he offers mortals in their own ultimate interest. Therefore, responds Micaiah, this very blindness of Ahab's is the sign of God's disfavor. The vision of the heavenly council is another instance of the *opus alienum,* a dramatic inspiration equivalent in meaning to the trenchant metaphor of the Romans: whom the gods wish to destroy they first drive mad. The gods, of course, are all-powerful. They can destroy any way they please. Most often, they please to let mortals encompass their own destruction: they give them enough rope to hang themselves. Consider this last metaphor. Only a person deprived of his reason, hell-bent on self-destruction, will use the freedom to act—the scope given him by a long leash—to fashion of the rope a noose for his own neck.

The appositeness of the metaphor must by its own logic—and by the logic of the story plot be lost on Ahab. It need not be lost on the college of YHWH's prophets, they who are betraying their trust by reducing themselves (and by implication their lord) to the status of a mortal monarch's yes-man. But it *is* lost on them. Their spokesman Zedekiah hears only one thing in Micaiah's subtle message. Micaiah is, directly or indirectly, calling his colleague a liar. He says, in effect, as he vents his rage in a blow to the cheek, "You call me liar—well, you're another!" The actual wording is in the form of a question, a challenge to which Micaiah may respond. He asks how Micaiah can prove that the lying spirit is in Zedekiah and not in himself. Micaiah's vision, since we have only his own word for it, proves no more than did Zedekiah's horns of iron. There can be no proof until the pudding is ready for the eating. You will know which of us was right, replies Micaiah, when the kettle you helped set on the fire boils over. We are left, thus, with a vision of the war's aftermath, when recriminations are rife over responsibility for the misbegotten campaign, when the dead king's heir appoints a commission of inquiry into the lapse on the part of his father's intelligence service.[16] Micaiah's image of Zedekiah's plight at that time evokes the lyrics of one of our folk songs: "O sinner-man, where you goin' to hide from?" That the sinner is YHWH's own representative, that he acts not alone but in company with an entire college of fellow prophets, is a telling indication of the value that Scripture places on the principle of majority rule in matters spiritual.

Poetical Review

Dialogue. The predominance of dialogue in a nonhistorical narrative, as con-strasted with its rarity in a narrative formally like **historiographic** account, is exemplified here. In the prophetic narrative (verses 1–21) dialogue constitutes two-thirds of the text; in the "historiographic" verses 29–38 dialogue is less than one-quarter; in the combined narrative dialogue constitutes about 55 percent of the text. Note also how crucial the dialogic element is in manipulation of varying **points of view** to develop the psychologies of the oracle-consulting personae, that is Ahab, Jehoshaphat, the chamberlain, and the contesting prophets.

Oblique expression. In **dialogue,** examples are the *opus alienum* in Micaiah's vision, Zedekiah's retort, and its violent accompaniment. The blow struck by the latter is reminiscent of the preacher's marginal note to himself on his sermon manuscript: "Argument weak here. Yell like hell!"

Character. Ahab is the courtly and courageous king for all his sinfulness and the fate it entailed.

Literary and metaliterary conventions. In our discussion, examples are factors in evaluation of author's intent in regard to both historicity, and the conventions of foretelling and divination, in **ancient literature,** especially as compared with the supposedly greater sophistication of guesswork in our own generation.

Historicity and fictionality. Compare in particular the weight of the internal literary considerations to the extra biblically available inscriptions from Assyria and their testimony.

Metaphor (versus literalness). Note the disposition of the blood of Ahab despite the historic datum of his honorable burial.

7

Tales of Elisha and Others

The compass of the stories featuring Elisha, and their length and intricacy, as well as the extensive exegesis of other biblical passages required in connection with these stories, preclude a treatment here of all of them. Hence, the selection of stories for discussion in this chapter is to a degree idiosyncratic, if not altogether arbitrary. Common to all of the stories I shall treat is their preoccupation with existential problems of the prophetic enterprise; that is to say, for every answer that these stories provide, many more questions are suggested. Like the stories we have studied, these, too, seem to be essentially intraprophetic phenomena: in their self-critical stance in regard to the prophetic office as well as to those who seek prophetic guidance, they display a dramatic objectivity, a restrained humor, and an artistic control that is as impressive as it is daring. Were the authors of these tales less skillful as artists or less committed to the centrality of Israel in the moral economy of human history and Israel's prophets in that people's eccentric march or dance, in step or out of step with the beat of the divine drum, these didactic narratives could be construed as jeers at both people and prophets, possibly as skeptical smiles at the very notion that the human world is regulated according to the laws of a divine economy.

Tale 1: A Curing of Waters

19.The men of the town said to Elisha, "Lo now, the site of this town is quite attractive, as my lord can see. The water, however, is bad and the area causes early bereavement." 20.He said, "Fetch me a new dish, put salt in it." They fetched it for him. 21.He went to the spring, the water source, cast salt into it, declaring, "Thus said YHWH: I have healed this water. No longer shall bereaving death come from here!" 22.Thus the water became wholesome—as it is to this day—according to the word that Elisha pronounced. (2 Kings 2:19–22)

Jericho is a delightful oasis, situated about ten miles north of the Dead Sea and five miles west of the Jordan. Set in high mountain tundra that receives a mean annual rainfall of four to eight inches, it owed its lushness then (as it does today) to one of the largest springs in Palestine. The rains, when they come, are often as heavy as they are of short duration; and the floodings, along with strong winds, have disintegrated much of the mud brick buildings that constituted the settlement of Jericho in biblical times. Archaeologists are unable, as a result, to determine the size of the first city west of the Jordan to fall to Joshua or of the settlement that apparently replaced it some centuries later. According to Joshua 6, the city was put to the ban, its population exterminated, and its wealth of metals given over the YHWH's sanctuary. Joshua then pronounced a curse: "Damned by YHWH's decree be he who undertakes to rebuild this town, Jericho. At the cost of his firstborn may he lay its foundation and at the cost of his younger son may he set up its doors!"

It is reported that the site was rebuilt in the time of King Ahab and that the curse was fulfilled according to the word of YHWH pronounced by Joshua (1 Kings 16:34). A certain Hiel, from the city of Bethel, laid its foundation "at the cost of his firstborn, Abiram, and set up its doors at the cost of his younger son, Segub." The words *at the cost of* in English or *at the price of* is how all the newer translations render the Hebrew preposition *b,* which has this meaning as well as the meanings of "in, with, for, through, by, at, etc." The older translations render the preposition as *in* or *with,* thus giving rise to the notion that the foundation stone was planted upon his son, literally—upon his corpse. This mistranslation is the one and only basis for the notion that there was in antiquity a rite warranting the name "foundation sacrifice." While the archaelogist's spade has uncovered numerous earthen pots containing the skeletons of young children, witness to a common practice of such burial of the very young under the packed-earth floors of homes, not a single such skeleton has ever been discovered under the foundation stones of a city's gates. A dictionary of the Bible published twenty years ago (IB) contains this comment, typical of consensus scholarship, on Hiel's founding of Jericho: "The loss of his sons may well refer to a foundation sacrifice after the Canaanite pattern. Actually, however, archaeologists do not find too many such foundation sacrifices." As far as I have been able to ascertain the *not too many* does duty for *zero.* Aside from the misleading nature of this expression (Is this overstatement or understatement?), one may scour tomes on Syro–Palestinian archaeology without finding a definition of *foundation sacrifice* or a single example of a "Canaanite pattern" of this practice.[1]

Here, then, is a parade example of a metaliterary convention, itself owing its existence to an attempt to reify or concretize a metaphor (Joshua's curse and its fulfilment), functioning in the interpretation of a **literary expression** in a closely related text. This last formulation is not quite accurate. For what the reification of the metaphor serves to do is to conceal from us the link between the stories of Joshua and Hiel featuring Jericho's lethal ecology and the story of Elisha's heralding of YHWH's decision to remove that lethal element.[2]

Keeping in mind, then, that Tale 1 follows immediately upon the return of the fifty guild prophets from Transjordan to report to Elisha *in Jericho* and ever so casually begins with the observation of these prophets ("the townsmen") that the site of "this town," though pleasant, is polluted and that the tracing of the contamination to the water and the premature deaths to the "ground" is a judgment by the town's inhabitants, unsupported by the narrator (in his own voice or Elisha's), let us review the story of Jericho before the restoration of its salubrity and fertility, which continues down "to our own day."

Jericho, a city of indeterminate size in Joshua's time, was the first to fall to the Israelites crossing the Jordan into Canaan. Symbolic of the fate awaiting the sinful populations of Canaan's idolatrous cities is the war of extermination waged against it.

An entire chapter, Joshua 2, is devoted to the exploits of the Israelite spies dispatched by Joshua to reconnoiter the weaknesses of the city's defenses. But when, in chapter 6, the city's wall collapses to give the Israelite army entry into the city, it is not due to the wall's being mined at a weak point by Joshua's engineers. It falls because of a miracle wrought by YHWH. Each day, for six days, a single circuit is made about the city; seven priests blowing seven rams' horns, parading YHWH's throne, the Ark of the Covenant, with the army marching as vanguard and rearguard. On the seventh day, seven such circuits are made, at which point Joshua signals the army to shout the war cry; and the wall crumbles. The victory is YHWH's, the spoil is YHWH's, the destruction is as complete as only YHWH's will can decree, and for over three centuries this lush site remains desolate—this also determined by YHWH's will as expressed in Joshua's curse. What kept settlers away from this rich oasis? Was it a poison in the life-giving waters of Jericho's mighty spring? A miasma of contaminating radioactivity that made the area a people killer, affecting children with special force? And was the death a function of YHWH's response to the curse uttered by Joshua, or was the story of the curse a function of the mysterious death-dealing power? We literary critics are fortunate not to have to tackle the historian's task of ascertaining the fact of YHWH's causality in history or, contrariwise, attributing to YHWH natural, accidental, and undeterminable effects. The denouement of the story is to show YHWH as always and ever the cause of all effects—poisonous and curative.

Resettlement of Jericho was undertaken by Hiel. He braved the curse and paid the price. But the settlement endured. How? The answer is that in Elisha's time, YHWH, having decided to terminate the sentence on Jericho's ecology, had the announcement made by his prophetic representative. And was the healing of the waters accomplished by human agency, by the wielding of natural means? No. A spring's waters could not in any case be affected at the point at which the waters gush forth, for the purifying antidote will be washed away. And the antidote is, of all things, salt, the very element that renders water brackish, unsuitable for drink or for irrigation. But salt from a new dish, symbolizing a new dispensation from Providence, discloses once again that the Power that can mysteriously turn life-giving waters into a death potion can use the death-symbolizing salt to restore the spring's life-giving wholesomeness.

Tale 2: Bears and Bugbears at Bethel

23. And he went up from thence [i.e., Elisha from Jericho] unto Bethel: and as he was going up by the way, there came forth little children out of the city, and mocked him, and said unto him, Go up, then bald head; go up, then bald head.

24. And he turned back, and looked on them, and cursed them in the name of the Lord. And there came forth two she bears out of the wood, and tare forty and two children of them. (2 Kings 2:23–24)

This two-verse story, from the Authorized Version, following hard upon the cure of Jericho's waters, is possibly—in its gratuitous attribution of malevolence and atrocity to Scripture's benevolent God—the most embarrassing tale in the Bible;[3] for the most literalistic of fundamentalist believers, as well as for the Bible critic who may see it as a vestigial witness to a primitive prophetic ethos, the question remains the same: Why—whether it actually happened so or was a tale handed down by forerunners of the great prophets—was it included by the editors of Scripture's canon? It is the enormity of the atrocity that militates against what Montgomery sees in it, "a *Bubenmärchen* to frighten the young into respect for their reverend elders." Loving parents resort to bogeymen and bugbears in such tales because the exaggerations and other fantastic elements, distancing the tale from even the child's perception of reality, are devices to protect the child from the trauma of terror. Even in fairy tales, the punishment of villians is reasonably proportionate to the nature of their crimes. If Cinderella's stepsisters were doomed to spend the rest of their lives in shoes two sizes too small for them, their fate might seem reasonable to youngsters who have never suffered the agonies of corns or bunions. But even these innocents would find it a bit much if the girls had their feet chopped off.

The extravagant punishment visited upon the jeering children in this tale is certainly the most grotesque of its features. But it is only one of many elements that would have struck its Israelite audience as incongruous. Consider the following: (1) Baldness is a natural and common phenomenon, not reserved for the old or the middle-aged and unlikely to have been perceived as a mark of shame anymore then than today. (2) The miracle of the murderous bears is so out of keeping with the nature of these shy beasts as to be almost ludicrous. Bears keep to themselves and will normally run from humans. Only ravenous hunger will occasionally drive one to snatch a lamb from the flock. In the Bible, the image of berserk rage is a she-bear whose cubs have been stolen from her; but unprovoked, bears are known for their gentleness and playfulness.[4] (3) The Hebrew verb used to characterize the bears' attack is equally incongruous. It does not mean "tear" (or even "mangle" or "maul," as it is rendered in the newer translations); it means "to cut a cleft, to split into two" as in splitting rails or logs for firewood. (See discussion on **word play** in chapter 1.) Even more incongruous is the *scene* that is evoked, either dozens of children standing fixed in place, waiting as if paralyzed to be struck down, or—if they took to flight—the two bears methodically pursuing them for the sheer lust of killing. And what was Elisha doing all the while—directing them to head off the few who seemed about to escape?

The unraveling of this riddle requires first a proper translation of the Hebrew, which means to render the original not literally but **idiomatically.** Consider, for example, the term *go up.* Hebrew verbs of motion are not, like our own for the most part, general; they more often specific and related to topography. Cities, the better to defend them, were perched on hilltops; hence, to make war against a city (Ramoth–Gilead, for example) is "to go up" to it. One never goes simply to or from Jerusalem, one goes up to it or down from it. And Hebrew has a separate and distinct verb for each of these movements. Elisha is ascending the slope of Bethel when he is accosted and insulted. Bethel is a city that boasts a royal shrine dedicated to YHWH, whose legitimacy is repeatedly attacked in the books of Kings and in the Writing Prophets and whose jealous priests do not take kindly to the carping of interloping prophets (see, for example, Amos 7:10–17). The children are telling Elisha to keep going, not to enter their city, which would entail stepping onto a level plateau; such as he are not welcome. "Baldy," as they refer to him, is not a reference to any hairless pate; it is a specific reference to the characteristic mark worn on the forehead by prophets, the cicatrix we discerned in chapter 20, where the prophet had need to conceal it. So it is the prophet *qua* prophet who is being told to move on. The prophet does move on without entering the city. He casts them a backward glance and (here again we have the verb translated "curse," which may mean only upbraid) probably denounces their disrespect for YHWH, whom he represents.[5] (Since he is skirting the city, the forest from which the bears emerge is another incongruity, for cities were not built in forests.)

The first word used to describe Elisha's tormentors is *na'ar*. This word may refer to a male from the age of eight to eighty. It is not a term of derogation but, like the word *'eved,* (servant slave), it reflects a status of subordination. (It is used for the attendant of Elijah's in 18:43 and 19:3 and for the palace guards of the provincial governors in 20:14–15.) A ranch hand is a *na'ar* to the foreman and the foreman is a *na'ar* to the ranch owner. This word is modified by the adjective *small,* which may also mean "young" and, as in English, "petty, unworthy, unimportant, mean, base, etc." It is in this last sense that the adjective appears here. The mockers of Elisha were not young children, they were "worthless oafs, hooligans, hoodlums." And when the bears wreak their depredation among them, the term used for them—*yeled* /(child)—appears (as in Genesis 4:23) in the sense of "stripling."[6] In our own Germanic tongue, consider German *knabe* (boy), English *knave* ("the soldier or jack ranking below the queen in the deck of playing cards; rogue, scoundrel").

A mob is notorious for the false courage it derives from its sense of solidarity. But divided, they become weak: an ax stroke or two will cleave a block of wood, sending chips flying off in all directions. Let us try our own hand at translating the story.

> From there he went on to Bethel. He was yet enroute, when some mean-spirited rascals came out from the city. They jeered at him, gibing, "Move on, scarface! Move on, scarface!" He wheeled around and fixed them with a stare. He denounced them, calling YHWH to witness. Then two bears erupted from some woods, broke them up—some forty-odd knaves, the lot.[7] Kings 2:24–25

Our story is more in the comic vein than the tragic. The imagery is humorously apposite some two-and-a-half millennia after the composition of the biblical vignette. An itinerant and meddlesome preacher is beset by a mob of young hoodlums. Green in years and greener in wisdom, brave in their solidarity but cowardly at heart, they are dispersed by two "Smokeys." A millennium-and-a-half ago, a playful discussion among the talmudic sages questioned whether either bears or forest existed at Bethel before Elisha invoked YHWH's aid. One opinion, still in use today to tag a story as fantasy or balderdash or metaphor is *lō' dubbīm welō' ya'ar* "[There were] neither bears nor forest."[8]

Tale 3: The Willful Ax Head

1.The guild prophets said to Elisha, "Lo, now, this site where we are—by your grace—settled is too confined for us. 2.Let us, pray, move on to the Jordan—we'll fetch ourselves from there a roof beam apiece and fashion there a site for our settlement." "Go," he said. 3.Thereupon one of them said, "Undertake, if you will, to accompany us." "I will indeed," he said. 4.Thus, he accompanied them. Arrived at the Jordan, they set to cutting timber. 5. One fellow was engaged in felling a [tree] beam—his ax head fell into the water—he cried aloud, "Alack, my lord, that was mine on loan!" 6."Where," asked the man of God, "did it fall?" He showed him the spot, brought the ax head afloat to the surface. 7."Lift it out," he said. So he reached out and retrieved it. (2 Kings 6:1–7)

Jericho and nearby Gilgal and the royal capital at Samaria are important settlements where prophetic schools or brotherhoods may have subsisted on the charity of the population or on the income from nonprophetic employment. For a school of prophets to found a settlement on the virginal bank of the Jordan makes as little sense as the complaint about population pressure.[9] The entire story seems designed to create a setting for the miraculous retrieval of the ax head. The role of Elisha as the presiding guru of a prophetic collegium is implied but never spelled out either here or elsewhere, featuring Elisha or any other prophetic personality. He readily assents to the colony's proposal to move to a new site, as though the matter touched him not at all. When the point of the request is pressed home to the normally perspicacious clairvoyant, that it is not his permission that is sought so much as his agreement to make the move himself, he is equally amenable to this proposal. Now arrived at river's edge the action and dialogue is telescoped, as though Elisha was immediately at hand to hear the cry of lament, which is addressed to him: "my lord/my master." All this, so that the many miraculous achievements of Elisha may be augmented by one more, causing iron to float. Unlike the prodigies effected by Moses and Aaron, Elijah, and Elisha himself to part the Jordan's waters, the act here is performed without orders from, or invocation of, the Deity. Perhaps this lent surface plausibility to the suggestion recorded by Montgomery that the feat was one of imitative magic. But how comes it that a master prophet in the biblical tradition, which wages unremitting war against magic, indulges in that very practice?

But whether the feat performed by Elisha was magical or miraculous, this interpretation of this story ignores a number of details in the narrative. (1) Elisha is referred to only twice other than by pronouns: in verse 1, being addressed by the prophets, he is simply "Elisha"; in verse 6, it is "the man of God" who asks where the ax head sank. (2) The lament of the ax wielder tells us that the ax was borrowed. What difference does it make to the wielder who will bear the loss in any case, whether he has to compensate the owner or replace it for himself? (3) This narrative features two verbs in regard to a tool wielded on wood (*gzr* "to cut down" and *hippîl* "to fell"); and while biblical Hebrew has another ten verbs for "cutting," the verb for Elisha's action on the wood he throws after the ax head appears only here in an active finite form.[10] (4) The man of God who can make iron float requires information on where the head fell into the water. How far from shore can an ax head fly? Why could the wielder not wade into the shallows after it? (5) The gratuitous command of Elisha to retrieve the ax head and the note of compliance by the prophet who had lost it.

If the intent of this story is not to relate a literal event, the alternative is symbolism and metaphor—specifically the metaphor of an ax head "flying off its handle." The image in our English idiom, focusing on the danger of an ax head flying an erratic arc, is applied to a person so enraged as to lose all control. But this is not the only metaphor for an ax head, fixed on its handle or loosed from it. Let us recall Isaiah's famous metaphor on the king of Assyria who, blind to the truth that he is merely YHWH's instrument, arrogates to himself unlimited power and the freedom to use it as he will.

> Shall the ax head vaunt itself as against
>   the one who hews with it?
> Shall the saw overween as against
>   the one who handles it?
> As if 'twere like for a staff
>   to handle the one who wields it!
>   for a shaft
>   to lift up [a creature] not made of wood! (Isaiah 10:15)

In our tale, as in both the English idiom and Isaiah's metaphor, the ax head represents *power*. In the English idiom the ax head represents the person, and its departure from the handle represents loss of control. In Isaiah's metaphor ax head and handle together are a person as instrument, wielded or handled by a power beyond him or it. In this tale of Elisha ax head, handle, and handler figure as personae interrelated by the plot in order to spin a parable.

The person proper is a prophet. His power instrument is the ax head, which he wields by means of its handle. But the ax is borrowed and, like the ax–king or king–ax of Assyria, has a consciousness of its own. In its consciousness the ax head is like a riding horse, its handle like a saddle attached to it, and the handler like the rider seated upon it. A horse reluctant to be ridden has been known to inflate its belly while the saddle is strapped round its girth; then, when the rider is seated, the horse will exhale, causing the loosened saddle to slip around and the rider to describe a centrifugal arc on his way to the ground. Other bits of equestrian lore are the mount who, acknowledging only one master, will attempt to dislodge a stranger from the saddle or the one who will resort to various kinds of equine mischief when it senses an inexperienced seat on its back or hold on its reins. So, too, our ax head. It knows that its handler is not its master, resists being wielded by him, and (treating handler and handle as one, like rider and saddle) it flies off the handle. The handle somehow no longer fits because the person who grips it is not fit to handle it. All this is grasped in a flash by Elisha. The ax head escaped one handle (a poor fit). He whittles a new handle, with prophetic knowledge of the shape required to fit the ax head, sends it flying to the spot where the ax head sank, and the ax head rises to the occasion, like a steed responding to its master's whistle. (Or should we say, like a lover to a true lover's call?).

The wielder of the ax was member, perhaps yet novice, of a prophetic band. The prophet's power is the power of prophecy, which comes from God and for which the mortal candidate must be first suitable and then suited. When the tyro prophet laments the loss of the power, the departure of the spirit, it is the man of God, master prophet, and magister of the prophetic school who enacts the lesson: he who would shape human character and behavior to conform to God's will must himself be shaped into vessel—worthy bearer and instrument–agent of that will.

Tale 4: A King Sacrifices His Son

In none of the three preceding tales is any effort made to locate the events in a historic setting. Even in Tale 1 the reference to the lifting of the curse laid on Jericho's rebuilders by Joshua is left to the reader's inference. The narrative we are about to examine (2 Kings 3:4–24) is quite different. Longer than these vignettes and, like 1 Kings 20 and 22, featuring kings at war, it is given a recognizable historic setting; and there is thus some excuse for the assumption of scholarship that history (*wie es eigentlich gewesen*) is what the narrator was about. And the more verisimilitudinous the setting (in terms, for example, of geographical and topographical detail, national boundaries, ethnic rivalries, political alliances, and royal characters) the more likely that it is history. In the case of Tale 4 the historical background provided by the Bible is supplemented, to a degree rare in Israel's history, by our possession of a fascinating extrabiblical document, the Inscription of Mesha, king of Moab.[11]

In thirty-four lines of text, most of it highly legible, Mesha tells us of Moab's oppression by Israel, to which he put an end when he succeeded his father on the throne. He mentions Omri by name, as well as a number of cities that he retook from Israel, most known to us from biblical texts. From one of these cities, Nebo, which he put to the *herem* ban he slew some seven thousand (perhaps the entire adult population) and captured the shrine vessels of YHWH, which he paraded in a victory procession before his own God, Chemosh. He also mentions by name the tribesman of Gad, who had from time immemorial dwelt in the town of Ataroth.

It seems rather ironic to us that the recovery of this non-Israelite monument has served to confirm the essentially historical *genre* of the biblical narratives relating to the lands east of the Jordan—ironic both because the biblical information on Israel's settlements there is very sparse and because the Moabite Stone contains details that seem to conflict with—if they do not contradict outright—the biblical information. The map showing the borders of tribal or national entities as reconstructed on the basis of biblical data will help us to gain some perspective on these problems. We note first that Moab proper is situated between Edom to the south and Reuben to the north with the streams Zered and Arnon as the respective boundaries. Reuben—and Gad to the north—are the Israelite tribes settled on the territory wrested from Sihon the Amorite in the time of Moses. But for Mesha's placement of Gadites in Ataroth "from of old," we should have no idea that by the middle of the ninth century Reuben, as a tribe, had long ceased to exist. Furthermore, Mesha in lines 1–2 identifies himself as the Daibonite (= Dibonite) and in line 28 recapitulates his mastery of his territory by the biblical formula (*mišmaʿat*) boasting that all of Daibon was obedient to his will. Inasmuch as Mesha succeeded his father on Moab's throne, this would indicate that the boundary of Moab was considerably to the north of the Arnon in the time of his father; it would also suggest that Dibon rather than Kir-hareseth, far to the south, was the royal capital. Some scholars assume, from the mention of Omri and "his son," who lost out to Mesha, that Omri conquered Moab. The Bible, however, has but one reference to Moab's reduction to tributary status, this in the time of David (2 Sam. 8:2); and the Hebrew term for Omri's "oppression" of Moab suggests exaction of tribute, not conquest. It would seem, then, that David's feudal supremacy over

Moab, a principality to the south of Reuben or Gad in his time, was intermittently sustained by the rulers of the Northern Kingdom, who were heirs to the Transjordan territories of the former United Kingdom. The beginning of the end, not only of Israelite hegemony but of settlement of Transjordanian territory, is thus marked by Mesha's victories and razing of Israelite centers. By the time that the stories of Elijah and Elisha were written down (if not composed) the east-of-Jordan territories whence they arose may have become legendary as areas once constituting part of Israel's patrimony.

If Omri, then, had not so much conquered Moab as halted its expansion northward and reimposed a yearly tribute, there would be no clear discrepancy between the biblical accounts—sparse as they are—and what we glean from the Moabite Stone. In the matter of the timing of Mesha's revolt, however, there would appear to be clear disagreement between the two sources. Mesha has Omri in possession of "the territory of Mehedeba wherein he was ensconced [all] his days and half the days of his son, [a total of] forty years, when Chemosh restored it in my days." Since the biblical data allot Omri a reign of twelve years and Ahaba a reign of twenty-two years for a total of thirty-four years, scholars fault the chronology of one or the other source. Another contradiction is the repeated biblical assertion that Mesha's revolt took place after the death of Ahab (2 Kings 1:1, 3:5) and the twice-appearing implication in the Moabite Stone (lines 6–8) that the revolt took place during the reign of Omri's son (Ahab).

These discrepancies do not require a questioning of the veracity of either account. Either or both of the accounts may be speaking loosely on the matter of the revolt's occurrence, since neither anticipated that the fixing of a historical moment would so occupy savants in the distant future. The forty years of Omride occupation of Mehedeba may be a round number, the *his son* in the inscription may refer to Omri's grandson. On the other hand, the revolt, from Mesha's point of view, may have begun in Ahab's reign (perhaps in the last six years of his reign when he was preoccupied with the coalition against Assyria) and have consisted merely in the withholding of the annual tribute. The actual advance northward may well have begun only with Ahab's death—this would be the purport of the biblical notices, which record no conquests by Moab—and have extended through the two-year reign of Ahaziah and into the twelve-year reign of Jehoram.

With this caveat in regard to approaching the narrative of a campaign against Moab as historiography, let us turn to the story itself:

> 9. The king of Israel, along with the king of Judah and the king of Edom, set out. They marched some seven days until there was no water for the army or the animals in their train. 10. "Woe!" cried the king of Israel. "YHWH has called out these three kings to hand them over to Moab!" 11. But Jehoshaphat said, "Is there no prophet of YHWH's near, through whom one may ask an oracle from YHWH?" One of the courtiers of Israel's king spoke up, "There is. Elisha ben Shaphat, he who poured water over the hands of Elijah." 12. Jehoshaphat said, "In him resides YHWH's oracle." So the king of Israel, Jehoshaphat, and the king of Edom went down to him. 13. Elisha said to the king of Israel, "What have we to do with one another! Go to the prophets of your father, the prophets of your mother!" The king of Israel pleaded with him, "Don't [speak so]—YHWH [seems to have] called out these three kings to hand them over to

Moab!'' 14.Elisha replied, ''By the life of yhwh–Hosts, in whose presence I have stood, were it not that I esteem Jehoshaphat, king of Judah, I would grant you neither glance nor notice. However—fetch me a harpist.'' As the harpist played, the hand of yhwh overcame him, 15.and he declared, ''Thus has yhwh spoken, 'This wadi will become pools, everywhere pools.' Yes, this is the meaning of what yhwh said: You shall feel no wind, see no rain, yet that wadi shall fill with water that you may drink— you, your cattle, and your animal mounts. 16.And so slight a thing, this, in yhwh's sight—he will thereto put Moab into your power. 17.You will wreck every fortified city, every choice town, you will fell all the fruit trees, you will stop up every water well, you will render barren and stone-littered every fertile field.'' 18.And so it came about, at morning, at the time of the regular offering, that water coursed down from the direction of Edom, the earth everywhere covered with it. 19.Meanwhile, the Moabites at large having heard of the kings' advance to war against them, they were rallied—from the youngest to wear the warbelt and up—and took up their positions on the border. 20.Early that morning, the sun's rising rays upon the water, the Moabites saw the water at a distance as red as blood. 21.''Blood there!'' they exclaimed, ''The kings have taken to their swords, their forces attacking one another. Now, Moab, to the plunder!'' But upon their reaching the Israelite force, Israel rose to the attack against Moab, and they were repulsed in a rout.

They [the Israelites] invaded and ravaged Moab. 22.The towns were laid in ruins, every fertile field strewn by the soldiery with stones, every water source they stopped up, every fruit tree they felled, until [only the capital] Kir-hareseth remained intact, every stone in place. And the slingers now mounted the attack and let fly against it.

23.The king of Moab, seeing that the battle was going against him, rallied seven hundred picked swordsmen, to hack a way through to the king of Edom. In this they failed. 24.Then did he take his son, the firstborn, who stood to succeed him on the throne, and offered him up atop the wall as a burnt offering. Then was there fierce wrath directed against Israel. They withdrew their attack upon him, and went back to the[ir own] land. (2 Kings 3:9–24)

The three verses preceding this story constitute a notice of Jehoram's coming to the throne and twelve-year reign and the judgment that though he continued in the offense of Jereboam ben Nebat so displeasing to yhwh, he departed from the practices of his mother and father, indeed doing away with a Baal pillar that his father had erected. Verses 4–5, which begin our story, are a parenthetic introduction to the beginning of Jehoram's campaign. The Hebrew of verse 4 (literally, ''King Jehoram set out *that day* from Samaria'') cannot refer to the preceding temporal notice of Ahab's death and Mesha's revolt, for two years' rule by his brother Ahaziah elapsed before Jehoram came to the throne. The ''that day'' or ''at that time'' of the verse is a reference not to time but to occasion. Jehoram's sallying forth from Samaria literally *followed* (it did not precede) his recruitment of Jehoshaphat; for the decision to attack from the south inaugurated the campaign.

The calling upon Jehoshaphat to join the campaign and other parallels to the story in 1 Kings 22 are instructive for the ends that these similar yet different narrative strategies serve. As in the previous story, Jehoshaphat is at one with Israel's king in agreeing to join in the campaign. Unlike the first account, there is no recourse to oracles to ascertain the favor or disfavor of yhwh in regard to the proposed war.

Instead, the dialogue would, at this point, seem to emphasize that the decision to attack Moab from Edom to its south is the responsibility of Jehoram and Jehoram alone. But why this decision? An examination of the map (p. 2), in particular the topography of the area, and the recollection of many biblical tales featuring fords across the Jordan at spots close to Jabesh–Gilead or Succoth–Penuel (fords that are less than a day's march from staging areas in the hill country of Manasseh and Ephraim), will make this decision seem absurd indeed. For this decision entails the logistical difficulties of moving the main Israelite force thrugh the highlands of Judah, descending to the rift of the Arabah, skirting the inhospitable shores of the Dead Sea and cutting into the wilderness of Edom to cross into Moabite territory at a ford of the Wadi Zered.[12] The response to this seeming absurdity is that an attack from the north had been precluded by Mesha's destruction of Israelite settlements there and his fortification of the strong points. The problem with this response is that while it makes sense for a reader today who has access to the Moabite Stone, such was not the case of the biblical readership. They knew nothing about the Moabite Stone, they were told nothing about the extirpation by Mesha of Israelites (whether Reubenites or Gadites), and they probably had as hazy a sense of the topography of the territories of ancient Edom and Moab as is owned today by the average Israeli Jerusalemite. The reader of this tale in biblical times had no map (political or topographical) to follow as he read this story.

All this foregoing discussion is in the interest of gauging the intent of this story's author in the matter (or degree) of history or fiction. As I have indicated, the judgment that the author's purpose is essentially historiographic is—ironically—reinforced by our recovery of what is unquestionably a *historical* document and, at that, an extrabiblical one, hence "proof" of the historicity of the biblical chapter despite discrepancies between the two documents—on the absurd assumption that a historical setting makes historical fiction any the less fiction, like the equally absurd assumption that a fiction cannot be essentially concerned with truth. In addition to the foregoing considerations, our biblical author has taken the trouble to include a signal to the careful reader that the story is not to be taken literally in the sense of *historically factual*. Whereas our story features three kings in alliance against Moab and twice pointedly refers to one of them as "the king of Edom," there are two "historical" notices—one preceding and one following our story—that in the time of Jehoshaphat there was no king of Edom![13] What is clear from these historical notices and implied in our story is that Edom was at the time of our story vassal to Judah as Jehoshaphat may have been close to vassalage status to Ahab and his son, Jehoram. The uses to which the historical setting are put are, however, only to be discerned in what may well be the fictional plot of an attack on Moab from the south. Let us consider now the plot elements that are so patently fictions.

First, there is the matter of the forces mustered by Jehoram not to punish the mighty Mesha who has extirpated Israel on the east side of the Jordan but to put down a rebellious vassal whose predecessors on the throne have paid tribute to Israel's monarchs since the time of David at least a-century-and-a-half ago. This long-endured subjection is altogether understandable when one looks at the map and keeps in mind the topography of the two sides of the Jordan, the fertility and population resources of Judah or Israel to the west, the Negevlike climate of most of Moab to the east of the Salt Sea. What does defy credibility is the need to muster confederate armies or to mount an attack on the soft underbelly of so unformidable a foe. And why should this huge force of three kings march for seven days when Moab's Kir-hareseth lies on the King's Highway a mere twenty miles from the Book Zered, Edom's northern frontier. And, supposing that Moab was so formidable a military power as to dictate a seven-days march to the east—in order to make a surprise crossing of this drought-dried wadi—how could the king of Edom permit an exhaustion of water supply on his own territory? Had this taken place on Moabite soil, we might have guessed that the Moabites were not surprised, indeed that they had time to poison or stop up every water well or cistern; but this is ruled out by the notice in verse 19 that Moab rallied to meet the allies "on its border."

We then have, in direct discourse, in the hyperbolic metaphor of the king of Israel's lament, the element we missed here when we compared it to the parallel story in 1 Kings 22—the recourse to oracular consultation of YHWH before the campaign's onset. The very mention of YHWH as a responsible factor in the war is the narrator's dramatic way of letting us know that YHWH had indeed been consulted. The hyperbole lies in how that responsibility is expressed; for YHWH has not instigated the war and cannot therefore have "called out the kings." It is bad enough that as in chapter 22, the YHWH prophets have interpreted his response as favorable. And (in sardonically ironic contrast with Micaiah ben Imla's interpretation of this favorable response in chapter 22) it is now the king himself who so interprets the favorable oracle: the exhaustion of the water supply is clear evidence that YHWH has lured him and his allies to their doom. As difficult as it is to imagine the king of Edom's being ignorant of the water holes in his own territory, it is even more difficult to believe that King Jehoram's quartermaster had not informed him when a retreat to the west was still possible that only a few days' water rations remained. But one does not—or should not—raise questions about verisimilitude in a didactic fiction. The king of Israel, unlike his father, who paid for his optimism with his life, is quick to despair. Not so King Jehoshaphat of Judah, who at this juncture asks, as he did at the planning stage of the war in chapter 22, whether YHWH cannot be petitioned for another oracle.

To what end? What question does Jehoshaphat propose to put to YHWH? Coming, as it does, after Jehoram's cry of despair, is there a note of humor (on the narrator's part) in putting the question into Jehoshaphat's mouth? Is he not made to appear like the patient who has just heard from his physician that he has an uncurable disease, only a few months to live; when asked if he has a last wish, he replies in the affirmative: he wants another opinion—a more favorable prognosis. If Jehoram is correct, YHWH has already spoken the doom of the allied kings. If he is wrong, then YHWH's favorable response is in place, and one need only wait and see how the present predicament is to be overcome. The formulation of Jehoshaphat's question is strange, to say the least. As, in chapter 22, he asked for another of YHWH's prophets to whom to put the question of whether or not to make war—this after four hundred of them have spoken as one—so now he asks whether there is not in the neighborhood a prophetic office through which a query may be sent to heaven. What would a prophet of YHWH's be doing here in this remote region of Edom's desert? Never mind. One of the courtiers of Israel's king—not, mind you, Edom's—just happens to know of one, Elisha ben Shaphat by name; and lest this eremite's credentials be questioned, his antecedents are vouched for: he is the one who served (the legendary) Elijah as bodyservant. And—what happy coincidence!—Jehoshaphat has already heard of this fellow, who attended that nemesis of Ahab's, the prophet Elijah. He can vouch for him personally: YHWH's word is [in?] him!'' ''And so they *went down* [that is, in all humility], to him: Israel's king, and Jehoshaphat and Edom's king.''

Not a word is said of the address to the prophet. The audience begins with Elisha singling out the king of Israel for abuse. As though he were Elijah himself—which, in a sense, we have seen him to be—he taunts Jehoram to consult the prophets of Ahab and Jezebel. The prophets of Jezebel were long ago slaughtered at the foot of Mount Carmel. And whose but YHWH's were the prophets of Ahab? Elisha's taunt of Jehoram is thus as hyperbolic and metaphoric as Jehoram's repeated complaint. ''Do not put me off,'' he pleads, in effect; the situation is desperate despite the favorable reply from YHWH to the question whether he should go to war against Moab! Elisha relents, not for the sake of Jehoram (guilty of what else other than being son to Ahab and Jezebel?) but out of deference to Jehoshaphat. And then, so unlike the Elijah whose successor he is and so like the professional ''sons of the prophets,'' he requires the playing of a musician to invoke a prophetic trance.

The prophetic oracle has two parts: what YHWH will do and what the Israelites will do. The first part is the transformation of the dry wadi into pools of water. This transformation will achieve two results: provision of drinking water for man and beast and the delivery of Moab into Israel's hands. The first result is obvious, the second less so. The Moabite forces take up defensive positions ''on their border.'' The water ''comes from the direction of Edom.'' Since there is no sign of rain or storm and since water flowing down channels from the mountains could only be, in that area, from the east to the west, the water can only come up from the ground. This is a commonplace in that part of the world. The dryest of these water beds support vegetation in summer's drought, and digging there will often expose water a few inches or a foot or two below the parched surface. So the water rose up from below. What, then, is the meaning of the water coming from the direction of Edom? The answer is in what follows the next morning. The attacking forces are facing north. The Moabites have taken up their defensive position facing south. The rising of the water occurred to the south of them, a bit to the front and north of the Israelite lines. The Moabites experienced none of the moisture phenomenon in their area. When the slanting rays of the rising sun fell upon shallow puddles of water where only parched earth was expected by the Moabites, they mistook the spots of sparkling redness for blood; assumed the allies had had a murderous falling out; and, breaking ranks, descended not as a disciplined force on the attack but as an unruly mob eager for easy loot. And so they were, as Elisha promised, delivered into Israel's hands.

The second part of the oracle is not a command of God's. It is a vision in which the prophet foresees the behavior of the Israelite attackers. And it is given prominence by its detailed foretelling and then by an almost word-for-word repetition when the vision is fulfilled. The attacking forces conduct a wanton campaign of destruction, the equivalent of a scorched earth strategy. To what end?

Everything falls to the invaders. The capital of Moab is surrounded. Its king gambles on one last desperate sally. He attacks with a picked force in the sector manned by the weakest of the allied soldiery, the contingent of Edom.[14] With the failure of this last attempt, the king of Moab knows that it is all over. He withdraws into his city and makes a last desperate appeal to his god. Atop his city's wall, in full sight of his relentless enemies, he erects an altar, slaughters his son, the heir to his throne, and burns him to ashes—an offering to his god Chemosh, who thus far had proved himself unable or unwilling to bring him succor. And where everything else had failed, this last despairing expedient proved efficacious. A ferocious anger fell upon Israel, which thereupon gave up the siege, returned home, never to trouble Moab again.

The anger against Israel was a manifestation of divine displeasure. Why does the narrator not tell us what form it took? Because he does not know? Because he does not care? Or because the omission of this detail is a device to alert us to the significance of the entire story? Never mind whether it was a plague, or what plague in particular. Our attention is not to be distracted from the fact. The sacrifice of Mesha's son was efficacious! Why? To recommend such an expedient to us when all else fails? Absurd! To praise the power of Moab's god Chemosh? Equally absurd. Then the power that responded to the sacrifice was YHWH's! Why did he so respond?

Another clue is in the apparent contradiction between Elisha's assurance that YHWH would deliver Moab into Israel's power and YHWH's failure, at the end to do so. If the ineluctable kernel of historic truth in this tale is Israel's failure, why would a propagandist of YHWH's transmit a tradition of a false oracle or invent it and put it into the mouth of the peerless prophet Elisha, whose very presence in Edom or Moab is a feat of conjuration? The answer, of course, is that the oracle *was* fulfilled in the defeat of the Moabites seduced to their destruction by the blood puddles on their border, this improbable event symbolic of YHWH's grace to which the victory owes. Israel's follow-up of that victory remains as the only factor to which the final failure may be traced. And once again, it is the Book of Deuteronomy that provides us with the psychology essential to the understanding of this event.

In this book in particular, a sensitivity to the finely balanced ecology of God's creation gives expression, in symbolic precepts, to imperatives of moral theology. In the hierarchy of nature the animal ranks above the vegetable, and the human above the animal. But human insensitivity vis-à-vis any of these can spill over into other categories with morally sinful consequences.[15] The precepts apposite to the war waged against Moab, a ravaging so savage that the beseiged king knows he cannot treat for terms of surrender with an Israelite king who carries out the threat that Ben Hadad only voiced (in 1 Kings 20), appear in Deuteronomy 20:10–20. In contrast to the wars of extirpation that Israel will (says Moses) have to wage against the inhabitants of Canaan, wars pursued beyond Israel's territories are subject to regulation. You must offer such cities terms of surrender, rendering them subject but sparing the inhabitants' lives. Only if they refuse terms and you reduce the city after a siege may you execute the male adults and take women and children captive. The conclusion of this passage is critical for our purpose, for its thrust is pointedly contrary to the Israelite conduct of the campaign against Moab.

19. Should you in battle for a city be forced to a long siege, you may not, brandishing your axe freely, destroy its trees—it is of their fruit you eat, and you may not fell them wantonly: Are trees in the open human, able to withdraw from you in a siege

defense? 20.Only those trees that you know are not productive of food may you
destroy by felling to build your siege-works with them, until that city is reduced that
has been waging war against you. (Deut. 20:19–20)

Trees are God's gift to mankind. To use them according to the use the creator intended
for them is proper. Trees that do not produce fruits edible by mankind may be used for
lumber; not so, fruit trees. To use them for lumber, even if they are in enemy territory,
betokens insensitivity to divine teleology, ingratitude, perhaps even rebellion. And
the account of Israel's war against Moab pointedly describes the destruction of fruit
trees and adds thereto the pointless and ruthless stopping up of wells and springs, the
ruining of the none-too-plentiful farmlands of the Transjordan steppe. And this in a
war against their Moabite cousins (who, along with the Ammonites, are descendants of Abraham's nephew, Lot). YHWH warned Moses that Israel might not pass
through Moab's territory, the land he himself had allotted to them, on their way to the
promised land: "Do not harass Moab, provoke no war with them." (Deut. 2:9).
    One piece of territory, east of the Jordan and north of Wadi Arnon and west of
Ammon, was in the time of Moses still in the possession of one of the primordial
native folk, unrelated to Israel and its cousins. The king of this area, Sihon the
Amorite, refused the request of Israel for a peaceable transit of his lands and forfeited
his kingdom as a consequence. Not originally conceived as part of the promised land,
this territory was settled by the Tribe of Reuben, which—faced with the contesting
claims of Moab to the south and Ammon to the east—early disappeared from history.
Part of that territory was claimed and held by the tribe of Gad, Omri ruled it as king of
Israel as far south as Wadi Arnon, and apparently laid Moab to the south under
tribute. And this was the status quo until the loss of the area in the time of Jehoram ben
Ahab, when its Israelite inhabitants were expelled or exterminated, this by the boast
of Mesha himself, inscribed on his stele.
    We can only guess at the degree of trauma experienced by Israel in this national
disaster. It must have been considerable. But nations do not forsake their gods in
defeat. Mesha himself attributes the defeat of his predecessors to the fact that the
Moabite god "Chemosh was angry with his land." And as he credits Chemosh with
his own triumphal restoration of ancestral lands, so Israel surely attributed its loss to
the displeasure of its God YHWH. Yet this entire chapter of Israelite history is lost from
biblical view, except for this story: the story of successful invasion of Moab, the
routing of Moab's forces in keeping with YHWH's promise, the subsequent ravaging
of Moab in contravention of Deuteronomy's prescription for the conventions of war,
the siege of the enemy capital, and—implicit in the foregoing—the offering of no
quarter to its penned-up king. The sacrifice of his firstborn atop the wall of Kir-
haresheth is Mesha's signal to his enemies that it is they who have driven him to this
despairing, extraordinary expedient. The barbarism of child sacrifice is the stuff of
legend; the barbarism that gives new life to legend is laid at the door not of the
bereaved father, whose loss of heir betokens his own loss of immortality, but of his
relentless enemy. And the verdict he implies is confirmed by Scripture. Israel is
responsible, and YHWH it is who holds her responsible. Inasmuch as Mesha tells us
only of his victories, and the Bible concedes the defeat of Israel, the entire account
may be a historical fiction, but a fiction that is not to be denied its truth. The facts of

history may be read and judged for value in terms of moral truth. Moral truth transcends the facts of history. Such, at least, is Scripture's view.

Poetical Review

Dialogue. Almost 50 percent of the narrative is given over to dialogue. Dialogue in Verses 7–8 serves to evoke parallels in the narrative of 1 Kings 22—Jehoshaphat and his connection with prophetic inquiry. So, too, Jehoram's repeated lament, in **figurative speech,** of YHWH's luring him to defeat and disaster. In the mouth of the anonymous courtier is the introduction, as if for the first time, of a YHWH prophet formerly in the service of Elijah. The two-part oracle of YHWH through Elisha promises YHWH's deliverance and victory and predicts how Jehoram will respond to that help.

Repetition. One example is Jehoram's lament. Repetition of word and act are seen in the prediction of Israel's savagery and the fulfillment of the prediction.

Plot. The exploitation of an invasion of Moab from the south to stresses YHWH's grace in the matter of water in an arid waste, providing life for Israel's forces and luring Moab to defeat. Note the savage warfare with which Israel responds, in disobedience to YHWH's will as expressed in Deuteronomy, and the response of YHWH to the desperate act to which his people have driven the Moabite king.

Character and point of view. Note the dramatic objectivity of the narrator: not a hint of approval or disapproval of either of the allied kings or of the Moabite king, who (as the narrator must have known) had written an end to Israelite settlements in TransJordan east of Judah. Consider also the **voice** of narrator, prophet, and kings of Judah and Israel.

History and fiction. The contrast between a historic *document* (the Moabite Stone) and a narrative (the biblical story) of questionable **genre. Metaliterary considerations** are apposite to the question of genre.

Literalness and Metaphor. Examples are the luring of Moabite forces to defeat and the grace of YHWH in supplying a victory, to which Israel may respond in impious ingratitude; and the sacrifice of a son by a desperate father. Would such an act have been deemed likely, in any **ancient** society, to win back the support of its own tutelary divinities?

Poetic Integrity, Unity of Scripture in the narratives in 1 and 2 Kings, in the "historical" narratives in both Pentateuch and Kings and the interface of divinely sanctioned precepts and didactic narrative.

Tale 5: A Father
Sacrifices His Daughter

The interpretation of Tale 4 as essentially a moralistic fiction removes the story of Mesha's sacrifice of his son from the list of attestations to the "historic" practice of immolation of the young in ancient times. The "normal" practice of the ancients of a rite unknown in our own time is a metaliterary convention and in my opinion is utterly erroneous. The small number of stories featuring child sacrifice is but one of the factors that led me long ago to this not-widely-shared opinion. Indeed, I was struck by how much of the evidence for such a practice derived from the Bible itself. We have seen in Tale 1 that there was no such thing as foundation sacrifice. In Tale 4 the child is not an infant (as least explicitly so), and the entire act may stand for a metaphor. If we look in Scripture for other stories of children being sacrificed by their parents, we shall find that there are only two. In one of these, the Binding of Isaac (which will be treated in a subsequent volume), no sacrifice occurs. The other one occurs at least two centuries before the time of the Mesha story, its locale the contested territory of Transjordan, indeed the Gilead area in the north, from which derive the two archetypal prophets Elijah and Elisha. Although no prophet is featured in this story (How significant is the role of Elisha in Tale 4?), its cumulative relevance for judging whether child sacrifice was a substantiated practice in ancient times (that is, a reliable metaliterary convention) or a literary theme variously deployed as metaphor is what prompts me to present the story here.

Judges 2 tells of a Gileadite born out of wedlock and denied a place in his father's family by his half-brothers. This outcast, Jephthah, goes north to the territory of Tob, east of the Sea of Galilee; there he achieves a reputation as chieftain of a raiding band. When the Gileadites are under pressure from the Ammonites to the east, the elders seek out the successful freebooter and offer the chief magistracy of Gilead if he will return to lead them in battle. After consulting an oracle of YHWH's, Jephthah agrees. There follows a detailed account of the claims and counterclaims of the two peoples to the territory; and when Ammon refuses to halt its incursions, Jephthah leads an invasion of Ammon.

> 30.Jephthah made this vow to YHWH, "If you deliver the Ammonites decisively into my hands, 31.then whatever first comes out of the doors of my house toward me on my successful return from the Ammonites—that will become YHWH's. I shall offer it up as a burnt offering."

> 32.Jephthah crossed over to the Ammonites to wage war against them. YHWH gave them over into his hands. 33.He defeated them from Aroer to the approach of Minnith—some twenty towns—and inflicted a great defeat as far as Abel-cheramim, until the Ammonites submitted to the Israelites.

> 34.When Jephthah arrived home in Mizpah, lo there was his daughter coming out to greet him with dances to the timbrel beat—she his precious child, other than her he had neither son nor daughter. 35.The moment he caught sight of her, he tore his clothes. "Oh woe, daughter mine, how you have brought me down to my knees, have become—you [the chiefest] of my troublers! For I—I have oped big my mouth to YHWH and dare not retract!"

> 36."Father," she said to him, "you have oped big your mouth to YHWH. Do to me according to what came out of your mouth—seeing that YHWH has given you vengeance of your enemies, the Ammonites. 37.Allow me this"—she said further to her father—"give me two months respite that I may go to the hills, sink down in lamentation, I and my companions, over my virginity intact." 38."Agreed," he said and allowed her two months[' respite]. So she wept over her virginity in the hill country. 39.After the two months she returned to her father, and he executed upon her the vow he had taken—and [or *thus*] she never had carnal knowledge of a man.

And she became a fixed rite in Israel: 40.year in, year out, the maidens of Israel would go to rehearse [the fate of] the daughter of Jephthah the Gileadite, four days in the year. (Judges 11:30–40)

As we have seen in the case of prophets, so in the case of the assorted magistrates, rulers, or governors (mistakenly rendered ''judges,'' a juridical term too restrictive to describe the office) who rose in the premonarchical period, not everyone who bore the title was an exemplar of virtue. Jephthah's illegitimacy in terms of birth is, however, unfortunate for him personally, not a black mark against him. The one less-than-admirable act on his part is the rash formulation of his vow, for which he pays dearly. The seriousness of the vow and its tragic consequence mask a note of silliness, almost of the comic, that is deliberately injected by the narrator.

The biblical attitude to vows comes through clearly in Deuteronomy:

22.Should you make a vow to your God YHWH, do not delay in paying it, for YHWH your God will exact it of you and you will be punished. 23.On the other hand, you commit no offense [risk no punishment] if you refrain from vowing. 24.Take care for what comes out of your mouth, fulfill what you have vowed to YHWH your God of your own free will, what you have uttered with your own mouth. (Deut. 23:22–24)

Vows are man's doing, not God's. By their very nature, they represent an attempt to bribe God. A quid pro quo is offered to Deity—as though there were anything that man can do for God that he cannot do for himself. But God will hold you to any promise that you make to him. Since the vow is not a formal contract, it does not normally stipulate the date by which the human must fulfill his promise once the condition set for God had been fulfilled. Procrastination would not explicitly constitute reneging, says the text; but God will hold you to what is implicit. Your payment falls due the moment the condition you set for God has been met. No one asked you to make a promise in return for your petition's being granted. So be careful before you utter such a promise; pick your words with care; and, having spoken, fulfill promptly the promise you volunteered.

The incongruities in the Jephthah story begin with the formulation of his vow. What he intends is that he will sacrifice the first animal upon which his eyes will alight on his return home. (This would constitute a rather picayune payment for the vast profit accruing to him by reason of his God-granted triumphs, but that is an incongruity in the very nature of the vow to God.) Had he so formulated his promise, there would have been no problem and no story. In order for him to trap himself by his promise, the author has to make him say what he does not intend, and in doing so the author indulges in an incongruous metaphor, literally, ''the comer out that comes out of my house's doors toward me.'' Animals come out of barn sheds and paddocks, not out of a house's doors.

The second incongruity is in his meeting with his daughter. She did not wait for him at the farm, there to do a private dance for him. She was among the celebrants gathered at the town of Mizpah to hail the conquering hero. As such, she is the first of his household upon whom his eyes fall. Normally, his catching sight of his beloved girl would have filled him with joy. Only much later would he have come to realize that his vow's expression had trapped him and doomed his daughter. But, for dramatic effect, recognition of her and of the vow's implication of her is instantaneous. He tears his clothes, the first token of mourning despair. But then the words he utters are hyperbolically oblique. His daughter has done what his enemies had failed to do, brought him, literally, to his knees. She, his chief delight, has become his chiefest agent of agony. Yet she, of course, has done nothing at all. Jephthah then continues with the explication, the actual reason for his plight, his utterance of the vow. What all the translations have missed is the force of the Hebrew verb governing *mouth*. It is not the normal word for "to open." It appears in Scripture about a dozen times altogether, almost always with *mouth,* in poetry or in elevated prose. The sense of "gaping, yawning" is clear, for example, in the context of Earth's opening her mouth to swallow the blood of her children or the bodies of the Israelites in rebellion against Moses.[16] In this context, then, Jephthah's confession is equivalent to our "Me and my big mouth!" And his daughter, who can only be doleful in her agreeing that the promise must be kept, agrees also that papa's opening of his mouth reflects little credit on his judgment.

The crowning incongruity is substantive, however, not stylistic.[17] It lies in the assumption that the vow must be kept explicitly, literally, despite the absence of any such intention on the part of him who made it. Silly as this assumption may seem to us, we should be free to assume that it would not have appeared silly to biblical Israel were it not for the passage on vowing in Deuteronomy 23. There, as we have seen, it is what is implicit in the vow that requires fulfillment. Whereas Deuteronomy stresses that God stands in no need of your promises but that you must nonetheless keep your word to him, the assumption of Jephthah attributes magical and mechanical power to the spoken word even when it is unrelated or, as here, diametrically opposed to the human and moral values of mankind at large and of Scripture in particular.

This raises another question. Did Jephthah actually sacrifice his daughter? Those who have raised this question before have been accused of being apologists for Scripture. Unable to accept human sacrifice as morally tolerable and embarrassed by Scripture's depicting one of its heroes committing such an abominable act (and not even taking him to task for it), these apologists try to deny the obvious. Verse 39 clearly reads, "He did to her the vow that he had taken." In raising this question anew I deny any apologetic concern. My interest is in literature and my concern is to grasp its meaning and discern its affirmations, not to judge their truth. My question, then, is not whether a historical chieftain did or did not in fact sacrifice his daughter. It is whether the logic of the story points to the consummation of the sacrifice. And if the text read, "and he offered [her] up as a burnt offering," as Jephthah's vow in verse 31 reads, that would be the end of it. But the text does not read so. After stating that Jephthah "executed upon her the vow he had taken," it immediately goes on to say that his daughter died a virgin. The verb for the sexual inactivity of the maiden may express the past tense (as, for example, King James: "and she knew no man") or the pluperfect tense (as in most modern translations: "she had never known a man"). To render it as pluperfect is also to hold the author guilty of a redundancy, for the maiden's virginity has been the whole point, twice repeated, of the two preceding verses. There would be no redundancy if the author's intent is that the fulfillment of the vow did not necessitate the actual sacrifice of the daughter but that it did eventuate in a life of spinsterhood; hence the lamentation for her tragic fate, to die a virgin, a metonym not for unviolated purity but rather for childlessness.

Let us consider the biblical context of the vow that Jephthah made and fulfilled. Jephtah says two things in his promise: that whatever greets him on his return will belong to YHWH (that is, become his property) and that he will offer it up as a burnt offering. This makes it clear that what he had in mind was the first animal he would see. But a human is also an animal, an animated creature; the Hebrew word for such an animate creature—human or beast—is *nefeš*. And it is the literal construction of his vow, rather than his narrower intent, that makes for this self-ensnaring device. The one passage that is apposite to the context of this vow is in Exodus.

> 11. When YHWH brings you into the land of the Canaanites—as he swore to you and to your fathers—and gives it to you, 12. you shall transfer to YHWH every first issue of the womb. That is, every first issue dropped by cattle you own—the males belong to YHWH. 13. Every first issue of the ass you are to redeem with a sheep, or if you are not minded to redeem it, you must break its neck. Every firstborn human among your sons you must redeem. (Exod. 13:11–13)

What is here legislated in regard to the firstborn male was voluntarily vowed by Jephthah—an animal to become YHWH's property. The animals here fall into three categories. The first is animals that may be eaten by Israelites and offered up as sacrifices; in regard to these the Israelite is given no choice, the animal must be handed over to YHWH. The second category is constituted of draft beasts, which may not be eaten or sacrificed; here the Israelite has a choice of redeeming it for himself by substituting a head from the flocks or destroying it. The third category is the human; here there is no choice. The human firstling, which may neither be sacrificed nor destroyed, must be redeemed.

In the light of this (regardless of what a historical Jephthah might actually have done in some dim barbarous past of Israel's) the Jephthah of our author's tale would have had no choice. He would have had to redeem his daughter by substituting a sacrificial victim. This conclusion, if confirmation be required, is borne out in 1 Samuel (14:1–45). King Saul pronounces an adjuration: any of his soldiers who ease up on the pursuit of the defeated Philistines before evening to taste a morsel of food is to be doomed. His own son, Jonathan, unaware of the adjuration, dips his staff into a honeycomb and tastes of the honey. Evening falls, the fast is broken, Saul seeks an oracle as to whether God will prosper a nighttime resumption of the attack, and God refuses an answer. The sacred lots are thrown to determine who has done something to incur divine displeasure. The lots point to Jonathan, and his inadvertent violation of the food ban is discovered. Saul swears that Jonathan must die. The army, pointing to Jonathan's responsibility for the victory, a sign that heaven was his associate in the day's battle, swear that no harm will come to him. The concluding verse says, "The army redeemed Jonathan, so he died not."

The redemption of Jephthah's daughter might have been regarded as total or partial. If the latter, then the substitution of an animal victim would have satisfied the second part of the promise in the vow, the offering up of an animal as a burnt offering. The first part of the promise, "[She] shall be YHWH's," would then have remained in force. Daughters in ancient Israel were regarded as the property of the father, a concept of which a vestige endures in the father "giving away" the bride in our own wedding rites. In the act of marriage a father gave up his daughter and her husband acquired her as his property. If Jephthah's daughter now belonged to YHWH, she could not be given or taken in marriage. By cloistering her off from contact with any male, Jephthah would have "executed upon her the vow he had taken."

The term *to go down* in association with weeping, which I have rendered "sink down in tears," appears elsewhere. Why the hills are the place for lamentation is not clear, although many another text features the sound of mourning coming from a hilltop. The one feature that our narrator is least likely to have invented is the yearly rites (four times a year or once for a four-day span) for Jephthah's daughter carried on by Israelite maidens.[18] The meaning of the word for their rites (rendered by me as "rehearse," by others as "lament, chant dirges") is also unclear. But whatever their manner or purpose, they are not mentioned elsewhere in the Bible and may have been restricted to Gilead or a small area of it. Rabbinic tradition records that in postbiblical times, certain days were occasions for the maidens to parade their charms to entice the young men into wooing them (a milder version of the Sadie Hawkins's Day celebrated in Dogpatch and at one time on sundry college campuses). Perhaps these rites for Jephthah's daughter were intended as a specific against the fate of spinster-hood that she suffered. One might surmise that it was the existence of some such custom—of no significant religious weight, yet harking back to a local pagan tradition—that became associated with the sad tale of the great chieftain's only daughter, which is treated with subtle lampoon in this tragicomedy. If this be judged idle conjecture, it is at least far less mischievous than the sober adducing of it as evidence for child sacrifice in the ancient world.

Tale 6: A Prophet Abets Regicide

7.Elisha came to Damascus. Now Ben Hadad, king of Aram was ill. Word came to him, to wit: "The man of God has come, all this way." 8.The king bade Hazael, "Take tribute in hand, go to receive the man of God, and make inquiry of YHWH through him, in this wise, 'Shall I recover from this illness?'" 9.Hazael went to receive him, taking tribute in hand, indeed the choice bounty of Damascus, freight of forty camels. Arrived, he stood before him at attention and declared, "Your liegeman Ben Hadad, king of Aram, has sent me to you, asking, 'Will I recover from this illness?'" 10.Elisha said to him, "Go, say [to him] 'No.' You will survive all right! YHWH, however, has revealed to me that he will die." 11.He made himself stony-faced, and maintained it so for an eternity. Then the man of God broke into tears. 12.Hazael said, "For what reason now does my lord weep?" He said, "Verily, I know what disaster you will inflict upon the Israelites: their fortresses you will put to the torch, their elite youth you will put to the sword, their infants you will smash, their women with child you will rip open." 13.Said Hazael, "Really now? What is this dog, your servant, that he should wreak so great a feat?" Elisha answered, "YHWH has revealed you to me as king over Aram!" 14.He left Elisha's presence. He came before his lord. He said to him, "What said to you Elisha?" Said he, "To me he said, You will survive all right." 15.On a following day, he [Hazael] took a compress, dipped it into water and spread it over his [Ben Hadad's] face. So did he die. And [so] Hazael succeeded him as king. (2 Kings 8:7–15)

This nine-verse narrative is a parade example of how Scripture's authors will exploit a historic event to create a fiction, a fiction that is rich in humorous subtleties and lampoon of the human even as (in this instance) it skirts the ever-perplexing theological problem of a benevolent God acting in history, who must therefore be somehow involved in, and somewhat responsible for, events whose malignant reality we find difficult to reconcile with his benevolence.

The historical kernel of this narrative is the death of the biblical Ben Hadad (or the Aramean Hadadezer)[19] and the usurpation of his throne by Hazael. Our knowledge of this historical tidbit is derived from an inscription of Shalmaneser III in which he reports the defeat of Hadadezer beside the Orontes River. "Hadadezer perished," he reports, without making it clear whether in, or subsequent to, that battle. He continues, "Hazael, a commoner [literally "son of nobody"], seized the throne, called up a numerous army and arose against me." What the Assyrian does not report is that the "son of a nobody" not only usurped the throne but vacated it by regicide. And, indeed, had Shalmaneser reported such an act of treacherous assassination, it would be a gullible historian who accepted such a report as fact, not because such an act is in itself unbelievable but because the conviction of ninety-nine out of a hundred people charged with murder creates no presumption of guilt for the one-hundred-and-first. Unless the ruler is killed in the course of an openly declared coup d'état, regicide—like most cases of murder within the family—is secretly committed and publicly denied. To accept this account of Ben Hadad's death as a historical certainty, its truth guaranteed by the presumption that God himself revealed this fact to the biblical narrator (a presumption neither explicit nor implicit in the biblical text), is, alas, analogous to the credulity of classical historians who repeat, as sober scholars, the tales of the twelve caesars preserved by Plutarch and Suetonius. These accounts of acts performed in bed and bedroom, in bath and bathroom, have the cumulative effect to persuade us that for a period of several centuries the throne of one of history's greatest empires was regularly occupied by a succession of tyrants who—whatever their other competencies or disqualifications—shared, almost without exception, a proclivity for rape, sodomy, incest, matricide, fratricide, sororicide, and other such things.

The question raised in the foregoing rumination is a literary-critical one but an essentially metaliterary consideration crucial to the divining of the poetical intent of the author: Is he intent on history or invention? On fact or fiction? And if fiction, what kind? Propagandistic, satirical, moralistic, ideological? Aside from dialogue's constituting half of the entire narrative, a proportion much more likely in fictional than in historical construction or reconstruction, there are quite a few features in substance (plot, an essentially metaliterary consideration here) and manner (epithets, formulations of event and dialogue) that point to the narrator's flagging his metaphoric intent. Thus, to begin with what would appear central to the plot or action of this story, namely, Elisha's role in determining the succession to the throne of Aram, we recall now that in Tale 3 of chapter 5, YHWH accepted Elijah's resignation from service and withdrawal from history on the condition that he first anoint Hazael as king of Aram,

Jehu ben Nimshi as king of Israel, and Elisha as prophet in his own stead. Elijah fulfills none of these conditions. He recruits but does not literally anoint Elisha. It is Elisha as Elijah's surrogate who fulfills the first two conditions; and in this he, too, does no literal anointing. In Tale 7, which follows, Jehu will be anointed by a surrogate for Elisha. And in this tale, it is clear, Elisha does not literally anoint Hazael. But even this circumscribed role assigned to Elisha in the affairs of Aram is far-fetched. An analogy would be a divine charge on Billy Sunday, actually executed for him by Billy Graham, to go to Moscow to encourage a minor functionary by the name of Gorbachev to oust Leonid Brezhnev from his post and to take his chair for himself.

But we are jumping ahead of our narrative's progression. Let us go back to the beginning. The story starts not with Elisha's setting out from one of his haunts in the Jordan valley or in Samaria (or in Edom's wilderness?) but arriving—apparently without plan or purpose—in the southernmost capital of Ben Hadad's Aramean hegemony. Equally fortuitous is the timing of his arrival: Ben Hadad is suffering a grave illness. The strangeness of Elisha's appearance in Aramean territory is suggested by the narrator's report of his arrival at Damascus and the repetition of that report in the dispatch to Ben Hadad. Where the narrator refers to Elisha by name and the place of his arrival as Damascus, the herald announces him by the honorific title, "man of God," and alludes to the place with the words "all the way here." That the *here* where the man of God has arrived is not the *here* where Ben Hadad lies bedridden will be brought home to us soon, when we discover that the rich payment for the Israelite's oracular services are conveyed to him by camel freight. In any case, the considerable physical distance between Israelite prophet and Aramean king functions to necessitate an embassy from king to prophet, an embassy that serves in several ways to express the exaggerated respect of pagan monarch for Israelite divine. The king does not have Elisha brought to him, he sends a reception committee to the prophet now within his borders. The committee comes with a petition (addressed, once again, to the "man of God"); and the chief emissary in addressing the prophet declares that the king of Aram who dispatched him is the prophet's "son," (that is, his vassal). In keeping with this courtly address of a client–king to his royal superior is the text's (twice) characterizing the gift to the prophet by the word *minḥā* "tribute" —a train or file of porters or pack animals laden with gifts. But is there a note of incongruity in the king's bidding Hazael to take the *minḥā* "in your hand" and the narrator's repetition of this otiose phrase ("in his hand") before he goes on to expand on the gift's actually constituting forty camel-loads (rather than, let us note, 80 or 120 ass-loads)?

The tribute dispatched by Ben Hadad to Elisha by the hand (or agency) of Hazael and delivered by that hand to Elisha is clearly a king's ransom—both in richness and function. As a fee for a divination it is outrageously high, but it is not so if the oracular question cloaks a metaphor for a petitioner's prayer. The situation of the questioner and the size of the honorarium recalls the joke of three individuals, each informed by a medical specialist that he is suffering from a terminal illness and has no more than three months to live. Each is asked to make a last wish. One, a roué, desires to end his days in the company of Paris's most inventive courtesan. The second, a pietist, wishes to establish endowments for masses and novenas to be read for his soul's salvation. The third is most modest in his wish: he would like to have another opinion.

It is not opinion that Ben Hadad is after, it is a different diagnosis and a more favorable prognosis. As in the parallel narrative (chapter 5, Tale 5) the recourse to a given oracle bespeaks a tribute of faith in the power behind that oracle. Hence the tribute sent by Ben Hadad to YHWH's servant, the man of God, is literally intended, for all its metaphoric disguise, as a ransom for his life. Ben Hadad hopes to secure from Elisha and his God the kind of cure that in an earlier chapter (2 Kings 5) he had successfully implored for his field marshal, the scabrous Na'aman. (In that case, as in this, the petitioner "takes in his hand" a munificent gift: ten talents of silver, six thousand gold pieces, and ten changes of raiment.)

Noteworthy, however, is the difference in formulation in verses 8 and 9. In the former verse Ben Hadad instructs Hazael to take a tribute in hand without specifying its amount; in the latter verse, it is Hazael who takes in hand the forty camel-loads. In both verses, however, the question put to Elisha is exactly the same. Is is possible that the identical formulation of the question points to Hazael's fidelity to his master's charge, while the difference in the specificity of the tribute was one that Hazael managed to convey to the man of God: "My master charged me to bring you a gift— just see how I have interpreted that charge." The attribution of such an act of supererogation on the messenger's part and Hazael's conveying of that act to the prophet, thereby seeking to displace his master as the paymaster and hence as recipient of the prophet's favor, is clearly not explicit in the text; and our reading such intent into the thrifty text might itself be judged as a supererogatory act of eisegesis but for the intricate play of plot and character, gapping in regard to motivation, and the four-fold appearance of the theme of "to be or not to be" or—as it is seemingly phrased in the Hebrew, "to live or not to live." As we noted in an earlier context the Hebrew verb *ḥyh* means not only "to live" (as opposed to its antonym, "to die"), it also may connote "to be healthy, to convalesce," even "to prosper." (This last was our rendering of the verb in Jezebel's report to her husband in 1 Kings 21:15. As against the normal translation "for Naboth is no longer alive, he is dead" [so, for example, JPS], which makes the second clause a pointless pleonasm, we rendered it, "Yes, Naboth prospers no longer, he is dead.") It is the multivalence of this verb's connotations that our narrator exploits, along with a few other devices such as ambiguous spelling and multiplicity of possible pronominal referrents, to challenge the close reader of the text to examine the options for reading the characters of Elisha and Hazael even as he portrays a Hazael standing in befuddlement before the prophet who has delivered an ambiguous oracle.

As we noted earlier, Ben Hadad, in his charge to Hazael, tells him to put this question (for him) to the man of God: literally, "Will I *live* from this illness?" In our translation we replaced *live* with *recover*. An equally acceptable translation, which, however, conceals the presence of the preposition *from,* is, "Will I *survive* this illness." When Hazael tells of his master's charge, he quotes these words exactly. Elisha's response however is complicated by the spelling of one syllable. He says to Hazael, "Go say,"—and the next syllable means "to him" or "no/not" (pronounced identically in either case). The orthography (*ketib*) indicates the latter, while the lectio (*qere*) indicates the former. The text, therefore, may unfold the following readings of Elisha's response:

1. Go say to him, "You will indeed survive [the illness]"—[the fact is], however, that YHWH has revealed to me that he will, indeed, die.
2. Go say [to him], "No [you won't die]. You will indeed recover"—[the fact is], however, that YHWH has revealed to me that he will, indeed, die.
3. Go say [to him], "No, you won't recover"—YHWH [you see] has revealed to me that he will, indeed, die.
4. Go say [to him], "No." Prosper *you* shall, [O Hazael]—YHWH has, however, revealed to me that *he* will, indeed, die.

Readings 1 and 2 are alike in that the message Elisha tells Hazael to return to Ben Hadad is a favorable response to Ben Hadad's implicit prayer for recovery, despite Elisha's disclosure to Hazael that this response will be misleading. The difference between the two is that the formulation of reading 2 is—in the final event—not only misleading but a lie; for Ben Hadad will, by reason of Hazael's treachery, be given no time to recover. The formulation of reading 1 is, by contrast, true: Ben Hadad will survive the illness, not, however, Hazael's ministrations. The formulations of the response in readings 3 and 4 are both true to the events as they turn out. Of these two, it is reading 4 that to my knowledge has never been raised as a possible reading in any commentary. In part this is due to the ignoring of the sense of "prosper" in the verb *ḥyh*, in part to the excessive trickiness it seems to impute to Elisha or the narrator in the abrupt switch of the person addressed. That this reading 4 is indeed one of the intended options is disclosed in verse 14.

In this verse, Ben Hadad inquires about the Israelite prophet's response to his inquiry (petition). This time he refers to him as simply "Elisha." This is a subtle touch. As long as the response of the formidable agent of YHWH's was an unknown, Ben Hadad referred to him with the deferential title, "man of God." Now the response has been given, the die (so to speak) has been cast, and the king who presses to learn what it was need no longer stand on the ceremony fitting for a supplicant. The care that the narrator thus shows in his choice of every word is now to be kept in mind when we examine his formulation of Ben Hadad's dialogue. He does not ask, "What answer did Elisha return *about me?*" He says, "What said *to you* Elisha?" This translation of mine, transposing subject and indirect object, is to suggest that the formulation of the question opens up to Hazael a response that is true—to reading 4— and clearly a deception of the king. Hazael might have quoted Elisha (in. verse 10) verbatim: "Elisha said to him [Hazael], "Go, say, to him [Ben Hadad]." Instead he responds, "To *me* he said," that is, not to me as your representative but to me in my own person. The king is thus lulled into security and relaxes his guard, and Hazael waits for the moment when he can speed the unwary king to an early death. Returning now to the four possible readings of Elisha's response in verse 10, we can understand now the first clause of verse 11. The narrator leaves us in the dark as to whether it is Elisha or Hazael who keeps his face expressionless; and well he might, for it makes sense for both subjects. Elisha, having delivered an ambiguous, multivalent oracle remains poker-faced while Hazael, also expressionless, tries to read in the prophet's face some clue as to his real meaning. The prophet breaks down first—into tears, of all things. This is of no help whatsoever to Hazael in his attempt to understand the prophet's response, so he breaks into the understandable question, asking why this prophet plenipotentiary has suddenly succumbed to this show of weakness. The prophet's answer to this question is yet another pointer to reading 4 as true to his intent: Hazael will prosper—indeed, will become king, victorious king of a victorious Aram.

By such devices of multivalence and ambiguity in both plot and dialogue (homophony of negative particle and pronominal pronoun, descriptive action with grammatical subject left unclear, petition masked as inquiry, bribe disguised as gift, and fear and ambition lurking under show of exaggerated deference) our narrator suggests traits of character even as he ambiguates (a coinage I owe to Meir Sternberg) character. This raises the further question, What and to what degree (if at all) is Elisha's prophetic vision as disclosed to Hazael a causal factor in Hazael's murder of his master and usurpation of the throne? To some extent this question may be part of a larger general question: When an action predicted in an oracle is fulfilled by act of someone privy to the oracle, is the claim for the truth of the oracle as vindicated by its fulfillment vitiated by the consideration that such an oracle was a self-fulfilling one?[20] Our question here may also be related to another theoretical one, as a purely literary consideration: What does a character's action consequent to receipt of a revelation disclose about his faith or lack of faith in oracles? We raised that question from one point of view in regard to Hamlet and his father's ghost. From another point of view we might formulate a question about Macbeth. Did he repose any faith in the witches' prophecy that he would become king of Scotland? Of course, one will respond: Did he not act upon that prophecy? Of course not, another will rejoin. His acting upon that prophecy reveals his lack of faith. Had he trusted the oracle he would have remained inactive, waited for it to come true; he certainly would not have rushed to murder the king under his own roof. The witches merely plant a seed in Macbeth's mind (a soil well-prepared to receive it); it is no accident that the seed of regicide (and other murders to come in its train) is planted by hags representing the powers of darkness.

The relevance of this analogy—question and answer—to our narrative becomes clear when we ask what the point of it is, why is it told in Scripture? Its purpose cannot be to praise the Arameans for their respect shown to Israel's God and his prophet; it is unlikely that it serves to show the influence of a true Israelite prophet over an Aramean general's mind. What is the likelihood of the author's having been unaware that Elisha, functioning as an agent provocateur in a regicidal conspiracy, is engaging in an activity more appropriate to Satan's minions than to God's?

The answer to these questions must lie in the exhange between Hazael and Elisha, which we have yet to investigate. The answer that Elisha returns to the query about his weeping reveals that it is not with relish that Elisha brings to Hazael the tidings of his future greatness. He is in Damascus to give the oracle not because of, but despite, his knowing that Hazael will perpetrate fearful atrocities upon Israel. If we were to take this prophecy at face value, we should have to list this Hazael as equal to the most hated opponents who ever warred against Israel.[21] The one notice about Hazael's dealings with Israel fails to convey such a picture of him.[22] On the other hand, this ambassador of Ben Hadad's, whose cupidity for his master's station has been so subtly hinted at, is—in his self-effacing humility—made to appear as the most loutish of Nazis in his characterizing the wreaking of havoc on Elisha's people as "so great a feat [as you speak of]."

Thus, by insensitivity of character, gaucherie of expression, Hazael is limned for us as traitor by ambition and criminal by perpetration. But this delineation is only a matter of manner. The matter of the tale itself has Elisha present in Damascus to suggest by indirection a devilish option to a scoundrel who needs no prodding. This presence is a metaphor for YHWH's prophet's participating in the work of putting upon a neighboring throne the most barbarous of conceivable tyrants. This participation— depicted merely as a seer's letting a lackey share a revelation—is actually in metaphoric fulfillment by Elijah's surrogate of a charge by YHWH to anoint, no less, this very person as king: "YHWH said to him [Elijah, at Horeb], 'Go! Retrace your tracks—to the steppe of Damascus. On arrival, you are to anoint Hazael as king over Aram'." (1 Kings 19:15).

Let those rest content with the judgement who will that these narratives in the books of Kings are a patchwork of snippets from a variety of traditions, oral and written, about the separate and separable cycles of Elijah and Elisha. Let them continue to apply their investigative talents to penetrate into the motives of a redactor or a series of redactors who can piece together a string of contradictions, of prophecies unfulfilled on the part of true prophets of him who is alone Author of the future as he is of the past. Nor is it our business to quarrel with devout religionists who can yet read as sober—nay, inspired—historiography, an account that depicts the God of Israel as the all-knowing Director of an ancient CIA, directing his undercover operatives in cabals to overthrow neighboring governments, in the interest of bringing to power more ruthless tyrants for the punishment of his own erring people. But for those who recognize that theodicy is an equally difficult task whether atrocities are counted in the hundreds or millions, this story of Elisha–Elijah is eloquent witness that the prophetic faith and theology of ancient Israel coexisted with the awareness of the challenge posited by history's atrocities to the existence of a benevolent omnipotent God. And this faith and theology endures (so says the visionary servant of YHWH) despite the challenge not of the atrocities of the past but of those that will yet surely occur.

Tale 7: A Prophet
Abets Treason in Israel

If the exegeses of the various narratives we have so far treated are not altogether off the mark, one partial conclusion we might safely draw is that there is a difference between the scriptural authors' and more recent authors' approaches to **character and characterization.** For the present I should like to emphasize this difference in regard to consistency or inconsistency of character as it appears in modern as opposed to biblical poetics. In the narrative fiction and drama of recent centuries, personae will by their spoken words and their acts move the plot in one direction or another; or, conversely, the actions plotted will serve as a showcase for the exhibition of the personae's characters. To be sure, a dynamic view of character will not rule out a singular act of courage on the part of a generally cowardly persona, or a lapse into probity on the part of a generally dishonest one. But on the whole, we moderns look for consistency in character and expect explanations for deviations from it. By contrast, the episodic nature and didactic purpose of biblical narrative will permit an Ahab to be a foolishly shortsighted impious scoundrel in one series of narratives and a courageous warrior and tactful diplomat in others; docile husband of Baal-championing Jezebel, imperturbable witness of the slaughter of YHWH prophets and Baal prophets alike, and respectful inquirer of YHWH's disposition via the channels of his prophetic college. Thus, too, the picture drawn of Ben Hadad as a bombastic, drunken lout we may see as a fiction consciously employed for his own purposes by the biblical author who probably knew (as we can now divine from Shalmaneser III's inscriptions) that as the capable head of an Aramean coalition, he fought off the incursions of the Assyrians over a period of eight to twelve years.

In Tale 6 we beheld a loutish regicide-to-be, Hazael, foreseen by Elisha as a conqueror of Israelite cities presiding over wartime atrocities. The atrocities are notably passed over in silence in 2 Kings 13:1–7, 10:32–33, and 13:22–25 (see note 22). It is likely, therefore, that the portrait drawn of Jehu ben Jehoshaphat ben Nimshi in 2 Kings, 9 and 10 is not designed in the interest of poetical consistency or of historical verisimilitude.[23] Assuming the factuality of the events—in a broad sense, if not in literal detail—as related in these chapters, Jehu founded upon the ruins of Ahab's house a dynasty that endured a full century—no small achievement this when one considers that the reigns of some of Israel's kings were numbered in months or even a week. This continuing royal succession is pictured in 2 Kings 10:30 as a promise of YHWH's to Jehu, reward for carrying out his will in his annihilation of the House of Ahab. Lest, however, we take this notice as a general approbation of Jehu, the verses immediately following are quick to condemn him for failure to follow YHWH's torah for continuing in the offense of Jereboam I. Apparently traceable to divine displeasure are the defeats inflicted by Hazael and his wresting of territories from Israel, particularly in the areas east of the Jordan from Bashan in the north to the Wadi Arnon in the south.

Indeed, for all the flatness of the narrator's statements that the ruin of Ahab's house conformed to the will of YHWH, for all of Jehu's quotations of YHWH's pronouncements against this line in (apparent) justification of his depredations, it is hard to picture the scriptural author actually gloating over Jehu's unslakable blood lust. The aging Queen Mother Jezebel, informed of her son's death at the hand of his treacherous general and of the approach of that regicide, is every bit the equal of Jehu and far more regal in courage and defiance as she prepares her toilette, as if readying herself for her groom rather than for her assassin. Her taunt, "Done well, have you, O Zimri, assassin of his master?"[24] reminds her executioner that his predecessor in regicide enjoyed the fruit of his crime for a single week. And the call by Jehu to Jezebel's eunuchs to defenestrate their queen reflects as ignobly upon him who calls for, as upon those who respond to, this shameful mode of execution.

Indeed, a careful reading of 2 Kings 9:14–24 will disclose that it is in no fair fight that Jehu slew his master, King Joram. In verse 15 Jehu, addressing the colonels who support him in a bid for the throne, warns them that no one is to be permitted to leave Ramoth–Gilead who may seek to bring report of the conspiracy hatched there to

Joram in Jezreel. So the device of arresting the heralds sent out to him from Jezreel by King Joram is a trick to force the king himself to come forth from behind the city's wall, a trick made explicit by Joram's belated recognition and call to his cousin, the King of Judah: "Trickery, Ahaziah!"[25] And Jehu does not flinch from the challenge presented by his master, who wheels about to escape the trap himself: he speeds the fatal arrow into the king's defenseless back.

And if Joram's descent from Ahab and Jezebel, including him in the issue doomed by YHWH to extirpation, is mitigation for Jehu's act of regicide, the same cannot be said for Jehu's hounding to his death Ahaziah, king of Judah, whose only complicity in the crimes of Israel's monarchy is that his mother Athaliah, taken to wife by good King Jehoshaphat, was daughter of Omri and sister to Ahab (2 Kings 8:18, 26). In a similiar vein we may question whether Jehu's slaughter of forty-two of King Ahaziah's kinsmen excited the applause of the biblical narrator (see 2 Kings 10:12–14) or, for that matter, whether he approves the seemingly dastardly way in which Jehu accomplished the death of Ahab's "seventy sons in Samaria" along with Ahab's "nobles, intimates, and priests" in Jezreel (10:1–11).

But the most morally questionable instance of Jehu's acts of violence is the one in which he seems to act out of zeal for YHWH's glory alone: his eradication from Israel of the Baal cult (2 Kings 10:18–28). Ahab's two sons who succeeded him on the throne are never cited as promoters of Baal worship. Ahaziah died (as we saw in 2 Kings 1), for sending to Baal-zebub of Ekron for an oracle; and King Joram, seeking an oracle from Elisha in the wilderness of Edom elicits from YHWH's prophet the sneer that he resort to the prophets of his father or mother (3:13). How did Baal worship make a comeback after Elijah's slaughterings on Mount Carmel? Where did the temple reserved to Baal stand, the temple to which every last one of his devotees came under Jehu's threat of death for noncompliers? What badge of identification did these Baalists possess who filled their god's shrine "from mouth to mouth"? And how did they manage to screen from entrance any YHWH worshiper who might have sought to intrude on this exclusive service to an alien deity? These and other incongruous details in this narrative of a final and conclusive extermination from YHWH's territory of a god who will reappear in YHWH's own city, Jerusalem, even in his own house on Zion's mount,[26] suggest a literary (metaphoric) and metahistorical intent behind all this seeming pro-YHWH frenzy of butchery. In view of this, the story of Jehu's anointment as king by a prophetic agent provocateur, the only story of Jehu in which a prophet of YHWH figures, must be told in the interest of some message other than a justification of Jehu's revolt, regicide, and regnal success.

The Anointing of Jehu

1. Elisha the prophet, now, summoned one of the prophetic confraternity. He said to him, "Strap on your belt, take this flask of oil in hand and go to Ramoth–Gilead. 2. When you arrive there, search out Jehu ben Jehoshaphat ben Nimshi. When you have done this, single him out from among his peers and take him to one of the innermost chambers. 3. Take up the flask of oil, pour some on his head and say, 'Thus said YHWH: I have anointed you king over Israel.' Then open the door and swiftly depart—not a moment delay!"

4.So the neophyte, the neophyte prophet, went to Ramoth–Gilead. 5. On arrival, there were the army commanders in session. He announced, "A word with you, commander!" Said Jehu, "With which one of us?" He replied, "With you, commander!" 6.Forthwith he went into an inner room, poured oil on his head and announced to him, "Thus said YHWH, God of Israel: I have anointed you king over YHWH's people, over Israel. 7.You are to strike down the House of Ahab, your liege lord, that I may thereby avenge the blood of my servants the prophets, the blood of all YHWH's servants spilled by Jezebel. 8.The House of Ahab in its entirety is to perish, so that I cut off every male of Ahab's line, near and far in Israel, 9. making the House of Ahab like the House of Jereboam ben Nebat and the House of Baasha ben Ahijah; 10.and as for Jezebel herself, the dogs are to devour her in Jezreel's plat, no one is to bury her!" He opened the door and made a hasty departure.

11.Jehu, now, came out to his lord's liegemen. They greeted him with the question, "Is all well? On what business came that crazy to you?" He replied, "Yourselves know the type and his ranting." 12."Not true," they said, "tell us plain!" "Oh," he said, "he went on and on; in brief, to me: 'Thus said YHWH: I have anointed you king over Israel.'" 13.Instantly they removed each one his robe and placed it at his feet, sounded a fanfare on the horns and made proclamation: Jehu is king! 14.Thus came about the conspiracy of Jehu ben Jehoshaphat ben Nimshi against Joram. (2 Kings 9:1–14)

There are, in this narrative, three plot perspectives or thematic elements that may be abstracted from the seamless whole for the purpose of ascertaining its kerygma(s). Let us call these elements *analytical foci*.

The most general focus (and the most obvious) is the relationship of this narrative's plot to the "historical" or metaphoric–historical context and, in this connection, to the prophetic cast of Baal prophets and YHWH prophets. This thematic cynosure appears in verses 6–10, the address of YHWH to Jehu, in the mouth of the prophet who has been sent to anoint him: the main purpose of YHWH in fixing on Jehu as king is to accomplish the end of Ahab's dynasty, a bloody end that is primarily in retribution for the blood of YHWH's prophets spilled by Jezebel, a theme we last heard sounded in 1 Kings 19, in Elijah's audience with YHWH at Horeb. I have stressed before the metaphoric nature of this theme, as opposed to a literalistic historical reading of it; but it will bear repetition here. After chapter 19, not a single Baal prophet makes an appearance, while YHWH's prophets (true and false) are ubiquitous. In chapter 20 and 22 in the case of Ahab and in 2 Kings 3 in the case of Jehoram ben Ahab all royal contact is with prophets of YHWH. (The assumption must be that the absence of Baal prophets will be traced to their elimination by Elijah after the confrontation on Mount Carmel. Perhaps, too, that is why Ahaziah ben Ahab in 2 Kings 1 must send all the way to Baal-zebub in Ekron for a non-YHWH oracle.) As the prophet of YHWH consulted by Jehoram in chapter 3 is Elisha himself, so too in chapters 5–7 the unnamed king associated with Elisha's prophetic activities can only be the same king, Jehoram. In addition to this primary crime for which YHWH wills retribution, there is the implicit reference to the crime against Naboth and his line in the command that Jezebel's corpse is to be exposed in "Jezreel's plot."

The second focus is the parallel of this narrative with that of Hazael's succession to the throne of Aram and more generally its relationship to the larger chain of association of YHWH, Elijah, Elisha, Hazael, and Jehu, which was presented to us first in 1 Kings 19. In that chapter, let us recall, a single prophet who, representing YHWH, was able to eliminate some 450 Baal prophets (plus 400 Ashera prophets)—all the cohorts of Jezebel—had for fear of Jezebel's threats fled south to ask his God to end his career and his life. We saw, furthermore, that the incongruity of fear of death prompting a recourse to suicide (an incongruity not altogether false to the human condition) was but a cloaking metaphor (and a signal of that metaphor) to introduce the theme of that hoped-for, longed-for final battle in which the proponents of false gods are defeated for all time. The failure of this hope for Elijah, his disappointment that his triumph on Mount Carmel had not turned out to be history's last chapter, led him to a despair not shared by his God. YHWH, not impatient for an end to history, accepted Elijah's resignation from service on three conditions, none of which Elijah fulfills literally. He does, however, recruit his successor-to-be Elisha, who in turn will as Elijah's vicar play a role in bringing Hazael to the throne of Aram and, through a neophyte prophet, fulfill the third condition, the anointing of Jehu ben Jehoshaphat ben Nimshi. Can we believe that our biblical author, so richly inventive in figurative plot and metaphoric denouements, did not intend for this Jehu to turn out to be, in one sense, a second Elijah, somehow succeeding where Elijah somehow failed? Though no prophet himself, this wily, sardonic, ruthless and bloodthirsty warrior becomes the agent of YHWH, not only in encompassing the doom oracle spoken by Elijah but in eliminating from Israel the Baal cult, even to its last adherent.

In this context, then, as in the case of the first focus, the metaphor of YHWH's ultimate responsibility as the lord of history exists in and alongside the specific death-dealing actions of human beings—good, bad and indifferent—who act knowingly, pretending to know, or not knowing at all that they are agents or tools of the God of history: the archer who aimed at an anonymous figure standing in a chariot and inflicted the wound fatal to King Ahab; the Hazael who smothered his master so that he might outdo Ben Hadad in ravaging Israel; and now Jehu, whose frenzied chariot drive is an identifying patent. His murderous rampage ends the careers of King Joram of Israel and King Ahaziah of Judah; Queen Jezebel and seventy of her husband's children and grandchildren; the guardians of Ahab's children whom he had tricked into slaughtering their royal wards; and the devotees of Baal whom, by guile and command, he rounded up to be butchered in the temple of their god. How well the epithet of a Hun king won from his traumatized victims—the Scourge of God—fits Jehu!

The third focus is the details of setting, cast of character, and dialogue in this story of a general stirred to regicide and usurpation, details not at all necessitated by the considerations raised in our two preceding foci. The setting of the story is Ramoth–Gilead. After his military associates have indicated to Jehu that they will support him in his bid for the throne, Jehu is made to say, "If it really be your wish [that I become king], let no one make a free exit from this city who might make for Jezreel to tell [of this plot]" (2 Kings 9:15). From this it would seem that Ramoth–Gilead is in Israelite hands, a border fortress being defended against Hazael's attempts to retake the city, which must, then, have been captured from Aram by Ahab before he sustained his fatal wound in the battle of the chariotry. Jehu is, in all likelihood, the commanding general of the defending forces now that King Joram has gone south to Jezreel to recover from wounds sustained in honorable battle. Why Elisha delegates the task of anointing Jehu instead of going to Ramoth–Gilead himself (as he did to Damascus to meet with Hazael) we are not told, but we can make a shrewd (and poetical) guess. If Elisha had not acted through a proxy, the narrator would have had to invent another device to make the point he contrives through the instructions that Elisha gives to the agent. These instructions, underlined by the information that the agent obeyed them scrupulously, stress the importance of the anointing's taking place in secret and the anointer's making a hasty exit the moment he completes his task. (Also instructive are the terms used to identify the agent. In verse 1, he is "one of the sons of the prophets"; in verse 4, he is "the *na‘ar,* the *na‘ar* prophet,*"* which expression the new translations fail to render faithfully.)[27] The *neophyte* prophet, arrives in the midst of a gathering of the military "brass." The dialogue between the tyro and the officers is an instance of **free direct discourse,** which will conceal from any reader unaware of this convention that the scene here is identical to that in Shaw's drama where a young messenger claiming to be an agent of divinity is put to the test. As Joan of Arc was able to pick out the modestly attired Dauphin from among his richly caparisoned courtiers, so here: the prophet asked to speak to the ranking officer, was challenged to pick him out from among his fellow officers and met the challenge with ease. His first success makes possible his interview in private, and he then departs as if he fears pursuit. When Jehu saunters out, his brother officers ask the purport of the prophet's mission. Mantic personalities are not always and everywhere accorded a uniform respect. As if to disguise their avidity to know what transpired in that mysterious tête-à-tête, Jehu's officers refer to the messenger not by his title, prophet, but by a characterization of that role drawn from experience with the less coherent exemplars of the profession: "that crazy." The use of this term, one indication of the bantering tone assumed by the questioners, opens the way for Jehu's evasive response: the rantings of such a one, as the question itself implied, are not to be taken seriously. But the bantering tone is dropped; the officers press on to extract a serious answer; and Jehu, pretending that he himself puts no credit in the crazy's message from YHWH's throne, offhandedly reports that decree. Now we understand why the young prophet had to leave so precipitously. Neither he nor Jehu could be sure how Jehu's colleagues would react to a proposal of mutiny. This problem of loyalty to one's king, a king the fourth in his line (the ascription of legitimacy is strengthened with the duration of dynastic succession), is suggested in verse 7: YHWH's charge to strike down the dynasty of Ahab specifically identifies the victim as "your liege lord." So Jehu, urged by his colleagues to trust them, discloses the prophet's message, yet in a manner that will not expose him to a charge of conspiracy. But he need not have worried. The officers are ready for mutiny, and Jehu is their candidate for the throne.

So, once again, the biblical author has taken a historical setting and historical personages and exploited both in a dramatic scene featuring imagined characters and dialogic invention to explore the problems of the mysterious way in which the hand of God stirs the brew in history's cauldrons, in particular the problems of the intervention in momentous affairs of the *activist clergy.* For this is exactly what the prophet represents. As contrasted with the cloistered brotherhoods or the priests content to minister in shrine or school for their regular prebends, the prophet is a cleric braving the highest heats of history's kitchen, interpreting the moves and the motives for the moves of the Hand that more than any other stirs the pot. The prophet may be an impressive clairvoyant or a raving hysteric. He may be the true voice of the true God, the lying voice of false gods, or the instrument of heaven to lure mortals into a course of self-destruction. When he dabbles in politics, the prophet may be a fish out of his depth, a roiler of the polity's peace, or the instigator of revolution and harbinger of a new political order. He often plays a dangerous game, exposing himself to risk ranging from ridicule to death. No wonder, then, that he needs a mark on his forehead, a sign that accept or reject his message as you will, you may not touch his person, for he stands under heaven's protection.[28]

The hyperbolic nature of the mass slaughter of prophets in the foregoing tales—be the perpetrator Jezebel, Elijah, or Jehu—becomes evident when we look elsewhere in Scripture for indications of prophets indicted or convicted on a capital charge. But to say that is not to assert that every element in a biblical narrative is pure fiction or that prophets never lost their lives for refusing to be silenced. Our next—and last— prophetic tale is an exemplary account of the existential quandary of a prophet and his public in a time of crisis. Taken from Jeremiah 26, it is not by the prophet but about him; it is not oracle but narrative; and its presence in a book of the Writing Prophets (like other narratives in these books) is an indication that the rabbinic division of the middle section of the Hebrew Scriptures into Neviim Rishonim and Neviim Ahronim (Early Prophets and Later Prophets) is poetically more sound than our, later division of it into History (Joshua through Kings) and Prophecy (Isaiah through Malachi).

Tale 8: A Prophet
Is Tried for Treason

1. At the beginning of the reign of Jehoiakim ben Josiah, king of Judah did this come to pass[29] by YHWH's direction: 2. "Thus has YHWH bidden: Take a stand in the court of YHWH's House and speak to the towns[men of] all Judah who come to worship in YHWH's House all the words that I charge you to speak to them. Omit not a syllable. 3. It may yet be that they will heed, turn back—each of them—from his evil pursuit, then may I relent of the evil [consequences] that I am considering wreaking upon them, yes, for the wickedness of their acts. 4. You are to say to them, 'Thus said YHWH: "If you will not heed me, following my oracular instruction, which I have set before you 5. (to heed the words of my servants, the prophets whom I dispatch to you, ever and ever dispatch though you have paid no heed). 6. then will I make this House like Shiloh, and this city itself will I make a [byword for] curse to all the nations of earth."

7. The priests and prophets and all the citizenry attended while Jeremiah spoke these words in YHWH's House. 8. Then when Jeremiah brought to a close all the address that YHWH had charged to be spoken to all the citizenry, the priests and prophets and all the citizenry put him under arrest, on this charge: "Die you must! 9. How have you dared prophesy in YHWH's name these very words, 'Like Shiloh will this House become, and this city laid in ruins without inhabitant!'" Thus did all the citizenry gang up on Jeremiah in YHWH's House.

10. When the high officials of Judah heard of these events, they proceeded from the king's palace up to YHWH's House. 11. The priests and prophets addressed the

high officials and the citizenry all: "A sentence of death is coming to this man, for his prophesying for this city that which you have heard with your own ears."

12.Then said Jeremiah to all the high officials and to the citizenry all, "YHWH did send me to prophesy concerning this House and this city the things you have heard. 13.Now then mend your ways and your deeds, heed the bidding of your God, YHWH; and YHWH will relent of the evil consequences he had pronounced against you. 14.As for me, I am in your power. Do with me as you deem proper and just. 15.Just so that you clearly recognize that if you put me to death, it is the stain of an innocent's blood that you take upon yourselves, even upon this city and its magistracy. For true it is that YHWH has dispatched me to you, to din these things into your ears."

16.[So it was that] the high officials and all the citizenry declared to the priests and prophets, "No sentence of death for this man: it is in the name of YHWH our God that he has spoken to us."

17.[The verdict was reached in this way:] Some of the nation's elders rose to address the full assembly in this wise, 18."Micaiah the Morashtite did prophesy in the days of Hezekiah, king of Judah. He addressed the entire citizenry of Judah, just so:

Thus has declared YHWH–hosts:
Zion as a field will be ploughed
Jerusalem shall be ruined piles
And the Temple Mount thick-forested terrace.

19.Did then Hezekiah, king of Judah, and all Judah [assembled] venture to execute him? Did he not rather show fear of YHWH and entreat YHWH's favor, so that YHWH relented of the evil consequences he had pronounced against them? We now are incurring fatal punishment."

20.[Citation was also made:] "Yet was there another one engaged in prophesying in YHWH's name, Uriah ben Shemaiah from Kiriath-jearim. He prophesied against this city and against this land very much as these words of Jeremiah. 21.When King Jehoiakim, together with his paladins and high officials all, heard report of his words, the king moved to execute him. Uriah, however, got wind of this and in fear took flight, reaching Egypt. 22.Then did King Jehoiakim send embassy to Egypt, Elnathan ben Achbor and dignitaries accompanying him to Egypt. 23.They extradited Uriah from Egypt, brought him to King Jehoiakim. He had him put to the sword and his carcass thrown into potter's field."

24.[Yet] was it the influence of Ahikam ben Shaphan, siding with Jeremiah, that kept him from delivery to the council for execution. (Jer. 26:1–24)

First, as to form: this chapter exploits the narrative technique of dividing the story into episodes, a synoptic episode relating the entire story in brief, followed by a resumptive episode in which additional details are provided. The failure to recognize this technique results (as may be seen by comparing my translation above with the best of those in the standard versions) in an assumption of a continuing chronological progression that, however, is at odds with the logical flow of events. This narrative division differs from most other instances of this technique in that the first episode, verses 1–16, is about twice as long as the resumptive episode, verses 17–24; normally the resumptive episode is longer than the synoptic one. If the normal ratio were regarded as a prescriptive norm, a reduction of this narrative to such a norm would be achievable by treating verses 1–6 as an introductory Episode A, followed by the synoptic Episode B and the resumptive Episode C. We might further note that in the synoptic Episode B, verses 12–15, the self-defense of Jeremiah is a detail, parallel to the details of the resumptive episodes in C, verses 17–19 and 20–24, and would normally appear alongside them in that episode. Let us see what impact the narrator achieves by ordering his material the way he does.

First there is the introduction, in which the **reliable narrator**[30] establishes the facts: YHWH did indeed lay his charge upon Jeremiah, a charge insistent that every detail of the message be proclaimed not just to the privy councilors of the realm but to the constituents of Judah at large, not just anywhere but in the precincts of the Temple whose continued existence is at hazard and that it emphasize the conditional nature of the threat as well as the reminder that this prophetic proclamation is not a novelty but the most recent of a long series of warnings ignored. The next sentence, verse 7, has the audience hearing "these words" spoken by Jeremiah, without actually having him repeat them—hence the assumption that the charge addressed to him by YHWH was indeed faithfully delivered by him.

Now to the personae in the proceedings. The audience in verse 7 is constituted of priests and prophets and the ʿam (the populus or burgess or citizenry enfranchised with the right of suffrage). All these participate in the indictment; that is to say, they act as a grand jury (not by a unanimous vote), and the charge in verse 8, "You must die," is free direct discourse, metonymic and hyperbolic. Strictly, it expresses a verdict arrived at, a sentence of death already passed; but what appears as a verdict is a charge and what appears as a sentence is an indication of the seriousness of the charge, a capital offense. So the summation of the action of the ʿam in verse 9 is not as normally rendered, that the people gathered or crowded around Jeremiah, a flat and otiose detail. It is rather, as we have rendered it, a verdict by the **reliable narrator** on the indictment itself: the charge was unjustified, the vote the result of an inflamed mob.

The trial itself begins only with the arrival in verse 10 of the kings' officials who sit as the empowered magistrates in cases involving capital punishment. From this point on, in verses 11, 12, and 16 there is a distinction between the priests and the prophets, who act as the prosecutors, and the high officials together with the ʿam, who act as judge and jury. The former address the latter, demanding (a verdict of guilty and) a sentence of death for Jeremiah for what he said. Jeremiah defends himself in verses 12–15, to the contents of which we shall return; then the officials and citizenry inform the priests and prophets that their demand is without merit. Thus, the inclusion of Jeremiah's argument in the first episode points to it as the decisive consideration that brought judges and jury to a verdict in Jeremiah's favor. This the the bottom line of the narrative, and to this bottom line we must revert to ascertain its narrowest kerygma.

But this bottom line was not so easily arrived at. In addition to Jeremiah's speech in his own defense, there were elders (different from the prophets and priests, the prosecutors), who were members of the deliberating ʿam and cited precedents in support of both prosecution and defense. Thus, the resumptive episode begins in verse 17 with the precedent of the canonical Micaiah (whose oracle we have in the

Book of Micah, but no mention of his having been subjected to a trial), who by his threat brought King Hezekiah and his parliament to reconciliation with YHWH and a grant of his pardon. Let us note that nowhere in the citation of this precedent is there a statement that Micaiah was ever charged with (or tried for) treason. The precedent is merely in the threat that he delivered in YHWH's name, a threat as dire as the one delivered by Jeremiah. Thus, the precedent points not merely to a rejection of the demand for the death penalty, it repudiates the indictment itself. The second precedent cited is introduced in verse 20 by *wᵉgam* (literally "and also"). This conjunctive phrase refers not to the subject of the following verb "to be" but rather to those other elders who argued against the purport of the Micaiah precedent. They cite the case of a prophet who only a few years before had been apparently slated for execution without a trial by King Jehoiakim, who went to the trouble of extraditing him from Egypt just so that he could kill him for a prophecy regarded as *lèse-majesté*. The resumptive episode concludes that for all the deliberations of the judges and the presentation of precedents in argument, the influence of one noble, the father of the Judean appointed to govern Jerusalem after its fall to the Chaldeans, proved decisive.

Returning now to the first episode—the charge against Jeremiah, Jeremiah's self-defense, and the verdict rendered in his favor—the accusation addressed in dialogue to Jeremiah seems to stress the point that to entertain the notion that YHWH might punish his people by allowing the destruction of his own House (a threat made by YHWH himself to Solomon in 1 Kings 9:6–9) is in and of itself heresy, blasphemy, and treason. The unlikeliness of such a presumption points to this address as **free direct discourse** and, at that, **oblique expression.** It has been well said that when the gods are threatened, it is the priests who tremble. Even the threat to the Temple from the God to whom it is dedicated is conditional, contingent upon the refusal of the city's magistrates to alter their ways and their policies. Hence, the utterance of the threat can, by definition, be neither heresy nor blasphemy. It is, however, treasonous to those who exercise power in the government of both state and temple. And for this offense, the "ins" are prepared to take Jeremiah's life. So much for the indictment. In his defense, Jeremiah addresses these very points: that YHWH's message is a gracious invitation to Judah's leaders to change their ways and to forestall execution of his threat. That theirs is the power to judge and execute he does not question. The important thing is that whether they choose to heed YHWH's call for change or not, that is to say, whether they credit his claim that his opposition to their policies is indeed from YHWH or not, his perception that he speaks not for himself but for YHWH renders him innocent. And this point is underscored in the plain meaning of the verdict given in Jeremiah's favor: "A sentence of death is not applicable to this man, for it is in the name of YHWH that he has spoken to us."

The most striking thing about this bottom-line verdict (which to the best of my knowledge has gone unremarked) is that the ground for acquittal is not that "this man," Jeremiah, has spoken the truth in the name of YHWH. The question as to whether Jeremiah has correctly heard or utterly mistaken YHWH's will is just not relevant. It is enough that Jeremiah spoke (as he pleaded in his own defense) under the compulsion that YHWH wished him so to speak. It is clear from the surrounding chapters that the magistrates of Jerusalem (*yōšᵉvēy*) did not alter their policies. On what ground then did Ahikam ben Shaphan win a verdict for acquittal? Not guilty, by reason of what? Obsession? Irresistible compulsion? Insanity? How much has changed in the two-and-a-half millennia that separate our society, our sense of right and wrong, and our claims for the freedom of the clergy to speak and act in matters political from the society and the quandary of the society who were judged by, and had to deliver a judgment upon, Micaiah the Morashtite, Uriah of Kiryath-jearim, and Jeremiah, the maverick priest from Anathoth? Somewhere the line has to be drawn between the politician, who must take responsibility for his own policies, and the vicar of God, whose proclamations become irrelevant if he shuns the political arena but who, in the final event, must take refuge in the defense (as he must abjure credit or responsibility for his awesome role) that he speaks not for himself but for God. And upon whom must the search for the right place to draw the line devolve if not upon those who seek for truth in revelation from the God of truth?

8

Poetical Résumé:
Signs Literal and Metaphoric,
Prescribed and Proscribed

A review of the material covered in the preceding chapters will yield an impressive catalogue of **compositional techniques and rhetorical devices** deployed in masterly fashion by the biblical author—perhaps all the more impressive for the unintrusive way in which the author can exploit these in the treatment of profound themes in the human condition, affecting all the while the naïveté of a simple, unsophisticated teller of tales. In terms of narrative structure there are the synoptic/resumptive pattern; multiple and complementary endings; the soritical episodic series, where the concluding lesson unites the subject of the first episode with the predicate of the last; parallel arrangement of prefatory subplot and culminating parable; and elliptical introduction and surprise ending, where action is resolved and ellipsis filled in. In terms of rhetoric there are free direct discourse; oblique speech; double entendre; hyperbole and metaphor (the latter compacted in a word or two, or extended into parable); sarcasm, irony, and humor, sometimes verging on comedy. All this is description. If we permit ourselves a step in the direction of appreciation, can one fail to admire the adroitness of the narrator's use of voice; his subtle juxtapositions of the victim as hero and the hero as butt and of Deity as stern disciplinarian and tenderly capricious parent; and the iron fatality of the opus alienum and the unhappy ending that serves but as prelude to the opening of a new chapter of opportunity or the reverse, the taste of victory turning to ashes in the mouth?

Our investigation thus far has focused on biblical narrative. It should go without saying that the bulk of nonnarrative material in Scriptures, that is, the poetry of Psalms and the Prophets, Canticles and Lamentations is deserving of, and probably amenable to, a systematic poetical treatment of the kind attempted here and by others for prose narrative. Clearly, a subsection of such a treatment would also focus on the function of poetry integrated into narrative (such as the psalms of Jonah and of Hannah). Indeed, it may be a serious lacuna in the poetics of narrative not to consider the functions of such poetic constituents as the Song by the Sea, Moses' songs at the end of Deuteronomy, Jacob's at the end of Genesis, and Deborah's in Judges 5. Yet there remains in Scripture a substantial amount of nonnarrative prose: formulations of moral and legal norms, ritualistic regulations, even descriptions that partake of the nature of recipes or blueprints. The Jewish tradition refers to these prescriptions, regulations, and prohibitions by the noun halakha or the adjective halakhic, terms that we may adopt for their usefulness. As we have seen in the case of quite a few narratives, the conventions of Israel's assumed norms, even when more honored in the breach than the observance, are a vital factor in the interpretation of a narrative's plot, characters, and denouement; and it is only by recourse to halakhic passages (primarily in the Pentateuch) that we can gain entry into this world of norms. There is, however, a converse of this observation that is of equal validity and significance. Most—almost all—of the halakhic declarations appear in the framework of a narrative setting; that is to say, the *do*s and *don't*s, the *must not*s and *should not*s, the challenges, the dares, the warnings of punishment, and the promises of reward are presented as dialogue, be the speaker divine or human. Perhaps as good an example as any is the case of Exodus 32–34, the story of the golden calf's worship and its aftermath. That narrative, as we saw in episode after episode, cannot be read except in its cultic, prescriptive setting, the cultic prescriptions surrounding it can be fully understood only in the framework of the narrative setting, and a twice-appearing catalogue of prescriptions provides the key to recognition of the kerygma, which is the raison d'être for the entire convoluted episodic narrative. Additionally, it was suggested by the logic of this narrative and further supported in others that idolatry (like apostasy) is a rubric not to be taken literally, that these, like so many other similar rubrics, are to be taken as metaphors.

In later volumes of this work I hope to demonstrate that the essential premises and goals of the halakhic prescriptions, sacred and secular, will only become apparent after these halakhot are subjected to an essentially poetical treatment (as opposed to a diachronic, developmental, legal–historical treatment). For the present, however, as a fitting close to this volume I should like to approach the subject of halakha-as-metaphor in connection with an exploration of the evidence in (and beyond) the Bible for the life setting of the marks or signs that we several times assumed were worn or borne on their foreheads by the prophets.

The Biblical Locus for
Tefillin and *Mezuzah*

In the course of a parting address to Israel, which is about to cross the Jordan under the leadership of Joshua, Moses exhorts his people in these words:

> 4.Hear, O Israel! YHWH is our God, YHWH alone. 5.You must love your God YHWH wholeheartedly, singlemindedly, and exclusively. 6.These words with which I charge you this day must be indelible in your mind. 7.Hone them fine for your children, speaking of them whether at home or abroad, at bedtime and rising time. 8.Bind them as a sign upon your hand; let them serve as amulet [on the forehead] between your eyes; 9.inscribe them on the doorposts of your home and of your [city] gates.'' (Deut. 6:4–9)

Moses begins with a proclamation of Israel's monolatry: unlike its pagan neighbors, it acknowledges but one God, YHWH by name. A translation affirming that a person known by a proper name ''is one'' is as meaningless of a deity as it would be of a human being. A discrete entity is not normally in danger of being taken for more than one or less than one. The assumption that the Hebrew word *'eḥad* means "one" in its every appearance is an example of the folly of literalness. This folly would appear obvious to every speaker of English were he to remember that *only* is "one-ly" and *alone* is "all-one." The endurance of this mistaken rendering is a tribute to the mischief that has been done to biblical meanings by the substitution of a common noun *lord*, rendered as a proper noun *the Lord*, for the ineffable name YHWH and also to an anachronistic assumption by theists of the biblical persuasion that Moses anticipated the unitarian-versus-trinitarian division.

As YHWH alone is to be worshiped, so is "love of him"—exclusive loyalty to him to be shown by adherence to his covenant—mediated to Israel by his prophet Moses. His will is always and everywhere to be kept in mind and taught to Israel's children. All this is clear. But how are "these words with which I charge you this day" to be bound as a sign upon the hand, as "frontlets" (our translation "amulet" is a guess) between the eyes and inscribed on the doorposts. How and why?

As early as a few centuries before the Christian era, these signs—interpreted as protective symbols—were embodied in rites, understood to have been prescribed by God in such passages as this one in Deuteronomy 6. A small case, containing on parchment passages from this chapter was affixed to the doorpost of Jewish homes and was given the name *mezuzah*, which is the biblical word for "doorpost." Hollow cubes of leather, containing verses from Exodus 13 and Deuteronomy 6, were attached to leather straps used to bind them on the arm and on the forehead. Called *tefillin* (probably a plural of abstraction of the word *tephīlā*, "prayer, intercession"), they are rendered in the New Testament by *phylacteries*, from the Greek stem for "protection," which gives us the medical term *prophylaxis*. These tefillin (as also the mezuzah) are still an important element in Jewish religious practice to this day. And there is no question that as these symbolize the protection of God today, so did the signs of Deuteronomy 6 symbolize such protection.[1]

The Sign on the Doorpost (*Mezuzah*)

21.Moses summoned the elders of Israel and said to them, "Draw [from your flocks] and pick out lambs for your families. Slaughter the Pesach sacrificial victim. 22.Then take a bunch of hyssop, dip it in the blood in the basin, and apply some of that blood from the basin to the lintel and the two doorposts. You are not, not a one of you, to go out of his house's doorway until morning. 23.When YHWH passes through to attack Egypt, he will note the blood on the lintel and on the two doorposts. Then will YHWH stand watch over that doorway and not allow the Destroyer to enter to attack your homes.

24.This you are to observe as a decree fixed for you and your descendants for all time. 25.That is, when you shall have entered the land which YHWH will give you, as he has promised, you shall observe this rite of homage. 26.And when your children ask you, 'What does this rite of homage mean to you?' 27.you shall reply, 'This is the Pesach sacrifice to YHWH, acknowledging that he stood watch over the homes of the Israelites when he attacked Egypt—our homes he saved.'" (Exod. 12:21–27)

Modern biblical scholarship is agreed that the Israelites borrowed a spring harvest festival, the seven-day Feast of Unleavened Bread, and attached to it a separate rite or festival, that of the Pesach—this on the strength of pure conjecture. Whatever the prehistory of this festival, the preceding verses in this chapter prescribe one seven-day festival that is ushered in by the paschal sacrifice, which is linked, by the application of its blood to lintel and doorposts, with the plague that struck down firstborn of humans and cattle but to whose ravages Israelite homes remained immune. The three consonants of the name of the sacrifice, *psh,* are also the consonants of the verb describing God's action in regard to the Israelites' homes. It has been translated as "pass over" (and has thus given a new and unfounded name to the paschal observance) without a shred of supporting evidence. With the vowels customary in Hebrew nouns betokening physical defects (such as blind, deaf, mute, and so on), it betokens one crippled in the legs but so seriously that he is at a *standstill.* An older distinction between two crippled states is still preserved in the idiom "the lame [who can limp] and the halt [who cannot]." The verb *psh* appears twice in the story of Elijah's duel on Mount Carmel. There the Israelites are, according to the standard translations, *"hopping* between two opinions" when they are actually immobilized by indecision, and the Baal prophets are *"limping* on the altar" when they are really fixed in place, "keeping watch." In Isaiah 31:5, the verb appears with three synonyms, "protecting, delivering, rescuing"—and that is what the verb means here. YHWH protected the homes marked by the blood as Israelitish.[2] A fourth synonym, *šmr* (keep, guard, preserve) appears in "That was a night of *watchkeeping* for YHWH—to liberate them from the land of Egypt—that, this very night, is YHWH's: watchkeeping over all the Israelites throughout the generations" (Exod. 12:42).

The Sign on the Hand

There is but one reference in Scripture (other than the sign on the hand in the context of the mark above the eyes) to a mark on the hand. In Isaiah 44 YHWH reassures collective Israel. Referring to his people as "Jacob my servant, Israel whom I have chosen," he tells them that they need not fear, he will bless them and their offspring. As for this numerous posterity,

> 5.One shall say, "It is to YHWH I belong." Another will call himself by the name Jacob. One shall inscribe his hand, "Belonging to YHWH." Another shall take the name Israel. (Isa. 44:5)

I shall reserve for later comment the significance of what is here written on the hand. An extrabiblical phenomenon may, however, constitute a bridge between writing on the hand and protective marks on the doorposts. In Mesopotamia, protective figures of various sizes and shapes were a commonplace at public entrances, such as city gates, palaces, and temples. Private homes often featured a niche by the doorway in which such figurines were housed. One such terracotta figurine of a protective spirit bears inscriptions on its two arms. On the right, the legend reads, "Come in, spirit of well being"; on the left, "Out, spirit of evil."

The Sign on the Forehead

Perhaps the most famous of protective signs is ''the mark of Cain.'' Often understood as the sign of a murderer, it is, to the contrary, a sign that he is under God's protection, a warning to others that despite his outcast status he may not be attacked with impunity. The text in Genesis 4 does not state explicitly where on his body God imposed the mark. It is clear, however, that the mark would have had to be readily visible to serve its purpose; and this, along with the context of the prophetic markings ''between the eyes,'' virtually guarantees that the place of the mark was, indeed, on the forehead.

In Ezekiel 9 the placement of a protective mark on the forehead is explicit. The prophet is shown, in a vision, seven ''men.'' Six of them are armed with clubs—their assignment is to traverse Jerusalem and destroy all who have committed or tolerated actions abominable to God. The seventh, belted with a scribe's writing case, is to precede the destroyers ''putting a mark on the foreheads of all those who groan in pain over the abominations committed within her [Jerusalem]''; these are to be spared. The word for ''mark'' here is *taw,* and the verb for imposing it derives from the same stem. The word *taw,* the name for the last letter in the Hebrew alphabet, means ''mark''; and that letter, in the old script is a cross, or our letter *X.* In the nonalphabetic cuneiform script of Mesopotamia, the logogram for *pallurtum* or *pillurtum*—which means ''cross, crossroads, mark on forehead''—is a cross. In recent memory, an illiterate would write an *X* (''his mark'') in place of his signature; and in our idiom it is still the sign *X* that ''marks the spot.''

Slaves in Mesopotamia wore characteristic marks of their status, and one idiom for manumission is ''the slave-mark is removed.'' Since the person who removes the slave-mark is a ''barber'' and the mark is ''barbered off,'' many scholars assume that the mark was a kind of tonsure, rather than a cutting into the flesh that left a surgical scar or cicatrix. This assumption overlooks the paramedical function of the ancient barber as both dentist and surgeon. (The diagonally striped pole before the barber shop is a formal convention for the symbol of healing, the intertwined serpent on a staff called the caduceus.) The term *baldspot* for the mark on the brow, together with the associated expressions for ''cuttings in the flesh''—along with the probability that such marks were more likely to be of a permanent nature—would seem to support our conclusion that these marks were such that they would not have been easily erased.

But whether as tonsures or incisions, several passages in the Pentateuch express the view that such marks are not compatible with the holy status of the servants of YHWH. Leviticus 21 begins with YHWH charging the Aaronide priests that except in the case of the closest of kin, they may not contract the ritual impurity that is occasioned by contact with a corpse. Verse 5 continues, ''They shall not make any baldspot on their heads, nor shall they shave off the tufts of their beards, nor shall they make incisions in their flesh.'' The following verse gives as the reason for the foregoing prohibitions that their service at YHWH's altar requires them to preserve their state of holiness to him.

Leviticus 19, addressed to the entire Israelite community, begins with the charge,

"You shall be holy, even as I, your God YHWH, am holy." Verses 27–28 prescribe, "You shall not trim off the locks of your scalp nor destroy your beard's tufts. You shall not make incisions in your flesh for a dead person, nor impose upon yourselves tatoo marks."

Deuteronomy 14, part of Moses' address to Israel in its entirety, begins, "You are children of YHWH your God. You are not to gash yourself nor make a bald-spot between your eyes for the dead. For you are a people holy to YHWH your God."

The function of the various markings on doorposts, hands, and foreheads are clearly protective. In the cases of the paschal blood on Israel's doorposts in Egypt, the mark of Cain, the marked foreheads of YHWH's faithful in Ezekiel (as presumably in the case of Elisha and the anonymous prophet of 1 Kings 20), they are means by which people may be identified as being under God's special care. The mark of identification, whatever its shape, stands for what is explicit in Isaiah 44: "[Belonging] to YHWH." Similarly in Zechariah 14:20, homage to YHWH will be so widespread that "even on the bells of horses [it will read]: Holy to YHWH." And on the holy diadem of pure gold, the frontal blossom bears the intaglio legend proclaiming its wearer, the high priest, as "Holy to YHWH" (Exod. 29:30).

Why, then, the prohibitions of such signs in Leviticus and Deuteronomy? The answer must lie in the nature of the markings that are proscribed. Whereas the wearing of a legend that proclaims one's loyalty to YHWH could hardly be objection-able, there must have been markings that did not betoken such loyalty or faith in YHWH's protecting power. And the specific context of the three prohibitions may provide the clue. All of the marks proscribed are in connection with a recent death, seemingly of a close relative. Of all the beliefs and practices that we are pleased to call superstitions—and many of these are based on misunderstandings or malicious distortions, "superstition" being a comfortable description for the other fellow's religion—the best documented and readily to be witnessed in our own time are those associated with death and the power of the spirits of the dead. The placing of flowers (or stones) on the gravesites of those we loved is a vestige of the food offerings placed there in ancient times. Deuteronomy 26 speaks of a sacred triennial tithe that is to be shared with the needy. This tithe of produce, according to verse 14, may not be consumed in a state of mourning or impurity; nor may any part of it be given to the dead. Thus, it is clear that Scripture does not proscribe such offerings to the family ghosts (this word in the sense of German *Geist*, "spirit") either on the grounds that they constituted superstition or ancestor worship.

Among the powers of the dead was that of seeing into the future—hence the various practices of necromancy, divining the future with the aid of mediums who could make contact with the spirits of the dead. Recourse to such consultations with the ancestral spirits is, however, consistently condemned in Scripture but not because these were regarded as superstitions, that is to say, fallacious beliefs. On the contrary (and this gives us an insight into the biblical view of magic of every kind) these practices were outlawed for the very reason that they were regarded as efficacious.[3] Magic, to the biblical mind, represented an arcane wisdom, an esoteric science. There are powers in the universe (and it is questionable whether there is any point in dividing them into natural or supernatural) into which one can tap, given the proper technique, to achieve one's end.

The biblical proscription of magical practice is based on reasoning similar to that which prohibited recourse to non-Israelite deities for oracles. To inquire of Baal-zebub of Ekron, for example, is to acknowledge that power. Worse than that (as reflected in the indignant question, "Is it for lack of a God in Israel that you send to inquire of him?") is that such recourse is an avoidance of YHWH.[4] 1 Samuel 28 teaches a lesson as to the futility of seeking a second opinion when YHWH's prognosis is reflected in his silence. King Saul, on the eve of a fateful battle with the Philistines, has consulted YHWH through the three legitimate avenues of inquiry: Urim and Thummim, dreams, and prophets. YHWH has not answered, the clearest sign of his disfavor. Saul, in desperation, although he has previously sought YHWH's favor by stamping out illegitimate diviners, now has recourse to the medium at En-Dor. She raises the spirit or ghost (the Hebrew word here is *'elōhīm*, "a deity, numen") of Samuel, who spells out the meaning of YHWH's silence: Saul is doomed. In like fashion, in Isaiah 8:16–20, the prophet announces that there will be no more oracles for a time from YHWH, "who is hiding his face from the house of Jacob." As for those who will then have recourse to the mediums, on the ground that "a people may inquire of its ghosts (*'elōhīm*) for oracles (*torah*), of the dead on behalf of the living," there is for them not a glimmer of hope.

Whatever the magical technique proscribed, the reason for its prohibition is not that it cannot work (it may indeed) but that it represents a failure to consult God's will, to implore his help. It is a recourse to a mechanical, soulless, power—which can be manipulated—instead of to the God whose operation on behalf of humans is contingent upon their conforming their morality to his will.

The identifying of certain people by marks as being under YHWH's protection for the benefit of YHWH's own destructive agents or for humans who may not recognize his protegés is anything but magical. Signs imposed to ward off demons or spirits, however, whose actions are presumed to be independent of God, are essentially magical. And it is such marks that are prohibited.

When, therefore, in Deuteronomy 6, Moses charges Israel with singleminded and exclusive dependence on YHWH, he stresses that it is in obedience to him that their protection lies. "These words with which I charge you this day," he says, are to serve as your protection! The concluding commandment, then, is not a literal command at all. It is in fact the very reverse. It is a daring metaphor. Put not your trust in such traditional protective markings but in your fidelity to YHWH's will.

The freedom with which the biblical author deploys metaphor may be remarked in the two parallel statements of God's sparing of the Israelite firstborn in Egypt, in Exodus 12. In verse 23, where Moses is addressing Israel, the words read, "When YHWH passes through to attack Egypt, he will note the blood on the lintel and on the two doorposts. Then will YHWH stand watch over the doorway and not allow the Destroyer to enter to attack your homes." In one sentence YHWH is divided, as it were, into two persons, himself and his destructive agent. It is YHWH himself who traverses Egypt to afflict it, who notes the blood on the houses and stands guard at each entrance. So his is the destructive power and his the protective role. The actual work of destruction is performed by the Destroyer, who is barred from Israelite homes by YHWH in person. In verses 12–13, where God is speaking to Moses, the words read, "I shall pass through the land of Egypt that night, strike down every firstborn in the land of Egypt, man and beast. So shall I execute sentence on all the gods of Egypt, even I, YHWH. And the blood on your houses shall serve as a sign on your behalf, that you are there; so that when I see the blood I may stand watch over you and there will be no destructive attack on you when I do strike in the land of Egypt." Here the destructive attack is impersonal, although YHWH protects in his own person. The striking metaphor in this passage lies in another element. The striking down of the firstborn is not a malevolence directed at the actual victims, it is a punishment of Egypt's gods, a judgment on Egypt's values, symbolizing the end of the privilege Egypt had arrogated to itself as the firstborn of heaven, the tolling of the bells for the death of an old order and the birth of a new.

Are we stretching the metaphor? Who can tell? The limit of a metaphor's stretch is its breaking point. A far greater danger, for interpreters, is the contraction of the metaphor to the point where, being literally construed, it disappears altogether. That is where the literalist interpreter opens the text to questions of such an order of naïveté that it would force a blush to his cheeks did he not ascribe the literalism to the text itself. To give but one or two examples of such questions: Did God really need a sign from the Israelites in order to identify their homes? Does not the marking of the homes with blood reveal that blood was chosen for its potent magical properties?

That the "command" to bind "these words" and to write them, in Deuteronomy 6, is a metaphor is conclusively confirmed in the three other passages where these markings appear in a prescriptive context. Two of these texts are in Exodus 13. The first ordains the paying of homage to YHWH through the observance of his Festival of Unleavened Bread. The second ordains the dedication of the firstborn male (born to every female) to YHWH: animals suitable for sacrifice must be given over to the sanctuary, unclean animals must be redeemed by the substitution of a sacrificial victim or destroyed, and human males must be redeemed. The observance of the Feast of Unleavened Bread, according to verse 9, "will serve you as a sign on your hand and as a mark [not, despite the translations, "a reminder"] between your eyes." The dedication of the firstborn males will similarly, according to verse 16, "serve you as a sign on your hand and as an amulet [a frontlet] between your eyes." The context in both excludes any literal interpretation.[5]

The third passage is Deuteronomy 11:13–21. First, there is the promise of seasonal rains and abundant harvests if Israel observes YHWH's commands. Then the threat of drought, famine, and death if Israel is seduced into worshiping other gods. Then the passage concludes:

> 18.Therefore, place these my words upon your hearts, at the core of your being, binding them as a sign upon your hands that they may serve as amulets [or *frontlets*] between your eyes. 19.And teach them to your children, speaking of them at home and abroad, at bedtime and rising time. 20.And inscribe them on the doorposts of your homes and your town gates 21.to the end that long may you and your children endure on the land that YHWH swore to your ancestors to give to them, as long as heaven endures over earth. (Deut. 11:18–21)

The opening words of verse 18 cannot be literally construed, for words are not written on hearts or minds or souls. As this command is a metaphor, so are the charges that follow. Of particular interest is verse 19, which harks back to the passage, Deuteronomy 6:4–9. Here as there, what is central to the charge is not merely keeping the words in mind but teaching them to one's children, honing them, like the repeated strokes of a blade against a whetstone. And if we turn back to Exodus 13, we shall see that this feature, too, appears there. Both those passages anticipate the entrance into, and the taking possession of, the promised land. "At that time," says verse 8 in regard to the observance of the Feast of Unleavened Bread, "you are to tell your son. . . ." And verse 14, in regard to the rite of the firstborn male, says, "And when in the future your son asks, 'What does this mean?' you shall say. . . ." Both these recitations to the future generation rehearse the great power deployed by YHWH to effect Israel's liberation from Egypt. We are compelled, therefore, to restate the force of the metaphor. Israel's protection lies not in inscribed markings or writings, not alone in obedience to God's word, nor even in observing the tributes of festival and firstborn rites but in the teaching, the inculcation in future generations, of the power and grace of YHWH. Exodus 13:9 makes this explicit, "that the torah of YHWH be ever on your lips [literally, *in your mouth*]."

The signs are a metaphor for protection; the rituals of festival and firstborn are a metaphor for homage and obedience; the ritual metaphors are a pretext, an occasion, for teaching posterity; and that teaching, carried out through the generations, is itself the mechanism for the preservation of Israel on the sacred soil promised and delivered by the God of freedom.

It will be helpful at this juncture to repeat some observations on the relationships between the literal and the metaphoric and between that dichotomy and the question of "truth" in the sense of trueness to the intention of an author. Language is essentially metaphoric. All words are symbols. Language is essentially speech. Written language is, therefore, a visual symbol for an oral symbol. Letters are written symbols of which written words are constituted. And the letters that represent vowel sounds and consonantal sounds are, like those separate sounds themselves, essentially meaningless. The very concept of the literal ("according to the letter") is essentially a metaphor and, strictly construed, essentially meaningless, at that; for if "the letter" is essentially meaningless, then so must be that which is "according to the letter." The word *literal* is, however, an elliptical expression; it stands for "literal sense" or "literal meaning," by which we intend the narrowest or strictest sense construction of a word or phrase, in contrast to the metaphoric or figurative sense, by which we intend a broader or freer construction. The categories of literal and metaphoric are, therefore, not absolutes. Each of these categories represents a range of more or less strict—or more or less free—constructions. And as the more literal is not closer to truth, so the less literal, or metaphoric, is not further from the truth.

The dynamic range between one end of the literal–figurative spectrum and the other is eloquent witness to the aliveness, the vitality, of language. Both the literal and the metaphoric assume lives of their own. A nail is by definition not alive; never having been alive, it should not, properly speaking, ever be qualified by the adjective "dead." And yet one expression for "dead" in its most literal sense is the simile "as dead as a doornail."[6] Or consider the metaphor "hit the nail on the head." One very much doubts that this expression is ever used in a literal sense. One does not remark on the obvious. Furthermore, to hit the nail on the head does not mean not to hit it on its flank. It means to hit it on the head *squarely,* that is to say, at a right angle, which is itself a metaphor for the *parallel* planes of the hammerhead and the nail head.

When is a horseshoe not a horseshoe? Well, it is most likely to be a horseshoe when it is an object of iron, curved and fixed to a horse's hoof. It is least likely to be a horseshoe when it exists in the mind or on paper as a particular kind of curve. But what is it when, having been cast loose from a hoof, it is hung over a doorway? It is still literally a horseshoe and metaphorically a symbol of good luck. But is it only a symbol of good luck, or is it a fetish—an inanimate object endowed with the power to effect good luck. And if it is the latter, is this due to its having once been attached to a hoof or would it have that power if it were hung straight from the blacksmith's forge or if it were but a shape painted on a surface. This is the kind of problem we face when we speak of ritual as metaphor. Ritual is, in a sense, metaphor in action or the acting out of metaphor. Hand shaking, hat tipping, hand kissing, bowing, kneeling, kowtowing—all are rituals, actions in token of—or metaphors for—respect.

How, now, shall we address Deuteronomy 6:8–9, "Bind them as a sign upon your hand, let them serve as amulet between your eyes; inscribe them on the doorposts of your home and your [city] gates"? Are we to assume that the command is literal and its content metaphoric? That is to say, the command reflects the will of God that we practice rites along the lines of the traditional Jewish mezuzah and tefillin? That the failure to do so will not only deprive us of their protective force but perhaps even invoke God's anger? Or shall we assume that the command itself is also a metaphor? That, indeed, associating it with the prohibitions in connection with the markings for the dead, we ought take it as a metaphor for a reverse command, a prohibition of putting any weight on such mechanical rituals, rituals that smack more of magic than of true religion?

The answer, in terms of what the author of this passage intended, is beyond us. Even were we to assume that one hand penned both these lines in Deuteronomy 6 and those prohibiting cuttings in the flesh and baldspots on the brow, the commands would not be self-contradictory; for the latter only refers to lacerations and the former has to do with signs in general; the latter prohibits only such signs in specific connection with mourning and, probably, the fear of the spirits of the dead; whereas the former uses signs in general as symbolizing the idea of protection. Hence, it is not clear whether such markings as the professional prophets might have worn would have been included in the proscriptions of marks made "for the dead." What is clear, however, or should be, is that the legitimacy or illegitimacy of marks or of any other ritual does not inhere in the wording of one or another biblical passage but depends on how the rituals are conceived by the person who practices them. If the rituals are conceived as technique, based on automatic functioning of will-less, soulless powers or laws, then their theoretical base is magic and their practice is contrary to the thrust of Scripture's teaching. If they are reminders of the power of God, affirmations of loyalty to him, then—whatever their possible roots in long-forgotten pagan practice —they are in consonance with Scriptural teaching. By the same token, a hallowed tradition, original in the worship of Israel's God, may have to be repudiated if it comes to be viewed as an effective fetish, unrelated to his moral will.

Take, for example, the prophet Jeremiah. In Jeremiah 3 he attacks Judah for worship at shrines that he regards as illegitimate, the rubric for these being "under every leafy tree." He looks forward to a time when YHWH will reclaim his rebellious children. The hallowed Ark of the Covenant, the portable Throne of YHWH, which dated back to Moses and the wilderness wanderings, he takes as a symbol of the worship of YHWH anywhere at all and subjects it to scorn.

> 16. And in those days, when you will abundantly increase—declares YHWH—people will no longer utter [the words], "the Ark of YHWH's covenant." It shall not come to mind. They shall not mention it or miss it or make another such. 17. At that time, they shall call Jerusalem "Throne of YHWH"; there shall all nations rally to it, to Jerusalem, to the name YHWH! (Jer. 3:16)

Thus the legitimacy of Jerusalem's Temple as the one and only shrine of YHWH is affirmed at the expense of the older symbol of YHWH's seat. Yet Jerusalem, too, could become a fetish, its inviolability regarded as assured by the fact of the Temple's presence there, rather than by Judah's living in conformity with God's will. And a few chapters later, Jeremiah, citing the destruction centuries before of YHWH's great shrine in Shiloh, suggests that this shrine, holy to YHWH, now treated as a fetish, will share the fate of Shiloh's sanctuary:

> 1. The word which came to Jeremiah from YHWH: 2. Stand at the gate of YHWH's House, proclaim there this word, "Hear the word of YHWH, all you of Judah who enter these gates to do homage to YHWH. 3. Thus said YHWH–Hosts, God of Israel: 'Improve your behavior, better your deeds, and I will give you permanent residence in this place. 4. [But] put no store in false slogans,⁷ such as, these [buildings] are YHWH's palace, YHWH's palace, YHWH's palace. 5. Not so! Only if you improve your behavior, better your deeds (if you deal justly one with another, 6. oppress not stranger, orphan, and widow, spill no innocent blood in this place, follow not other gods, to your own hurt) 7. will I give you permanent residence in this place, in the land which I gave to your ancestors for all time. 8. But see now, you are setting store by false slogans, of no efficacy! 9. Will you engage in theft, murder, adultery, false swearing, offering incense to Baal, following other gods you have not known [that is, to whom you are beholden for nothing]? 10. And will you then enter, stand in my presence, in this House, which bears my name, and declare, "We're saved," only to go on with such abominations? 11. An outlaws' lair? Is that, in your eyes, what this House, which bears my name, has become? Yes, I can see, too' declares YHWH."

> 12. "Just go now to that place of mine in Shiloh, where I first fixed an abode for my name, and see what I did to it—on account of the wickedness of my people Israel!" (Jer. 7:1–12)

Unless human nature has changed radically in the past two-and-a-half millennia, it is highly probable that the percentage of murderers and adulterers in the Temple throng addressed by Jeremiah was not too far different from the percentage of those malefactors in attendance at Sabbath worship in today's churches, mosques, and synagogues. The difference between then and now would seem to lie in the preachers. The ministers of today are either more politic or less given to hyperbole and metaphor. The "offering of incense to the Baal" is certainly a metaphor; for even if "the objects made for Baal" in 2 Kings 23 are to be taken literally, they were removed from the Temple of YHWH by the pious King Josiah early in Jeremiah's ministry. And if the human condition has not altered appreciably in regard to concerns moral and cultic, the address of prophet then as of minister today was not to the atheist or professional criminal but to the pious and the devout. They have one view of their righteousness, and their preachers have another. They put much store in the trappings and literalisms of their established churches, and their preachers—well, perhaps, I had better let the matter rest. Jeremiah, after all, saved his choicest condemnations for his rivals in the priestly and prophetic ministries.

We have not yet exhausted the subject of baldspots in connection with rites attending death. Scholars have, to their own satisfaction, variously dated the priestly writings (such as Leviticus) and the Book of Deuteronomy. However they may differ on the former, they are all agreed that Deuteronomy was known to Jeremiah and certainly by the prophet who postdates him, Ezekiel. Yet there are at least nine occurrences in the Writing Prophets of bald marks in connection with such other mourning rites as the donning of sackcloth and gashing of the flesh (Amos 8:10; Micah 1:16; Isaiah 15:2, 22:12; Jeremiah 16:6, 41:5, 47:5; Ezekiel 7:18, 27:31). Two citations will suffice. In Jeremiah 16 a prophecy directed against Judah and Jerusalem foresees death on such a scale that there will be few left to lament or bury; corpses will decay in the open, food for the scavengers of field and sky. YHWH forbids the prophet to participate in mourning or offering condolence to the survivors, who have forfeited his favor: "Great and small in this land shall die, they shall receive no burial, lamentation will not be made for them, nor will anyone gash himself or make a baldspot for them" (verse 6). Ezekiel 7 is a prophecy of doom on all four corners of the land: sword, pestilence, and famine will ravage town and countryside. Of those surviving it is predicted, "They shall gird on sackcloth, horror will envelope them, humiliation [be written] on every face and on every head a baldspot" (verse 18).

These late and casual references to a continuing mourning practice, betraying no sense that the practice itself should be condemned as contrary to the will of God, is not evidence that the prophets were unaware of formulations of prohibition nor of these prophets' rejection of such presumed prohibition. It is more likely that the prophets understood the prohibition as it was intended—as precept, not as law, to be honored for the meaning behind it and not in rigid or slavish adherence to its literal formulation. The mark on the forehead is a metaphor for protection. Its proscription in certain contexts (for example, in the implicit rejection of sign in favor of obedience in Deuteronomy 6:7–8) is also, essentially, a metaphor. But metaphors have lives of their own, often assuming meanings considerably different from that contained in their first appearance. There would, therefore, be something absurd both in insisting on the literal context of a metaphor and in proscribing a literal interpretation in the name of its metaphoric message—for example, abjuring the use in religious cere-mony of tefillin or mezuzah on the grounds that the metaphoric intent of Deuteronomy 6:7–8 is that anything suggesting an apotropaic mark is forbidden or, conversely, understanding that passage as denying God's protection to those who fail to "bind them upon [the] hand" or "inscribe them on [their] private and public entrance ways."

In stressing this last conclusion, I may be seen as straying from the primary concern of philologian or poetician (to ascertain the original and "plain meaning" of a text) out of concern not to give offense to those who have made doctrinal denominational commitments to practices seen as required (or repudiated) by that text. To such a perception I would respond with the reminder that the **scholar** of **poetics** must take care not to mistake *an* interpretation of his or hers as *the* interpretation, excluding all others; that, further, the most objective of philologians, while dealing with cultural precipitates, cannot reproduce a predicted precipitate in a test tube. But there is more than this at stake for the enterprise of poetical analysis, especially of Scriptural texts. Just as dogmatic readings of Scripture have often closed off this book, or much of it, from the attention of serious searchers for the good and the beautiful, so have the supercilious conclusions of ''scientific'' philologians turned away many a clear-sighted religionist from an avenue of interpretation that would enrich and deepen a text to which he is so committed as to preclude possibility of further justification.

There is yet, I believe, another and important consideration: the dictum put forward by Oscar Wilde denying the applicability of aesthetic judgment to any artifact produced for any practical use, that is, other than for sheer enjoyment. Whether or not Wilde was merely being provocative, there is a school of aestheticians or a current in aesthetic theory that moves in the direction of sundering form from substance, art from life, morality from aesthetics—a splitting of the nuclear hendiadys *kalos k'agathos*. Even were such a bifurcating enterprise feasible elsewhere, it would be absurd in the case of Scripture, the literature in service of theology and morality. Here, to overlook substance or kerygma is to reduce its value to zero, just as to overlook its artistry is to mistake the kerygma. When Scripture ceases to be relevant to our experience of life, it is as artistically trivial as it has been declared meta-physically and ethically pointless. The Bible has proven that it can survive a failure of aesthetic appreciation, but it is doubtful that it could survive as an artistic creation shorn of a capability for compelling assent to its moral affirmations and its mordant yet never morbid assessment of the human condition. If preachers have often done violence to its teachings by subjecting them to a slavish literalism, if humanistic scholars have so rarified the metaphors as to virtually empty them of relevance for contemporary conduct (thus turning metaphor into its own kind of self-contained literalism), we shall inevitably repeat the one error or the other unless we keep ourselves open to what others have found in it most precious for their lives.

To live according to the spirit is a meaningless enterprise without the letters in which the spirit is defined. To live according to the letter, it would seem, is often to do violence to the spirit. Literalism at its extremest would appear to be the most common form of idolatry (idolatry construed metaphorically, to be sure). The sword of language is double-edged. It can cut for benefit or harm, with its literal edge or its figurative one. Perhaps true prophecy is the capacity to know when to wield the blade and with which edge.

9

Retrospect and Prospect

In the Preface I stated that the existence of a poetical grammar for Scripture was a hypothesis to be demonstrated, that I would sketch some of the elements of, and the problems attending, the constituents of such a grammar in part I and illustrate them in the exegetical essays (part II). How successfully have my discussions vindicated the hypothesized grammar. Specifically, to what degree have the features in part I been indeed illustrated in part II?

Figures of Speech

Chapter 8, focusing on metaphor and its ubiquity in Scripture, particularly illustrates **figures of speech.** The ubiquity of metaphor, the preponderance of the highly figurative over the more literal, in every one of the narratives studied can hardly be overstated. In Chapter 8 the exploration of metaphor focused on texts of a prescriptive, rather than narrative, nature. (We had already seen the interweaving of regulatory norms in narrative and vice versa in such texts as Solomon's Judgment, the aftermath of the golden calf, and the Israelite invasion of Moab. Now we were able to discern the essential concordance of cultic regulations, prescriptive norms, and parenetic exhortations with the kerygmas developed in the narrative genre.) I concluded (as I had at a number of other junctures) that the harmony of the prescriptive with the narrative was evidence for the poetic unity, or single authorial voice, of Scripture.

As didactic fictive narrative is to history, as the figurative is to the literal, so is the prescriptive or parenetic to the biblical literature that modern scholarship calls "law" or "legal," whether in the domain of the political, the religious, or the moral. In contrast with (and probably in opposition to) the genetic treatments of legal corpora called "codes" and "priestly legislation," the way is now open to an essentially poetic address to the material subsumed under these headings. And as legal texts from the cuneiform cultures of ancient Israel's ambience have been adduced to inform on ancient Israel's legal system, it may well be that a poetic address to biblical "law" will stimulate a poetic address to "law" or "codes" recorded in cuneiform script.

Foci of Literary Analysis

Foci of literary analyses are the elements on which the modern literary critic focuses when examining a narrative for analysis and explication. In Part I, I provided a sketchy description of these foci as employed in an address to modern fiction and expanded upon their peculiarities in biblical narrative.

In regard to points of view, the biblical author is a master of shifting them. By concealing, disclosing, or ambiguating the identity of the "viewer" as author or narrator, he can exploit the assumption of the narrator's reliability—most of the time. At others, he may suggest the possibility that the narrator is less than totally reliable or reliable as to honesty but endowed with less than omniscience. One cannot exhaust all the possible strategies available to an author in the matters of to what extent and how he may play with his narrator's less-than-total reliability. For one thing, many a strategy may be yet exploited. For another, many strategies may already have been deployed without having been recognized as a case in point of narratorial unreliability. For example, the narrator who in his own voice seems to expound on the wisdom and wit of his hero yet reports a direct discourse of that hero in which he discloses himself as a garrulous fool would be an example of a not-totally-reliable narrator. In similar manner the storyteller of Tale 1, Episode 2 in chapter 6 or Tale 6 in chapter 7 are subtle and brilliant exploitations of the unreliable narrator strategy.

I shall return to this theme shortly in another connection. For the present, note that point-of-view mastery is especially evident in that chief of narrative dramatic techniques, dialogue. And in respect to this feature it is questionable whether there is any other narrative tradition, ancient or modern, in which there is a display of such artistic control. Not a line or a word in dialogue is superfluous. Whatever a character says or thinks (internal dialogue) and how this is formulated play a vital role in the development of the denouement and the cipher in which the kerygma is encoded. And matching this artistic feature of thrift in dialogue is that of free direct discourse, whereby the narrator inserts into a character's dialogue a clue as to how that dialogue is to be understood by the reader.

Repetition, on the other hand, which very often strikes us as a hapless device or a disfigurement in some artistic creations, can never be so characterized in biblical narrative. Every instance of repetition of incident, action, or speech, in narrative or dialogue, has proven to be a signal for a closer reading, alerting reader to a subtler communication than the one anticipated or a subtler nuance of that communication.

One category of repetition (or rather of seeming repetitiveness) unique to Scripture (we found in the Pentateuch, Jonah, the "histories," and the Writing Prophets) is the synoptic/resumptive-expansive technique. Unique to Scripture and a uniquely powerful strategy to achieve in narrative the parabolic ends of this ideological and ideational literature, the limited number of narratives studied precludes at this time an appreciation of its many possible variations.

Finally, there is one particular feature—the consistency of character development—in which biblical narrative art seems to fall short of the standards upheld in modern fiction. But here, too, the deficiency is only seeming. For the characters or personae in biblical narrative are means to the author's goals rather than (as in today's fiction) ends to be realized in and for themselves. Thus, a biblical persona, figuring in a series of plots or actions, can for all his flatness as a type (e.g., king, prophet, rebel, loyalist, heroic warrior, master criminal) be round in a variety of subtype features such as any flesh-and-blood creature is all too capable of displaying.

Form and Content, Poetics and Exegesis

It will be useful here to review the differences between my approach to the poetics of biblical narrative in Part II and that of most others who have published significant investigations on this phenomenon. Some of the latter engage in literary commentaries on whole books of the Bible, with the resulting preponderance of analysis that a line-by-line examination entails. The rest begin with questions of poetical categories, techniques, strategies, and foci of interpretation and proceed to examine narrative instances in which these characteristics are featured. My approach in Part II is to address one entire narrative unit at a time and poetical features in the order that they unfold in the narrative's progression. This difference in approach eventuates in a number of features in my analysis that do not appear (or not with any great degree of prominence) in theirs:

1. a culminating of each analysis in a synthetic conclusion, that is, in the formulation of that narrative's kerygma;
2. frequent recourse to reconsideration of metaliterary assumptions underlying traditional or modern interpretations of a text's kerygma;
3. frequent recourse to halakhic texts for the elucidation of Scripture's mind-set in regard to conceptual constructs, moral values, and social or cultural institutions in the interest of buttressing or undermining one or another proposed kerygma;
4. an assumption, as a hypothesis to be vindicated, of the unitary nature of Scripture (or of one authorial voice behind formulations by different writers across stretches of space and time), a necessary concomitant of items 2 and 3; and
5. a predisposition to give the Hebrew text and its Masoretic mediators the benefit of the doubt, on grounds of the reasoning behind the principle *Lectio difficilior praestat.*

The foregoing is by way of description, not invidious comparison. The great weight that I place on the exegetical investigations against formal and theoretical questions of poetics derives, I believe, from this conviction: the ultimate aim of all literary scholarship must be the achievement of a deep and comprehensive understanding of the message or messages that the literature seeks to convey. The scaffolding exists for the edifice, not the other way around. The goal of the critic is not the display of his tools or of the skill with which they are wielded: in the case of Scripture it is exegesis, grasp of the kerygma.

In this connection it may be not amiss to note that scholars—academicians in particular—are so often seized up by the give-and-take of methodological debate as to lose sight that method exists only in relation to objective. A colleague of mine in biblical studies, who reads every journal devoted to poetics, was heard to remark on my golden calf essay,[1] "It is thorough exegesis, but in what respect is it literary criticism?" Can one take issue with Adele Berlin's formulation?

> Many introductions to poetics make a sharp distinction between poetics and interpretation and I have done likewise . . . in order to show how the two differ. But it must be emphasized that the two have a symbiotic relationshp. "The relation between poetics and interpretation is one of complementarity par excellence . . . . Interpretation both precedes and follows poetics" (Todorov). . . . Poetics is useless in isolation; knowing the compositional rules of a text is of use only if we want to read the text. The contribution of poetics is to be found, in the words of Jonathan Culler, in its "attempt to specify how we go about making sense of texts, what are the interpretive operations on which literature itself, as an institution, is based." . . . In simpler words, poetics makes us aware of how texts achieve their meaning. Poetics aid interpretation. If we know *how* texts mean, we are in a better position to discover *what* a particular text means.[2]

Interpretation: Categories and Factors

Categories of interpretation and **factors in interpretation** concern the literal and historiographical. Unlike the foci of literary analysis, which must figure large in the analysis of narrative or drama of any time period, these categories or factors in interpretation are vague constructs of questionable legitimacy applied (often unconsciously) to literature from antiquity. While contributing nothing toward a laying bare of a text's compositional rules, they often dictate a vector of criticism that obscures the outlines of both poetics and interpretation. In contemporary literature, for all the proliferation of historical fiction (novels or romances), fictionalized history, and historiography in which imaginative reconstruction is inversely proportionate to data, we have little difficulty determining whether the intent behind the text is essentially fictional or historiographic—hence the absence in modern introductions to poetics of these metaliterary considerations. But in the study of the literature reaching us from remoter times, our assessment of the limits that dead history's hand had imposed on an author's freedom may so skew our critical vision as to nullify the testimony of these purely literary foci, which are such reliable guides to the literary designs of our contemporaries.

In this connection I would cite with appreciation and approval Robert Alter's characterization of the difference between "the Bible's sacred history" and that of "modern historiography":

> There is . . . a whole spectrum of relations to history in the sundry biblical narrative . . . *but none of these involves the sense of being bound to documentable facts that characterizes history in its modern acceptation.* It is often asserted that the biblical writer is bound instead to the fixed materials, whether oral or written, that tradition has transmitted to him. This is a claim difficult to verify or refute because we have no real way of knowing what were the precise contents of Hebrew tradition around the beginning of the first millennium B.C.E. A close inspection, however, of

the texts that have been passed down to us may lead to a certain *degree of skepticism about this scholarly notion of the tyrannical authority of ancient tradition,* may lead us, in fact to conclude that the writers exercised a good deal of artistic freedom in articulating the traditions at their disposal. As odd as it may sound at first, I would contend that prose fiction is the best general rubric for describing biblical narrative. Or to be more precise, and to borrow a key term from Herbert Schneider's . . . *Sacred Discontent,* we can speak of the Bible as *historicized* prose fiction.[3] (emphasis mine except the last)

Yet the borrowing of the adjective *historicized* by a student of literature to qualify the rubric "prose fiction"—understandable as it is in the context of a world that makes fiction synonymous with falsity and history synonymous with truth—is, nevertheless, less than helpful. Nor does it ward off, as we shall see in a moment, an attack by Meir Sternberg in defence of the Bible as history.

Let us first note, however, that Alter is arguing against two kinds of bondage applied to biblical narrative. The first is to make the biblical authors thralls, as it were, to the taskmaster historiography ("history in its modern acceptation" or, in von Ranke's so recent, so oversimplified, so celebrated formulation, *wie es eigentlich gewesen*): God himself had revealed to them the factual record of a past, a revelation not theirs to alter by addition or subtraction. The second bondage is not theological and is, if less objective, yet more scientific: it is the geneticist-imposed "tyrannical authority of ancient tradition" (also known as Jahwist, Elohist, Deuteronomist, Priestly).

Sternberg argues that the historian is denied not invented material "but the privilege and at will the flaunting of free invention." He continues:

> To develop chronicle into history, after all, the historian must supply a great many missing links—causal connections, national drives, personal motives and characteristics—and the imaginative gap-filling will remain acceptable as long as it operates within the limits of whatever counts as the rules of evidence. Therefore, to demonstrate that the biblical writers allowed themselves some latitude in the treatment of their materials is to demonstrate nothing much about their genre, unless that exercise of invention appeals and amounts to the total license of fictionalizing rather than to the constrained license of historicizing.[4]

The point that Sternberg is making here may be readily conceded. Narrative fiction and historical narration cannot be *formally* distinguished from one another—if, that is, one does not discriminate between the history of von Ranke's characterization (the narration of what actually occurred) and the history imagined by a modest and controlled narrator limiting himself, say, to what he believes a historical character was likely to have said in certain circumstances. But is it then possible to retrieve the "teleology" of a narrative communication "to control its typology" without reference to anything beyond what is contained in the communication itself? Sternberg thinks it is and offers as a "miniature illustration" the narrative of David's affair with Bathsheba as against Nathan's parabolic tale of the poor man's ewe–lamb in 2 Samuel 12: "Accordingly, as one purpose gives way to another—seeking redress for an anonymous sufferer to passing sentence on the king—the tale transforms from the history of an injustice to a fictional parable of injustice." We may stipulate that the parable is a fiction. But what presses us to accept as history rather than as freely composed fiction the story that Nathan did indeed so decoy the conscience of the king and did thereafter so upbraid him. Indeed, if we *had* royal correspondence retrieved from an excavated Jerusalem palace testifying of a king's promise to a wife named Bathsheba, comforting her over the death of her newborn son, would we be any closer to judging as history the entire tale of David's betrayal of a loyal captain and his consequent favoring of the faithless and usurped wife even in the matter of the succession?

I suspect that Sternberg would have no trouble responding to this question along the lines of a later statement of his ("The problem arises from the misidentification of history writing with historical truth.") But since for him "narrative fiction and historical narrative" are lines of discourse that can be plotted by poetical methods in the one biblical literature, he must nevertheless pose the critical question and supply his poetic answer: "So does the Bible belong to the historical or the fictional genre? The mist enveloping the question once dissipated to reveal its communicative bearing, the answer becomes obvious. Of course the narrative is historiographic, inevitably so considering its teleology and incredibly so considering its time and environment. Everything points in this direction."[5] But what is critical to my purpose is not Sternberg's discovery that the Bible creates (in its time) "a new mode and rhetoric of historiography" but the conclusion in which this discovery culminates:

> The new mode and rhetoric of historiography are themselves means to an end that any hint of invention would put in danger. Were the narrative written or read as fiction, then God would turn from the lord of history into a creature of the imagination, with the most disastrous results. The shape of time, the rationale of monotheism, the foundations of conduct, the national sense of identity, the very right to the land of Israel and the hope of deliverance to come: all hang in the generic balance. Hence the Bible's determination to sanctify and compel literal belief in the past.[6]

The cogency—no, the compelling force—of this argument for the literal historiographic intent of Scripture's narrative is not subject to question for some readers, indeed, for most readers. For the overwhelming majority of students of the Bible both today and in centuries past—naive believers and scientific skeptics, source analysts and holistic appreciators of the-Bible-as-literature alike—the Bible asks to be taken as literal and historiographic. But this is not so for all readers. And in identifying himself with the majority, Sternberg has overlooked a metaliterary question of crucial import for the poetic enterprise: What audience or readership was the biblical author addressing? More precisely, for a readership of what level of sophistication was the biblical author writing? Here, as in the instance of the Judgment of Solomon (see chapter 2), Sternberg has overlooked the possibility of a narrative composed to be read and appreciated on several levels. And presuming on the one level he has decided upon, he has generalized on the basis of this one postulated level on the literal and historiographic intent of all biblical narrative. The issue that I am raising here is not of the rightness or wrongness of Sternberg's position but the foreclosing of many options for poetic research on the strength of the demonstration of one possible narrative strategy, brilliantly elucidated and the more dangerous for its dazzling effects on our vision.

Let me illustrate the problem with particular reference to the narratives in the Bible's first book, narratives that because of their placement condition our psychology for the rest of Scripture. The Book of Genesis consists for the most part of stories written for an audience who are related to the chief personae in the stories as descendants (genetically or by adoption) and for whom every event in the ancestral past must be understood to have profound significance. By token and in consequence of this relationship of audience to personae and of the linkage of the audience's existential circumstances and vicissitudes to the events in which those ancestral personae were featured, these stories cannot but be a certain kind of story: history; for a real audience cannot have fictional ancestors, and fictive ancestors can only have fictive descendants. I am, and we are; therefore, they were. And furthermore, we are what we are, in large measure, because they were what they were. The immediacy for us of this ever-breathing, ever-palpitating, ever-living past—a past of which my grandfather of ten generations ago may be more real and present for me than he who sired me on my mother—is reflected and expressed in my self-consciousness as a Judahite, an Israelite, a child of Abraham by Sarah. And the import of that birth of mine so many centuries ago is brought home to me when the gates of my city are opened on market days to admit to its *suqs* those half-tamed half-brothers of Isaac: Ishmael, Hagar's son and camel-riding Midian, Aunt Keturah's son. And if, suppressing memory, I am ever so fatuous as to ask what's in a name or what is it to be a Jew (Judean), the psalm-singing Levites will remind to me on my next Temple visit that steppe-dwelling irreconcilable Edom still looks longingly at the richer fields that Papa Jacob wrested with providential help from Uncle Esau, elder brother by a minute or two.

There is in the foregoing picture an acceptance of the Genesis stories as an historical reality that in no way gainsays their essentially imaginative nature, nor the preponderance in them of metaphor and symbolism over information and other literalisms and that, furthermore, does not lead to their categorization as historiography in that term's modern acceptance. My acceptance of the patriarchs as my ancestors is a function of my reading them not just as protagonists in a story but as the heroes of the story. As reader, I claim them as my forebears whether I be "born-again" Christian or secularist Jew, whatever genetic pool I am heir to; for like all great artistic narrative, these stories elicit a "willing suspension of disbelief." And in this instance, particularly, the rewards of identifying with the heroes are immeasurable, inasmuch as these heroes are the instruments of the kind of God that I would want to exist, a God benevolent to humankind, friendly to my deepest and most intimate aspirations, and at the same time the lord of history.

But to accept this foregoing picture as literal historiography is to ignore that it bears all the hallmarks of fiction—ideological fiction to be sure, but fiction nonetheless. And so, while as a reader I accept the metaphors of my kinsmen, the children of Esau and Ishmael as the survivors of uncles who lost out in sibling rivalry with my father, as a scholar I must question whether my neighbors of Edom to the south or my Bedouin cousins of the steppes to the east and north share or ever shared a consciousness of relatedness to ancestors named Esau or Ishmael. And as a literary critic, a student of poetics, I must question whether the Bible wants me to accept literally (that is to say, as biographical history) the sottishness of Isaac, the bias and knavery of Rebecca, the supineness of Jacob, the filially pious passivity of Esau, all in the interest of supporting my faith that God will continue me in possession of the land wrested for me by such ancestral antics and postures. Rebecca's awareness of retribution, Jacob's lying half-truths to his father, Isaac's tremulousness on the threshold of awareness, and Esau's anguished protest that one brother's blessing need entail the other brother's curse—all spelled out by the narrator in action and dialogue—may as well bespeak the wonder of the Israelite theologian and moralist at the good fortune that made him and not his Edomite neighbors the darling of a destiny he did nothing to earn, him and not his cousins the chosen of a God of morality who yet manages time and again to overlook his chosen's unworthiness.

That the majority of readers, ancient and modern, laypersons and scholars, have assumed that biblical narrative is essentially history or at least intended to be accepted as such is testimony that for all the hallmarks of fiction in it, the biblical narrative exhibits certain features that seem to be utterly pointless except to reinforce the impression of detailed and accurate records upon which the biblical authors must have drawn. To cite Sternberg:

> Methodologically speaking . . . the Bible is even the first to anticipate the appeal to the surviving record of the past that characterizes modern history-telling. Such relics abound on the narrative surface itself, appearing as facts to be interpreted and brought into pattern. Recall how often customs are elucidated, ancient names and current sayings traced back to their origins, monuments and fiats assigned a concrete reason as well as a slot in history, persons and places and pedigrees specified beyond immediate needs, written records like the Book of Yashar or the royal annals explicitly invoked.[7]

As concerns the historicity or intended historicity of the stories in Genesis, it is the genealogies and the chronological data which seem to assure such categorization; lists and time sequences that in specifying political, ethnic, and geographic eponyms and chronological ordering fix us and our neighbors in time and place, in a divinely guided past that is warrant for our present and promise for our future. Indeed, there are those—among them scholars bound to the objectivity of the scientific–historical methodology—who will see in much of this historical narrative a propagandistic or self-serving fiction, designed to rebut the claims of encroaching cousins and to rekindle the patriotic fervor in Judean breasts. Interpretations of literature, as we stressed earlier, are subject to rating according to criteria of persuasiveness, not of truth or falsity. If a body of literature is to be approached on the presumption that it is essentially history (as is the case with biblical historians of the scientific school), the question for those championing this approach is, How does one differentiate the fictional, the tendentious, the propagandistic from that which is to be taken as a reliable record? If, for example, any incident or group of incidents in the career of King David may be judged as historically unreliable, why might we not regard as of one cloth the entire story of a shepherd lad or brigand chieftain who, by divine grace united a fractious group of tribes and established an empire stretching from Egypt to Syria? If, on the other hand, the stories in Genesis are, for all their verisimilitudinous settings and personae, essentially works of the imagination, designed to promulgate truths about God (theology), human nature (anthropology), and the link between the two (morality and sanctity), what is the purpose of genealogies and chronologies featuring names and longevities of people whose lives were snuffed out in a cataclysmic deluge or of eponymous ancestors who, according to another genealogy, had been sired ten generations before that time?

The geneticist approach to these problems—of pointless inclusions of certain kinds of data, of contradictions within the data themselves, of hallmarks of fiction and hallmarks of historiography—is to explain how these got into the text; that is to say, problems of coherence and consistency are recognized but are not reconciled. Their lack of coherence and consistency reflects their multiple provenance; and the redactor's responsibility for the hodgepodge is mitigated by the plea of pious compulsion, namely, to preserve writings from the past whose sanctity (if not authority) is guaranteed somehow by their antiquity. By contrast, the poetical approach views the literature as a whole. Nothing "got" into the text: whatever is there is the text. And the question must be, Why did the author (or the editor–compiler) make it what it is? What purpose—artistic, ideological, propagandistic, or kerygmatic—is served by vectors pointing in different directions and geopolitical genealogies featuring the same eponyms in different ancestral lines, by pointless chronological notices, or by numerological items that are patently absurd? And, let us remember, the success of the poetic enterprise must be judged by the persuasiveness of the poetic answers to the poetic questions.

In the case of the prophetic narratives that we have studied, narratives confined (with one or two exceptions) to the monarchical period of the Divided Kingdom, we had occasion to note how absurdly exaggerated numbers and quantifications, as well as implausible turns in development of plot, served to signal that the stories were to be read for their moralistic insights, not for their historical heuristic value. This characterization applies, as well, to the nonprophetic stories discussed in chapter 2 on judgments pronounced by Solomon and David and, in the case of the latter on how the way for that judgment was prepared by including here and there, in earlier chapters of 2 Samuel, seemingly otiose details about a single surviving scion of the house of Saul. We saw further that much that we thought we knew about the history of the monarchical period derived completely from the stories and that there is good reason to believe that the biblical author may have invented battles and wars and sundry campaigns, giving victory now to Israel, now to Aram, now to Moab, and again to Israel, as he backgrounded his theological and moralistic fables with the contentions of states striving for empire or independence. If the rest of the events in the "histories" of the books of Samuel and Kings are of a piece with these, then we have before us neither historicized fiction nor historical fiction nor fictional history. We have what Sternberg calls *ideological literature*. And the ideological narratives reflect a freedom on the author's part to invent plot and characters while investing both with the stamp of verisimilitudinous history. It is a nice question, which I cannot pursue here, just what the views of the ancients (Mesopotamian, Israelite, Greek, and Roman) were on historiography. But it must be clear that neither pietistic ideological compulsion nor collector's compulsion to preserve antiquarian traditions accounts for the introduction of two different Davids to Saul's court and parallels the long narrative of one of these (the conquest of the Philistine giant) with a chronicler's note (2 Sam. 21:19) that it was a Bethlehemite (like David) by the name of Elhanan who defeated the very same "Goliath of Gath whose spear-shaft was as thick as a weaver's beam." In general, let us note, verisimilitudinousness is a strategy to promote the persuasiveness of the narrative, even as narrative as a didactic tool is a strategy to drive home the lesson, and not to delude the reader as to the story's historicity. The girl who counted her eggs before they could hatch is more convincing for having lived and dreamed and dropped the eggs before they could be sold to buy a milch cow than a hypothetical lass who might have done the same thing. But having achieved his end by this narrative-as-history strategy, the biblical author must leave clues—to some of his readers at least—that historiography (and, at that, simpleton historiography) is not what he is about. So, having told at charming and delicious length (replete with dialogue featuring boasts and threats and callow youth's confidence that Israel's God who never has let him down will never do so) the story of how the stripling shepherd came to King Saul's notice (1 Sam. 17:1–58), the author brackets this story with an alternate version (1 Sam. 16:14–23, 18:10–15) of the warrior–minstrel who was recruited as the king's personal Orpheus.

Those in the audience who will, in any generation, read these chapters on the most naive and literal level will not note (nor, if they do note, will they linger over) the discrepancies in the description of David: in his coming to court, being spirited back somehow to Bethlehem so that he can be sent back to the king's battle camp, to be introduced to a king he has been serving for some time, who yet does not know him and upon asking his identity of his chief commander learns that he doesn't know him either. But for those initiates who have plumbed many an earlier biblical story for its varied metaphors, nuanced asides (in the mouths of narrator and characters) and multiple levels of meaning, this will be but another instance of a story whose broad outline was a matter of tradition (or, better, *traditions*). And what matter if one version or another were closer to the event *wie es eigentlich gewesen*? Whatever the version, the message is that character determines fate, not the other way around. And, let us note, the device of the less-than-reliable narrator is tailor-made for the fictive treatment of such traditions.

So it is, then, in the case of the author who introduces the two different David's I have just discussed. Indeed, the more the storyteller conveys an impression of reliable historicity in a tale whose historicity is dubious, the more he plays the role of unreliable narrator. But let us consider one more instance in Scripture, in which the *author* signals the narrator's unreliability or very possible unreliability even while he confesses that the actual truth of the events he is reporting can be known only to God.

In 2 Kings 1, a claim seems to be made three times that King David has sworn that Solomon would succeed him on the throne (twice by Nathan, once by Bathsheba), a claim then confirmed by David himself. Actually, Nathan never makes the claim to firsthand knowledge of such an oath: the first time, he is advising Bathsheba to claim to the king that he had so sworn to her; the second time, addressing the king, he expresses astonishment that the king might possibly have designated Adonijah while to Nathan and other loyal servitors specifying no successor at all. Now while the narrator contrives the impression that Adonijah has jumped the gun, so to speak—is celebrating his succession as a fait accompli while father David yet lives, a careful reading of the text will disclose that that is not so. Only Nathan describes the feast, and even he makes no claim that Adonijah has seized the throne: it is the invitees to the sacrificial feast who hail a succession that they regard as predetermined. And the specific mention is verse 41 of Joab as one of the feasters is another way in which the narrator indicates that the celebration is not a coronation. That shrewd general and veteran survivor of court intrigue would not have participated in a coup d'état while leaving the palace and a dying king in the hands of courtiers known to be loyal to a younger prince. No. The coup d'état is engineered by Solomon's minions and is reported to the feasters by Jonathan, son of Abiathar the priest; this latter is among the guests who hear with horror that Zadok the priest has anointed Solomon to general acclamation. Jonathan's report that it was at David's command that Solomon was anointed carries no authority whatsoever: he is reporting what has been proclaimed. But is (or rather was) the narrator who reports the command of David to Zadok and Nathan and Benaiah in a better position to know than was Jonathan ben Abiathar? Are we to take it for granted that he is an omniscient narrator? Or does his seeming omniscience cloak an assumption so obviously legitimate. After all, would priest (Zadok) and prophet (Nathan) be capable of a lie? We are left free to wonder what answer Joab might have given to this question, he who had never dreamed that David might be cozened (even in his senility?) into gainsaying the universal expectation that the senior surviving prince would come into the kingship. Or what would Abiathar have given for the credibility of Zadok, his unpedigreed rival to hierophancy? And, if we may permit ourselves one more question, is the narrator of 2 Kings 2, who relates David's deathbed charge to Solomon to arrange executions for Joab and Shimei ben Gera a propagandist upholding Solomon's innocence of bloodshed, or must he be seen as recipient of a revelation? What question of religion or morality hangs on the issue of Solomon's ruth or of the legitimacy of his succession to the throne? The facts of history speak for themselves—Solomon did succeed to throne and empire—and if God is the lord of history, is that a theological tenet always and everywhere to be taken literally? Or is it sometimes—perhaps more often than not—a metaphor for what he should be, and certainly would be if we recognized him as such.

Let us now return to Sternberg, who argues the historiographic nature of biblical narrative and also gives that narrative credit as constituting a (or the) "new mode and rhetoric of historiography." Alter, too, sees in the biblical prose fiction an innovative literary genre. He takes up Shemaryahu Talmon's suggestion that "the ancient Hebrew writers purposefully nurtured and developed prose narration to take the place of the epic genre which by its content was intimately bound up with the world of paganism, and appears to have had a special standing in the polytheistic cults."[8] He then elaborates on "the new medium" in a passage that I cite in full for its eloquent description of what I intend by my label "theological or moralistic fiction."

> What is crucial for the literary understanding of the Bible is that this reflex away from the polytheistic genre had powerfully constructed consequences in the new medium which the ancient Hebrew writers fashioned for their monotheistic purposes. Prose narration, affording writers a remarkable range and flexibility in the means of presentation, could be utilized to liberate fictional personages from the fixed choreography of timeless events and thus could transform storytelling from ritual rehearsal to the delineation of the wayward paths of human freedom, the quirks and contradictions of men and women seen as moral agents and complex centers of motive and feeling.[9]

I, too, am impressed with the uniqueness of biblical prose narrative, particularly in its temporal and geographic setting. And for all my disagreement with Sternberg's classifying it in the historiographic rather than the fictional genre, I am at one with him in this judgment:

> As regards cultural value, temporal scope, and persuasive strategy, this art of narrative has no parallel in ancient times. Alone among Orientals and Greeks, it addresses a people defined in terms of their past and commanded to keep its memory alive. . . . It also uniquely internalizes its own rules of communication, whereby the remembrance of the past devolves on the present and determines the future. . . . This cultural imperative . . . explains how there suddenly emerged a people [in the words of Herbert Butterfield] "more obsessed with than any other nation that has ever existed. . . . It was this historical memory that made Israel a people"; why "they stand alone among the people of the ancient world in having the story of their beginnings and their primitive state as clear as this in their folk-memory," in creating the history of a nation and even of humanity itself.[10]

Just so; but if the biblical narrative is history, it merely *records* the history, whereas the genius of its imaginative feat lies in its "creating the history of a nation and even of humanity itself." And in a very real sense, since a people's history or a race's history is semantically congruent and inextricable from its meaning or significance, the biblical narrative creates the people of Israel as it creates humanity.

Now if we keep in mind this awesome uniqueness of a narrative fiction that creates not only the history of Israel and humanity but the very ontology of this people and this race and projects the historic vectors onto the graph where imaginary numbers plot the realization of their united destiny (thus rounding out this "salvation history"), we shall have to acknowledge that this feat is the feat of *metaphor*—metaphor writ large and metaphor writ small; small metaphors in series building up to a greater metaphor; metaphors like a nest of dolls independent of, yet containing or capable of containing, one another in a travesty almost of gravidity; metaphor that gives meaning to the inanition of inert facts and, like God's breath, gives life to lumpish clay; metaphors of fiction and of history, which alike take up units of animated entities or conglomerations of them and, transforming them into the corporate fictions of family, society, nation, and state, make out of inorganic atoms and molecules the organic and hyperorganic entities that populate the realms of ethics and metaphysics. In the case of Genesis, Scripture begins with the Metaphor of Metaphors whom we call God. (In the Jewish mystical tradition one of the names for God, probably expressing his ineffability in a one-word metaphor, is Kibhyakhol, "As-it-were" or "So-to-speak.") It focuses on the creation of an inspirited clay collectivity, a metaphor termed *ha-Adam,* and, after sketching him in several disappointing incarnations, begins history anew with a flesh-and-blood individual destined to be your father and mine in a collectivity in which the other incarnated collectivities will ultimately find their salvation.

In recalling this let us recall the observation that **the literal and the metaphoric** constitute a spectrum, not a contradiction. The biblical authors invented the people of Israel as the instrument of God and thus transformed a factual historical entity (an agglomeration of peoples constituting a nation of sorts on the land bridge between Asia and Africa) into a normative metaphysical one. These authors did what they did in the interest of recruiting an entire population into the service of the God of history, who is also God of the ancestors. And since such a recruitment effort required that the *how* of the past and the shape of the future be accessible to the various levels of the population according to their capacity to grasp meaning on different stretches of the literal–metaphoric spectrum, the challenge to their inventiveness must be seen to have been awesome. And awesome, indeed, is the literary monument that witnesses that they were equal to the challenge. Imbedded in the narrative as skeleton in the flesh are the genealogies that both structure the narrative as "history" (literal truth) and simultaneously, by discrepancies in names and absurdly unrealistic chronologies, signal to the sophisticated reader, "All this is metaphor! The truths presented here are not the bare bones of facts bleaching in the bright sunlight. They are the metaphors of meaning flagged by unverisimilitudinous plots and oddly disarticulated structures."

It is a matter of wonder to many that there are yet today so many millions of fundamentalist creationists and their ilk (numbering among them quite a few holding doctorates in philosophy and science), who take the Genesis accounts as literal and accurate history. Nonfundamental biblicists reject, for the most part, the claim of these narratives to constituting accurate history, yet cling to the assumption that the authors intended them to be taken literally. They bolster this last assumption by pointing to "historic records" preserved in genealogies and explain away the discrepancies by recourse to geneticist analysis that attributes them to different "sources." Poeticists—except when they are seduced into speculating on the degree of historicity by their concern for fixing the literature's genre—are essentially not concerned with this question, and concentrate on the message that the narrative aims to communicate. As literary critics, they have no trouble at all with metaphor nor with reading the literature as being predominantly figurative. Why, then, do they so often keep company with fundamentalists and geneticist Bible scholars in ascribing literalist intention to narratives that would never be taken so if authored in our own time?

The answer, I submit, lies in the metaliterary convention that assumes a primitivity and naïveté for ancient authors as compared (implicitly) to the modernity and sophistication of the modern reader. As against the traditional approaches to the biblical texts (which may be characterized as primitivistic, historiographic, literalistic, and genetistic), my exegetical essays champion a view of the author–editor or author–editors as sophisticated, ideology-oriented, and philosophically inventive—hence figurative in expression and untrammeled by sanctified traditions from their society's past. Basic to this assumption is another one, namely, that neither the human condition nor human perceptions of it have changed in the last five millennia. As in intellectual capacity or artistic talent the ancient mind is not inferior to the modern one, so the relationship between the ancient mind and ours is one of continuity, not divergence.

Let us consider one biblical metaphor accepted as history or as intended history by most readers, including biblicists, today—a God who has determined to bring on to the stage of history the descendants of one man to the end that they may pursue on their ancestral soil the role of an entire nation in faithful service to that God. In order to allow time for this family to grow in size to national proportions, to prepare for them the promised land of which the sinful inhabitants must be given time to earn the full measure of their doom, this God deposits some seventy kindred souls into the midst of a mighty kingdom. There, over a period of 430 years, he warehouses this family, which multiplies into a nation of millions until it may erupt into history as the conquerors of the land scouted by a handful of ancestors (during which time a national or ethnic consciousness and solidarity will be preserved without a homeland, without a political consititution, without a distinctive religion or established clergy); and all the while, these multitudes will maintain not only their single ancestral identity but also their clan and tribal divisions, immune to assimilation to one another or to the native population!

Considering the constructs with which modern historians operate, considering the constructs by which ancient historians may have operated (however they may have differed from those of our time), is it conceivable that such a plot could have been woven and worn by people who could not tell apart the figurative from the factual, the metaphoric from the historical?

The inventive literary genius of Scripture's authors will continue to fail of proper appreciation until the constituent narratives receive explication along poetic lines and are related to one another as an organic whole, until the narrative and prescriptive texts are seen in their fully integrated complementarity, and until we have a clearer picture of how propositional statement can be expressed in lyric verse and flights of the poetic imagination be captured in cultic regulations. All this is to say that regardless of the variety of literary genres available to the biblical authors, the shape and substance of Scripture constitute a new genre, without precedent and probably inimitable: an ideological corpus blending story and structure, the novel and the historic, the real and the ideal, the poetic and the prosaic, the comic and the tragic, the mundane and the metaphysical into a literary achievement whose design we are only beginning to plumb.

The foregoing appreciation is not intended as a pious paean in praise of a literature that has been derided as primitive and childish, intolerant and Pollyanna-ish, dogmatic and priest-ridden, philosophically naive and metaphysically narrow. The survival of this corpus produced in an ancient tongue by a tiny and obscure people as the foundation of the religious and cultural vision of half the world speaks for itself. What I intend by it is an expression of gratitude to the literary explorers of my time who have opened a new chapter in the study of this achievement, to express thanks that I may be of their company, and to stress that lusty as these beginnings may be, they are still just that—beginnings. One danger to the development of the grammar of biblical poetics is the foreclosing of options, techniques, and strategies by the application of rigid categories of genre. Another is the blindness to the use of structures (which we are not accustomed to see as literary forms) either to flag the presence of metaphor or constituting metaphor in themselves. A third danger is the denial of a high level of sophistication to Scripture's authors (or intended audience), and to the authors of pagan antiquity as well, whose writings are adduced for comparative or contrastive purposes. Yet another danger is the humorlessness with which scholarship approaches its own critical enterprise, equating the objectivity of science with an intolerance of the quirkiness of the sense of humor, which will then fail of appreciation as an ingredient in the repertory of Scriptural authors or, indeed, of their near-contemporaries in eastern or classical antiquity.

There is often operative among students and scholars the notion that comic and tragic are polar opposites, that humor is a hallmark of the comic and entertaining and somehow antithetic or uncongenial to the serious and edifying. Perhaps a reflection of this is to be seen in Edwin M. Good's pioneering foray into literary appreciation; for in his *Irony in the Old Testament*[11] he either closes his eyes to nonsatiric forms of humor or attempts to force them into a Procrustean definition of irony. Generally speaking, we are more open to the possibility of humor in ancient texts if the butt of the jest is human rather than divine (losing sight, often, that a mockery of the Olympians may be only jest at the expense of the mortals who created them) and its tone grim and sardonic rather than gentle, tolerant, or lovingly playful. But the broad range of the spectrum of humor must be allowed for.[12] For all the ideological seriousness of the biblical authors, we cannot close off their freedom to indulge in the whimsy that metaphor often invites or entails. Surely, the authors of Genesis and Exodus had to be conscious of their own playfulness as they spun the tales of ancestors slow to mate and even tardier in procreation, yet spawning the many ancestors of tribes and polities once they had broken through the fertility barrier. And as surely as they were aware of the whimsy in Israel's growing to national proportions in Egypt's warehouse, so must they have smiled as they mocked the claims to historical precedence symbolized in the genealogies and chronologies that telescope all prehistory into twenty generations and Israel's history from Abraham to David into another fifteen.

Life and history are, as we suggested in chapter 4, often equated with a reality denied to fiction and its creatures. I would invoke another kind of reality in assessing the artistry of Scripture. Biblical Hebrew, to the best of my knowledge, has no word for the abstraction expressed in our word *art*. And even in modern Hebrew there is no alternative to the loan word *humor*. Artistry, however, is one of the meanings of *ḥokhmā* (wisdom), and the closest approximation to the humorous is the root *ṣhq* (smile, jeer, laugh). I fancy that if a humorless historian were to query one of the biblical authors on the preposterousness of his chronology, the answer would be along the lines of an inversion of a Latin saying: *Vita longa, ars brevis.*

NOTES

Abbreviations

ABN	Robert Alter, *The Art of Biblical Narrative*
ANET	Ancient Near Eastern Texts
AV	Authorized Version
HUCA	*Hebrew Union College Annual*
IB	*The Interpreter's Bible*
ICC	*The International Critical Commentary*
IDB	*The Interpreter's Dictionary of the Bible*
JPS	Jewish Publication Society (English Bible Version)
JSOT	*Journal for the Study of the Old Testament*
NEB	New English Bible
PBN	Meir Sternberg, *The Poetics of Biblical Narrative*
PI	Adele Berlin, *Poetics and Interpretation of Biblical Narrative*
RSV	Revised Standard Edition
SBL	Society of Biblical Literature (monograph series)

Chapter 1

1. *Oral literature* is, if not a self-contradiction, at least an oxymoron. The very existence of the expression owes to a presumption that long and intricate compositions such as epics and sagas—which have come down to us from antiquity in written form—existed before their reduction to writing, were transmitted by the mnemonic talents of a bardic tradition, were indeed composed without the help of the technology we call writing. The presumption, however, remains moot. At best, we may trace back a claim that an ancient literature was *based on* oral tradition; but the claim that a *fixed text* of a composition was strung together and preserved in memory by author or authors without first being either written or dictated is an absurdity. Do we suppose that Homer himself could have told the *Illiad twice in the same words?* The very term *literature* connotes a *text fixed in writing.*

2. Thus, there are poetical elements of great importance for other literatures that receive little or no notice here, and other poetical elements accorded a prominence out of all proportion to their significance elsewhere. See, in this regard, the Preface. This latter will be the case, for example, in regard to *free direct discourse,* which is almost a universal element in poetics and in regard to the *synoptic/resumptive-expansive* technique, which may be unique to Scripture.

3. Although both Robert Alter and Adele Berlin devote far more attention to character than to setting, neither ignores it altogether. Thus, Alter writes, "As elsewhere in biblical narrative, the revelation of character is effected with striking artistic economy: the specification of external circumstances, setting, and gesture is held to a bare minimum" (ABN, 37). Further on, he writes, "Biblical narrative offers us, after all, nothing in the way of minute analysis of motive or detailed rendering of mental processes; whatever indications we may be vouchsafed of feeling, attitude, or intention are rather minimal; and we are given only the barest hints about

physical appearance, the tics and gestures, the dress and implements of the characters, the material milieu in which they enact their destinies " (p. 114). Berlin writes, "It has often been said that the Bible rarely describes its characters. . . . Thus when we are given some detail about a character's appearance or dress, it is usually because this information is needed for the plot. . . . So there is description, even physical description. But what is lacking in the Bible is the kind of detailed physical or physiological description of characters that creates a visual image for the reader. We may know that Bathsheba was beautiful, but we have no idea what she looked like" (PI, 34).

4. Alter, ABN, chap. 6. Berlin, PI, chap. 2.

5. PI, 23–24.

6. ABN, 115.

7. If they cannot be shown so to function, their presence represents a challenge to the text as amenable to a poetical treatment and reinforces the credibility of the source-analytic approach. If they can, however, source analysis in regard to such "literary events" becomes otiose, a solution to a problem that does not exist.

8. See n. 7 and also Alter's treatment of Genesis 38 in the context of the Joseph story (ABN, 3–11). I have no sense of any repudiation of my revered teacher in expressing agreement with Alter's judgment that "Speiser's failure to see its [the Judah–Tamar story's] intimate connections through motif and theme with the Joseph story suggests the limitations of conventional biblical scholarship even at its best" (p. 4).

9. Berlin devotes an entire chapter to this element as it operates in biblical texts. Under this heading she includes such elements as dialogue and interior monologue, which I shall treat as a separate focus. This feature of literary analysis is admirably deployed by Lyle Eslinger in "Viewpoints and Point of View in 1 Samuel 8–12," JSOT 26(1983): 61–76 to the discrediting of source analysis and the vindication of these chapters as a unitary text.

10. PI, 38–40 and 64–72, respectively.

11. ABN, chap. 4.

12. The running of direct into indirect discourse and vice versa (parallel to strict versus free direct discourse) may be seen in the Hebrew expressions indicating oath or adjuration: in direct discourse, "He said, 'May God break my bones *if* I do this'." In indirect discourse, "He swore *that* he *would not* do this." Hence the *if* (*'īm*) of direct discourse is equal to the promise *I shall not* (*kī lō'*) and *if not* (*'īm lō'*) is equal to the promise *I shall* (*kī*). Yet in oath formulae reported in indirect discourse the promise that he would do something appears as *if not* (*'īm lō'*) instead of *that* (*kī*). The substitution of a formula for direct discourse for a particle introducing indirect discourse is balanced by an inverse phenomenon. The "May God break my bones"— the disaster invoked—is in direct discourse formulated as "May God do such-and-such and worse" (*kō yaᶜaśē vᵉkō yōsīf*), which would provoke laughter if uttered by a character on the stage.

13. Such instances of repetitive narrative events, revealing slight or serious inconsistencies and even contradictions, have been and are one of the main problematics addressed by source analysis. The surface plausibility of charging the two treatments to two different sources (or authors), while it clears one author of the charge of self-contradiction (a logical sin if not demonstrably an aesthetic one), falls away when one queries the sense of resolving one absurdity at the expense of positing another one: if it is not plausible that one author give two conflicting versions of an event, why is it plausible that an editor (whose task it is to improve an author's work) should parade his ineptness by attaching two such versions and passing them off as a unitary composition?

14. An example that comes easily to mind is the numbering in Genesis 2 of verse 4, combining and thereby connecting syntactically the notice ending one chapter of narrative with the beginning of the next chapter.

15. The function of the perfect tense as indicating completed action in a future time, that is, as a future perfect (as well as a present perfect and pluperfect), is often missed by translators of the biblical text. See, for example, the subtleties of this usage of the perfect in Psalm 126 in conjunction with the imperfects around it.

16. The sudden disappearance of waw-consecutive from literary Hebrew may (although I consider it unlikely) be due to a long time-gap between the last of the biblical books and the products of Tannatic times. The use of constructions other than the waw-consecutive in narrative might be support for my speculation. See, for example, Genesis 15:1 or 37:2. The latter, another example to add to n. 14 above, is in its second part a pure nominal sentence with subject preceding verb, while the former features the verb before the subject, as is normal in biblical Hebrew when the verse starts with an adverbial clause. Long overdue for research is the use in direct discourse of waw-conjunctive with the imperfect for action in the future alongside waw-conversive with the perfect tense. For example, in Moses' direct discourse in Deuteronomy, he consistently uses the waw-conversive with the perfect. By contrast, in God's direct discourse to Abraham in Genesis 12 the verbs for God's future actions are all in the imperfect in the first person with waw-conjunctive, then there is an abrupt shift to waw-conversive with the perfect for the future action of the third-person "families of the earth." In Genesis 26, God, again addressing Abraham, uses waw-conjunctive with imperfect for his future action (verse 3) then switches to waw-conversive with the perfect for the rest of his discourse.

17. For the reader's convenience I present the passage in translation. I use *now* as a marker indicating the beginning of a flashback:

There was a certain man . . . by name Elkanah. . . . 2.*now* this fellow had two wives, one named Grace (Hannah) and the other named Jewel (Penina); Penina had children but Grace had none. 3.*Now* this man would go up periodically to do obeisance, to make sacrifice to YHWH-hosts in Shiloh, there where the two sons of Eli, Hophni and Phineas, were priests to YHWH. 4.On a certain day Elkanah made sacrifice. *Now* he would give [meat] portions to wife Jewel and to all her sons and daughters. 5.But to Grace he would give one portion; though he preferred Grace, yet YHWH had closed tight her womb. 6.And her co-wife would ride her mercilessly with intent to make her grumble over YHWH's having placed bar across her womb. 7.Yes, so would he do year after year in YHWH's edifice whenever she [Grace] went up, so [also] would she [Jewel] ride her— but she [Grace] wept and would not eat.

Note the length of the flashback between verse 4's "On a certain day Elkanah made sacrifice" and the completion of the sentence in narrative time in verse 7, "but she wept and would not eat." Note also that although the flashback is a parenthetical narrative, she who would not eat is unidentified and must be understood as Grace—an understanding only made possible by the information supplied in the flashback.

18. Again, I must draw the reader's attention to this fact, that the synoptic/resumptive can only be a device of literature—written narrative. Examine any instance of this episodic technique in my translations and it will become clear that in an oral presentation the narrator would have to utter the words I supply in square brackets, or else their equivalents.

19. The waw is a copula, a connective particle, with a remarkable range of functions whether it connects clauses in waw-conversive fashion or in waw-conjunctional fashion (with the perfect tense the waw-conversive functions, in regard to punctuation and vowels, exactly as does the waw-conjunctive), or members of a series of nouns. (See, for the last item, Genesis 1:14, "and they shall serve for time markers, that is, of days and years." The waw connecting *'ōtōt* and *mō^adīm* is expressive of a **hendiadys**, the waw connecting *yāmīm* and *shānīm* iš expressive of a merism, and the waw connecting the two couplets is a deictic that also renders the first couplet into a construct.) The waw can, of course, be a conjunction; but it may also be a disjunction or have disjunctive force (as in any clause in which it can be rendered by *although*). It can even introduce a protasis with the sense of "if" (*'īm*) just as it regularly introduces the

apodosis with the sense of "then." For a discussion of the supposed disjunction '*ō* (or) as a deictic, see my essentially poetical analysis of Numbers 5:11–31 in HUCA 46(1975): 62–63.

20. It will not have escaped the alert reader that the paranthetic and/or flashback device, subordinating this insertion to the narrative time of the waw-conversive formulation, is hypotactic construction while the parallel episodes of the synoptic/resumptive represent parataxis. The characterization *waw-conversive* as opposed to *waw-consecutive* owes, of course, to the recognition that the opening of a scene or chapter with this construction is in no way to be related to what precedes (which may, indeed, be nothing). This alone reinforces my argument against the rendering of waw by *and*. The testimony of the absurdity of rendering waw by *and* when it introduces a resumptive episode is, I believe, conclusive.

21. Berlin, PI, 73–79, and 105. Alter devotes an entire chapter ("The Techniques of Repetition," ABN, 88–113) to this subject. He manages to cover so much territory in this brief space and offers such a variety of solutions, ranging from the highly theoretical to the specifically interpretive, that I shall forego comment at this point. I hope to make extensive reference to many of his interpretations in the essays that follow. In anticipation of my frequent disagreements and of a vigorous defense of my own interpretations as against Alter's, I would in all fairness acknowledge my debt to him. His is the pioneering effort. If he seems to me to rush to interpretation and to offer broad theoretical formulations that I find less than convincing, mine is the advantage of being able to compare the trails I have been blazing as alternate routes to the same objective with those he has already mapped.

22. Perhaps this explains, in part, why the only systematic treatments of phenomena that we classify as political, social, philosophical, theological, and the like are to be found in the guise or disguise of narrative prose or poetry. The propositional essay or treatise, less vivid and spontaneous than the narrative but much more direct and unambiguous, is the form in which we are accustomed to see nonfictional subjects. Perhaps herein lies the reason for our propensity to assume that ancient narrative is basically fiction. On the other hand and (as I shall argue), equally off the mark, is our assumption that there exists in early antiquity a narrative form that is not essentially fictional but is, rather, expositional of reality, namely, history.

23. Of relevance to my argument here is the cautionary note on which Berlin begins her book.

> [A]bove all, we must keep in mind that narrative is a *form of representation*. Abraham in Genesis is not a real person any more than a painting of an apple is a real fruit. This is not a judgment on the existence of a historical Abraham anymore than it is a statement about the existence of apples. It is just that we should not confuse a historical individual with his narrative presentation. (PI, 13)

Biblicists would do well to ponder Alter's view of some of the standard presumptions in contemporary Bible scholarship.

> The case of the Bible's sacred history, however, is rather different from that of modern historiography. There is . . . a whole spectrum of relations to history in the sundry biblical narratives . . . but none of these involves the sense of being bound to documentable facts that characterizes history in its modern acceptation. It is often asserted that the biblical writer is bound instead to the fixed materials, whether oral or written, that tradition had transmitted to him. . . . A close inspection, however, of the text . . . may lead to a certain degree of skepticism about this scholarly notion of the tyrannical authority of ancient tradition, may lead us, in fact, to conclude that the writers exercised a good deal of artistic freedom in articulating the traditions at their disposal. (ABN, 24)

24. *Antiquity* is, of course, a relative term. My use of it here is in relation to any period, society or language as far back as Sumer five millennia ago or as recent as Elizabethan England. The crucial consideration is the sense of a large gap between the mind-sets of that past culture and of our own and between the idioms of these cultures, be those idioms of different linguistic

families or of one family over a period that witnessed great or significant linguistic change.

25. Herbert Brichto, "On Faith and Revelation in the Bible," HUCA 39(1968): 37.

26. Stephen Jay Gould, *Time's Arrow, Time's Cycle—a Study in Myth and Metaphor* (Cambridge: Harvard University Press, 1987).

27. *Literal*, in the sense of "according to the letter," is from the Latin *littera*, which can mean both "letter" in the sense of the alphabetical character and "word" (*verbum*). But even if we amend the sense to "according to the word," the statement about the lack of meaning in the word is largely true; for, as Saussure has made us aware, words—while they may stand for a number of objects or concepts as referents—do not have specific meaning in isolation. Meaning is a function not only of the word itself but of the environment in which it appears.

28. I would stress here that my coinage of the latter category is due only to its being a nonliterary convention that is presumed in literary interpretation and not at all to the possible or likely roots of such conventions, as I discern them, in imaginative literature.

29. See 1 Samuel 24:1–22. The entire episode (and its contrast with its doublet in 1 Samuel 26:1–25) cries out for exegesis along the lines of poetical analysis that I am espousing. Can anyone miss the spirit of fun in the picture of Saul entering a privy, which just happens to be a cave in whose recesses David and his men are hiding out? And what about the picture of David sneaking up on the squatting king to snip off a piece of the royal robe? The dialogue in which the king "with his pants down" is cited as the opportunity that yhwh has promised David? The notice of David's remorse for the act that, resorted to in self-vindication, nevertheless exposed the king to ridicule? The careful reader will note that without recognition of the synoptic/ resumptive narrative technique, the order of the events is pointlessly confused.

30. See Herbert Brichto, *Problem of "Curse" in the Hebrew Bible*, SBL Monograph Series, vol. 13 (Philadelphia: SBL, 1963), 141, n. 56.

31. Ibid., 10.

32. Ibid., 29, n. 14.

33. In the real world, which is by definition a unity made up of differentiations, there can be no contradictories. (Even matter and antimatter are concepts in physics illustrative of the rule that two things cannot occupy the same place at the same time.) But in the conceptual world, the world of ideas, every idea or term for an idea is an abstraction from reality and exists only by virtue of the context from which it is abstracted. Hence, antonyms exist only in the context of one another, such as *peace* and *war:* truce is an example of a kind of peace that stresses the context of a continuing war.

34. Herbert Brichto, "On Faith and Revelation" HUCA 39(1968): 38–41 with special attention to n. 3. Today, in consideration of the feature of divergence which is central to obliqueness as against the convergence which is characteristic of metaphor, I would be tempted to dub this rhetorical figure *asymptotic metaphor*. But the phrase is cumbersome; *asymptosis* by itself smacks of a disease; and the borrowing from mathematics, while heightening the pretentiousness of the coinage, contributes little (as is the case with so many labels) to an appreciation of the phenomenon.

Chapter 2

1. Sternberg's analysis is in *The Poetics of Biblical Narrative* (Bloomington: Indiana University Press, 1985) 166–170. This volume appeared only after many of the essays as well as much of the introductory chapters of the present volume were written. Some of the material in it was familiar to me from the essays published in Hebrew, although I must confess that my first notice of these owes to Alter's reference to Sternberg in the Judah–Tamar article in *Commentary*, December 1975.

264 NOTES

This volume of Sternberg's includes (pp. 482–492) a shorter version of an article first published in HUCA 54 (1984): 45–82. It and my own essay on Exodus 32–34 in the same issue (pp. 1–44; see also chap. 4) reveal some of the similarities and difference in our respective disposition of poetical analytics.

2. The translation here is an attempt to convey what in the Hebrew I take to be an archaizing usage. The imperfect tense with the adverb *'āz*, here (as in Exod. 15:1) cannot be rendered by any of the regular forces of the imperfect (present, future, past durative). Hence, it probably is a recourse by the author to the older preterite, which had served for past tense until it was superseded by the postpositive perfect tense (while the form that had previously been that of the preterite also now served only for the afformative tense). (Note the same usage in 1 Kings 8:1 and, in contrast, the appearance of *'āz* with the perfect in 1 Kings 8:12.) A similar archaizing usage may lie behind the verb *wayyīšarnā* with a feminine subject in 1 Samuel 6:12, for the afformative taw almost certainly would not have displaced the yod in the third-person feminine as long as the preterite held the field. See also as a possible frozen use of the preterite the regular appearance of the imperfect with the adverbial *ṭerem*.

3. The use of *bī*, a particle with precative force, introducing an address by a subject to a monarch or a subordinate to a superior, bespeaks a protestation of sincerity on the part of the speaker, a prayer for a gracious hearing. As such it can hardly be anything but a frozen form of the introductional element in the oath formula, equal to *beḥayyai*, "By my life," "May my life be forfeit if. . . ." So, too, in the case of the same usage in verse 26. Here, the force of the oath formula in the mouth of the mother pleading for the life of her child suggests her readiness to trade her life for her son's. See the discussion below of the significance of the king's calling for a sword. For the meaning of the "By the life of . . ." in oath formulas, see my *Problem of "Curse" in the Hebrew Bible*, SBL Monograph Series, vol. 13 (Philadelphia: SBL, 1963) 150–156.

4. See n. 3.

5. Sternberg's interpretation of the king's "verbatim repetition of the mother's outcry" is worthy of careful consideration. My own translation, points up another aspect of the meaning-laden repetition. My rendering of the word *wayyō'mer* by the parenthetic "in command" is to bring out the pregnant ambiguity of the Hebrew, the quotation of the mother's words constituting the command. The signaling of this ambiguity (or better, ambi*valent* double entendre may be the author's reason for having the mother employ the peremptory negative *'al* while the king's repetition features the *lō* of general negation.

6. The translation of *lifney* by "at the king's behest" is more meaningful here than the equally possible sense of "before [or *in the presence of*] the king." The latter rendering in virtually all modern translations is less a matter of consensus than one of cultural lag in Bible study: the voluntative force of *lifney* was, along with many other fine points of Biblical Hebrew semantics, introduced by E. A. Speiser in his Anchor Bible translation. See his *Genesis* (Doubleday, 1964), 51, n. 11.

7. See n. 2. Aside from the use of the preterite here (when the option of the perfect tense was available to the author) the use of *'āz* to introduce the narrative, in a clearly intentional avoidance of normal paratactic usage, creates a caesura between the general and introductory notice of the two women's coming "to the king" and their appearance "before him" in the second clause of verse 16. The deliberation with which the author chooses his tenses is illustrated in my discussion below of the parenthetic nature of the last (nominal) clause in verse 27. In reference to this last item, I would draw the attention of the reader to the second clause of verse 22. Here, too, we have a nominal clause ("the other one then saying . . .") in contrast with the paratactic syntax of the beginning of verse 17 and the beginning of verse 22. Of further interest is that the second woman's response in verse 22, for all its formal paratactic correspondence to the speech of the first woman, is, unlike the latter, addressed not to the king

but to the first woman. This subtle twist in the dialogue makes it possible for the king to quote the two women with exactness; it also prepares us for the second-person address by this same woman to her rival in the parenthesis in verse 26.

8. The biblical evidence for this last sentence will be presented shortly. For the king's validation of his legitimacy in terms of his administration of justice, see my essay, "The Biblical Base of Western Democracy," in *Jews in Free Society,* ed. Edward Goldman, (Cincinnati: Hebrew Union College Press, 1978) 115–127.

9. The Hebrew for *whether . . . or* is w^e . . . w^e (see chap. 1, n. 19; see also the identical law in briefer phrasing in Exod. 21:16). The disjunctive force of the second waw is generally recognized (e.g., AV, RSV, JPS). The New English Bible, in its idiosyncratic fashion treats the second waw as a conjunction. My characterization of the New English Bible as idiosyncratic is not due to its preference for idiomatic sense over literalness but for its frequent departure from the plain Hebrew text whenever the translator feels that he could have improved on the Hebrew original. For example, in our story, the translator will not permit the false mother to ignore the king and address her rival. The Hebrew "Neither yours nor mine will he be" is rendered, "Let neither of us have it."

10. See Deut. 19:16–21 and n. 11.

11. Deut. 13:6, 12; 17:7, 12f.; 19:19f.; 21:21; 22:21, 23f.; 24:7.

12. See my "On Faith and Revelation in the Bible," HUCA 39(1968): 47–49. On the general issue of the monarch's obligation to provide a deputed judiciary, and specifically on the aging King David's vulnerability on this score, see 2 Samuel 15:1–6 for the single criticism raised by Absalom of his father's performance in the royal office.

Chapter 3

1. Brevard S. Childs, *Introduction to the Old Testament As Scripture* (Philadelphia: Fortress Press, 1979) 419. Wherever possible, in connection with these exegetical essays, I shall limit myself to one or two summary treatments for bibliographic citation. In regard to this chapter, the above essay will be cited simply as "Childs"; a second, more detailed and—in my opinion—most judicious exegetical treatment (cited hereinafter as "Landes")) is George M. Landes, "The Kerygma of the Book of Jonah," *Interpretation* 21 no. 1 (January 1967): 3–31.

2. Childs, 419.

3. Ibid., 419–22.

4. Landes, 15–16: "Consequently, the form, content, and location of the psalm show it fits quite suitably within its present context, and in harmony with the succession of events immediately preceding it."

5. What makes exile worse than death (which comes to all men) is its constituting denial of felicity in the afterlife. See my "Kin, Cult, Land, and Afterlife," HUCA 44 (1973): 1–54. The fate of being denied fellowship with kin in this life and the afterlife, within one's native borders or without them, is invoked in threats by the use of *'ārûr*. See my *Problem of "Curse" in the Hebrew Bible,* SBL Monograph, vol. 13 (Philadelphia: SBL, 1963), 77–96.

6. IB ad loc.: "Jonah is represented as thinking that by a flight from the territory of Israel he will be able to escape from the presence of Israel's God. The idea was common in the ancient Near East that the power and presence of a god was localized in the territory inhabited by his worshipers (1 Sam. 26:19; 2 Kings 5:17). Jonah does not know that all lands belong to God."

Apart from the fact that there is not a scintilla of evidence for such a belief in Israel or Israel's neighbors, it would still have to have been earlier than the traditions of a God who summoned Abram from Mesopotamia and manifested his literal presence in Egypt (Exod. 8:18). The author of the Introduction and Exegesis in this IB article on Jonah is not troubled by this consideration (nor were his predecessors) in dating the book in the late post-Exilic period. Landes, noting that Jonah "is aware of the impossibility of evading YHWH's overarching power and rule (as 1:9 implies)," resorts to an explanation of Jonah's flight that in no way eases the problems of his seeking to escape that "overarching power": Jonah thinks "to escape Yahweh's *cultic* presence where the prophetic oracle is vouchsafed." Somehow, apparently, that "overarching power" of Israel's God does not include the ability to manifest his oracular presence wherever he pleases. Landes thinks to strengthen this interpretation of flight from the "cultic presence" by the observation that "only two of Israel's prophets ever explicitly received YHWH's oracle outside the boundaries of the Promised Land: Elijah at Mt. Horeb . . . and Ezekiel by the River Chebar in Babylonia." How many instances does he want? And if Horeb is outside the Promised land for Elijah, was it not so also for Moses and his people? My intention in these remarks is not to hector my colleagues but to illustrate how far and to what lengths even gifted scholars will be driven, even to creating metaliterary conventions about ancient Israel's thought system, reifying poetical constructs as realia of ancient thought and practice and, no matter how silly they are, justifying them by relegating them to an earlier (read *primitive*) period.

7. Thus in the two instances cited by IB (n. 6). 2 Samuel 26:19 is a doubly oblique expression: those who incited Saul against his loyal vassal, David, in driving David from the land where alone sacrifices to YHWH may be offered (saying, in effect, worship other gods) have deprived YHWH of a worshiper and thus incurred his wrath (see n. 5 on the term *'ārūr* in this context.) In 2 Kings 5:17 an Aramean general, we are told, thinks he can have his theological cake and eat it, too, by continuing to serve his human master on non-Israelite soil while making offerings to YHWH on a sandbox covered with soil imported from YHWH's sacred territory in the south.

8. The verb *ḥiššᵉbā* with the ship as subject cannot be translated "threatened to be [broken up]" (NEB) or "was in danger of" (JPS), or "was like to be" (AV). *A Translators Handbook on the Book of Jonah* by G. Price and E. A. Nida (Stuttgart: United Bible Societies, 1978) reports, "The verb translated *was in danger* normally has the meaning 'think, plan,' and nowhere else in the old Testament is it used, as here, with an inanimate subject. Moffatt, in fact, goes so far as to say 'the ship thought she would be broken'; but such personification misrepresents the mind of the author"—the mind of an author who in 3:8 will tell of a royal decree that man and cattle alike are to don sackcloth and call out to the gods with all vigor! See p. 76

9. This term of address, in the mouths of the seamen, totally omitted by the New English Bible and equal to calling Jonah "Mr. No-Name," is a delicious touch that is obscured in the almost-correct Jewish Publication Society rendering "you who have brought this misfortune upon us." That initial *you* is missing in the Hebrew. Had the author intended a second-person address, the Hebrew would have read *ba'ᵃšer-lᵉkā*. By the near duplication of the indefinite *bᵉšelmī* of the interrogative clause in verse 7 in the definite *ba'ᵃšer lᵉmī* in verse 8, the author draws us to more concentrated focus on the four questions that the seamen put to Jonah. Having already named him, the one question they do not put to him is "What is your name?" See 1 Samuel 17:55–58 for an analogously delightful touch centering on a name. An individual's identity lies first in his patronymic—hence Saul's question to Abner about the youngster who has volunteered to face Goliath, "Whose son is he?" This is followed by Abner's oath denying knowledge of the lad's identity—ought not a general to know the name of a volunteer he has permitted into the king's presence?—and the king's pointed command, "You ask whose son he is!" When the nameless champion is brought into Saul's presence with the giant's head in hand, the king then asks—as if he only now deigns to inquire—"Whose son are you, lad?" And David responds (omitting his given name), "Son of your subject, Jesse the

Bethlehemite''—touching and ironic modesty on the part of a stripling hero destined to succeed his interlocutor on the throne.

10. This is not the place for a full discussion of the *'Ibrī/Ḫabiru* connection. The one common denominator of the two is that each is a member of what sociologists call an *outgroup*. (The ending of *'ibrī* is no more a gentilic than it is in the ordinal numbers or in adjectives like *ḥopšī* and *'ᵃrīrī*.) The Israelites in Egypt, like their fathers before them and their descendants afterward, are never identified by this term except on the part of, or in relation to, outsiders who have no reason to know them by their patronymic associations. As in Egypt or in Canaan before and after the sojourn in Egypt, the term *'ibrīm* may include, but may not be limited to, the Abrahamitic line and, indeed, may refer to non-Israelites in contrast with this gentilic group. Such is clearly the case in 1 Samuel 13:3–4, 6–7, and 14:21–22, where the two groups are clearly separate entities. The last mention of *'ibrīm* in historical narrative is when the clearly non-Philistine contingent in Achish's regiment is so referred to by the confederate Philistine lords. The *'ibrīm* never again appear as a group or group designation. The term in the singular (masculine or feminine) is only and always in connection with a status of bound service (indenture, distrainee, or the like): the context of Exodus 21:2 is that of a nonchattel bondsman, for *'ebed* alone may be "hired servant" or "chattel slave"; while in "your/his kinsman the *'ibrī*" (Deut. 15:12; Jer. 34:14) the term stands for the bondservant even without the term *'ebed*, and the full expression for "bound servant" of either sex appears with the adjectival *'ibrī* term in both genders.

11. The failure to look for an artistic intent in the author's ordering of time sequence will result in translations that ignore the plain meaning of the Hebrew as in the New English Bible or the Jewish Publication Society. Thus, "What can you have done wrong," they asked (NEB) when they already knew that he was trying to escape or "And when the men learned" (JPS) for *kī yāde'ū*.

12. See Solomon's prayer, 1 Kings 8, particularly verses 29–53 for the two terminals: the "house" built by Solomon in Jerusalem and "the heavens—your place of residence." Jerusalem's terminal is the *channel* of communication that Israelites may exploit from outside of Jerusalem to speed their prayers to heaven—the *derek* in verses 44, 48.

13. See Josh. 22:9–34.

14. Thus the descent motif (verse 7 *yāradtī,*) as in 1:3 (twice), 1:6; the repeated reference to YHWH's earthly seat in verses 5 and 8 with the particularly ironic juxtaposition in verse 5 of the prophet (now fleeing from YHWH's Presence) when he had thought himself ousted from that Presence yet was confident that he would be restored to it; the promise of sacrifice and mention of vows in verse 10, tying in with the sacrifice and vows of the seamen in 1:16, and all this in the threat-to-life imagery of drowning, present in other psalms but never so consistently developed as here. In respect to this last point, my image of Jonah's psalm as a mosaic of stones copied from elsewhere in the Psalter, or Childs's characterization of it as "a veritable catena of traditional phrases from the Psalter" (p. 423), while true as far as they go, do not do justice to the exploitation of so many borrowings for one richly developed context. Whereas the first half of verse 3 is almost verbatim that of Psalm 120:1, only here and never elsewhere does a cry for help come from *"the belly* of Sheol. The verb for "cast" in verse 4 is the same as in Psalm 102:11 but only here with "depth(s)" (*mᵉṣūlā*) into "the heart of seas" (Prov. 30:19). The verb for "surround" *sbb* appears often elsewhere but only here with "current" (*nahar*) as subject (but see Ps. 88:18). The second half of verse 4 is lifted from Psalm 42:8, but no other verse in this last psalm features water imagery. The rare verb *'pp* appears in two other poems (2 Sam. 22:5/Ps. 18:6 and Ps. 116:3) but only here in verse 6 with *water* as subject. Only here does seaweed wreath a head like a turban. Finally, the bases of mountains in oceanic depths, borrowed from Psalm 46:3 is featured only here in verse 7 in the experience of a human sinking

into the vortex of death. The terms for worshipers of nongods in verse 9 (duplicating almost verbatim those in Psalm 31:7) is not, of course, strange in the mouth of a monotheistic prophet whose one complaint about his God is the point of the whole story.

The length of the preceding argument renders inadvisable a detailed comment on Landes's conclusions in "The Kerygma of the Book of Jonah" (see n. 1), subtitled "The Contextual Interpretation of the Jonah Psalm." I would note for my readers, however, that despite his conclusion that the psalm is borrowed in its entirety by the author of the Book of Jonah, he does affirm that "the form, content, and location of the psalm show it fits quite suitably within its present context, and in harmony with the succession of events immediately preceding it." Why not, then entertain the possibility of the story's author being the psalm's author as well?

15. See pp. 18–19.

16. It may be worth noting that in 1:10 the author achieves the same dramatic effect as here by *gapping* (and thereby compressing) the greatness of Jonah's God as Jonah declares him and the response of the sailors to this declared Power, which is then followed by the hypotactic parenthetic (almost flashback) construction. By comparing instances of the parenthetic "resumption" with instances of the synoptic/resumptive technique, we shall be able to arrive at a fuller appreciation of the possibilities opened up by the latter.

17. The humor, let us not forget, is vital to the author's purpose. It begins with the absurdity of Jonah's flight from God, continues with his sleeping through a tempest and the contrast of his behavior with that of the pagan sailors and with his spending three days and nights in the fish's belly and, in that close room, giving thanks for a former deliverance to the God from whom he is fleeing. Now the absurdity wells up again in the attribution of human characteristics to the animals—all the foregoing items (and especially the last) building up to the last two words of the book, "and much cattle." The Jewish Publication Society, therefore, is less than faithful to the Hebrew when in verse 8, it renders the reflexive *wᵉyitkassū* as a passive, "they shall be covered." The following verbs refer to these animals as well as to humans: all alike are *to cry* to Heaven and *to repent*. There is nothing anomalous in the use of the distributive *'īsh* "each [man]," but the primary meaning "man" should evoke a smile, as should the concluding image in this verse. The author could have stopped with the turning away "from their evil ways," but he continues with the pleonastic "and from their lawless doings" (literally, "the lawlessness in their hands"). The Hebrew idiom is unexceptionable, except that instead of *hands* the synonym *palms* (*kap*, literally, "hollow") is used; this substitution is thus another incongruous touch; the flat underside of a hoof excluding the functions of holding, wielding, and doing of the human hand.

18. My translation of this verse, while correct, fails to convey the artful simplicity— childishness, almost—of the Hebrew; for the author again works to use the most basic of vocabulary, to employ the word *big* again. The Hebrew reads literally, "It was bad to Jonah a big badness." The last two words of this verse *wayyihar lō* are taken by the New English Bible, as by all the older translations, as betokening "anger." The sense of anger, however, is present in this verb only when its subject is *'ap*. In the impersonal use, as here, it connotes a wide range of emotional agitation.

19. See the end of n. 18 above for the meaning of the impersonal *hrh lᵉ;* I presume that the Jewish Publication Society's rendering of the expression by *grieve* is witness to agreement on this score. Aside from the question of anger, the rendering of the adverbial infinitive of *hiph'il ṭūb* (as though it were a second-person finite tense) cannot be justified. In the use, however, of the Hebrew term for "good" in the sense of "very greatly," compare the English **hendiadys** "good and mad" and idioms to baffle a foreigner such as "as good as dead" or "A miss is as good as a mile."

20. This plain meaning of the Hebrew is missed by virtually all translations. Note in the

discussion that follows how, in this resumptive-expansive episode, the interior dialogue—expressed first in free indirect discourse, then in a free direct discourse gloss of the preceding clause—focuses on Jonah's inner state and, in contrast with the synoptic episode, does not put it into the context of a prayer; in this resumptive episode God breaks in, unasked, on Jonah's musings. The subtlety of the rhetoric is not lost on Landes. He writes, "In his distress he does not cry to Yahweh for help. Feeling totally helpless and alone, with no external source to assist him to the death he so much desires, he directs his last request, not to Yahweh, but to himself: he asks his *nephesh* ('life principle, self') to die (4:8)" (p. 27).

21. Except for a review of the author's deployment of the two terms YHWH and (*ha*)*'elōhīm*. Apart from the sailors, who immediately recognize the lordship of YHWH from the moment he is introduced to them by Jonah, the term for deity (deities, the Deity) in connection with pagans is consistently the impersonal common noun *'elōhīm*. In the case of Jonah the term for the Deity is consistently the proper noun YHWH until chapter 4. In chapter 4, all of it dealing with Jonah, YHWH is the term for God throughout the synoptic episode (verses 1–4). In the rest of the chapter, the resumptive-expansive episode, the term for the Deity is *'elōhīm* except for two exceptions: (1) the final address to Jonah is by YHWH as in chapters 1–3, and (2) in the introductory notice of the resumptive episode (verses 6–7) the subject of the verb is the hyphenation YHWH–*'elōhīm*, a compound subject that never appears in the Bible except in Genesis (2:4–3:23). Here is a clue to the poetical function of the terms for Deity, which led to the documentary hypothesis. The full resolution of this riddle will have to wait on the companion volume.

22. In this instance of the synoptic/resumptive the second episode is not merely complementary to the first. It is complementary in the sense that the bottom line of the first episode is that YHWH does not care why his prophet is upset, while the second episode explains *why* he does not care: the prophet is selfish and self-centered, not just to the point of insensitivity to the plight of fellow creatures but to the point of blindness to God's investment in creation and to the whole meaning of the prophetic enterprise, to reconcile creation with its Creator. To appreciate the marvelous effects made possible by the narrative strategy of the synoptic/resumptive, consider that if the story ended with the bottom line of the synoptic episode (as it does, in terms of narrative time), we should be left without the focused indictment of the prophet's unprophetic stance and if the sense of the resumptive episode were placed in its proper narrative time position, the bottom line of the synoptic episode would be childishly pathetic. This narrative strategy yields two powerful rhetorical climaxes: the I-could-not-care-less of master to errand boy and then the sneer of the master at his errand boy's absurd sense of self-importance, a self-importance that owes to his election to an office that he is too proud to fulfill. And so the absurdity of the prophet's flight in response to the call of 1:1 comes full circle in the absurdity characterized in the bathos of the story's last words, "and cattle, much."

23. In addition, the king's councilors are "his big ones" in 3:7, and there is the *big* in 3:1 *gᵉdōlā lē'lōhīm* (great to the gods) as characterization of the city of Nineveh. My translation, "awesome," picks up the suggestion of the supernatural in the term *'elōhīm* when used in a construct or other adjectival context (as in Gen. 1:2 *rūᵃḥ 'elōhīm*, Ps. 36:7 *kᵉharᵉrē 'ēl*; see also Hab. 3:6 *harᵉrē 'ad* and *gibʿōt 'ōlām*). There is also the appearance of *big* in 3:5.

24. Thirty years after this suggestion was made in class (Sumerian 1), I give credit to my teacher, Samuel Noah Kramer, whose playful delight in deciphering and interpreting ancient texts is brought back to memory by the spirit that breathes in almost every line of the tragicomedy we are here interpreting.

25. Or with a form of *ḥṭ'* replacing *'āwōn*. This opening of a window into biblical Hebrew took place a quarter of a century ago when I traveled back to New York from the enchantment of E. A. Speiser's classes at the University of Pennsylvania. (I often learned as much from Muffs on these train rides as I did in Philadelphia.) It appears, with grateful attribution, in my *Problem*

of "Curse" in the Hebrew Bible, SBL Monograph Series, vol. 13 (Philadelphia: SBL, 1963), 208. The far-reaching import of this window is suggested in my essay, "Kin, Cult, Land, and Afterlife," HUCA 44(1973): 36, particularly n. 57, second paragraph. For the different force of *nś* '*wn/ḥṭ*' with human subjects, see my article, "The Case of the *Sota*," HUCA 46(1975): 63.

26. It is of interest that certain experiences and activities will be expressed in different linguistic families by analogous terms or images—thus Hebrew *spr*, "to count" in the ground-stem and "recount, tell" in the D-stem (cf. German *zahl, erzählen;* French *compte, conte*). So also comprehension and understanding (German *auffassen, begreiffen, verstehen;* French, *comprendre, entendre;* Hebrew *tps, 'md 'l*). So in the case of the idiom under discussion, the notion of bearing or carrying, which is the primary sense of *nś*, appears independently in English in the sense of a creditor "carrying" his debtor rather than "dropping" him. For all this correspondence in thought and idiom, English translators continue to miss the meaning.

27. In Psalms 86:15, 103:8, and 145:8 the *raḥūm* comes first; in 111:4, 112:4, and 116:5 the *ḥannūn* comes first. Even in Exodus 33:19 the verb *ḥnn* precedes the verb *rḥm*. The point made here is that regardless of when this juxtaposition first appeared, according to Scripture's own narrative chronology the first use of the adjectives cited was in Exodus 34:6.

28. In regard to this suggestion as to the verse's insertion in the "historical" Book of Kings and the purpose of this insertion, it is important to stress that I am not suggesting that this verse's insertion is evidence for or against any early or late dating for either the Book of Jonah or all or part of the Book of Kings. This is not an exercise in the kind of cutting and pasting indulged in by source critics. It is rather in keeping with my thesis (as it will be developed in the essays on the stories from the Book of Kings) that factual historicity is far from the main concern of the so-called historical books and that they, together with narratives from outside the Book of Kings (or Samuel or Chronicles, for that matter), are constitutive of the one authorial voice that also blends legal and perceptive formulations in the Pentateuch into narrative contexts in those books—formulations that also find exemplification in narrative and in the oracles of the Writing Prophets and narratives of their forerunners.

Chapter 4

1. At about the same time as publication of the present essay in HUCA 44 (1983), there appeared R. W. L. Moberly's intensive and comprehensive study, *At the Mountain of God: Story and Theology in Exodus 32–34*. JSOT Supplement Series, vol. 22 (Sheffield, UK: JSOT, 1983). No more mature and judicious a study of a single biblical narrative has to my knowledge and judgment, appeared in recent times. Moberly is au courant with much of the poetical approach to biblical narrative and manages a remarkable balance between this approach and respectful attention to source-critical and tradition history methodologies. That balance is also reflected in his bibliography as well as in his discussions (see, particularly, his chapters 3 and 4) of form criticism, genre, oral tradition, and other related issues, such as aetiology and cult legend. Of the scholarly works on our subject Moberly notes, "That Ex. 32–34 as a coherent narrative might be significant either historically or theologically is not raised as a possible option in any of these works. It is the [general] belief . . . that Ex. 32–34 is a complex of fragmentary and conflicting traditions whose present combination makes little attempt to conceal their diversity" (p. 12). Despite this "impressive" consensus, he goes on to identify one of his aims as to investigate "whether there is not in the text as it now stands a greater degree of unity, both literary and theological, than is usually allowed" (p. 13).

2. Moberly recognizes this: "The people's request for *'elōhīm* on the grounds that Moses has now disappeared is notable in that it implies that the *'elōhīm* are a replacement, in some

sense, for Moses.'' He continues with citing as support the use of *hiphil* '*lh* for the act of leading out (Israel) from Egypt attributed to Moses in 32:1, 7 and to the calf in 32:4, 8 (p. 46). On the following page and in n. 15 he refers to the problematic parallel of the verbs *hwṣy'* in Exodus 20:2 and this use of *h'lh*. He apparently fails to realize that the former is used exclusively with God as subject while the latter has the ''intermediary agent''—Moses or calf—as subject. They are so used for good reason: the verb *yṣ'* has (in addition to the meanings discussed among **idiomatic expressions**) the sense of ''to go free, be liberated from service'' (as in Exod. 21:2–5, 7, 11). The *yᵉṣī'ā* from Egypt is thus not merely an ex-(h)odos but a liberation from bondage. This sense, therefore, of the liberating activity is reserved for the liberating Deity and denied to his agent, real or putative.

In regard to Moberly's use of a plural verb with *'ᵉlōhīm* wherever my own translation in verse 4 renders the Hebrew demonstrative as a singular (*this*) for all its plural meaning in the original, the plural of majesty (a subcategory of the plural of abstraction) must always be rendered in terms of the nuance it carries in any given context. In this one, since there is only one calf it would be sheer silliness to render it by ''these are your god.'' The capitalization of the last word is, in this context, moot.

3. Our translation ''engraved it with a stylus'' for *wayyāṣar 'ōtō baḥereṭ* is closest to the Authorized Version's ''fashioned it with a graving tool.'' The *ḥereṭ* as a pointed instrument appears in Isaiah 8:1, where it is used with the verb *ktb* (to write). The verb *yṣr* (''to fashion, shape'' as in the working of clay to make pottery) is inappropriate with a pointed tool and would, furthermore, be vocalized as *wayyīṣer*—hence our rendering the verb along the sense of Akkadian *eṣēru* (to draw, make a mark). The Jerusalem Bible's ''in a mould, melted the metal down,'' and the Jewish Publication Society's ''cast in a mould'' reflect what translators presume the text must have meant in order to produce an *'ēgel massēkā* (a cast bull) but cannot be justified by attestation for either the verb or noun here. The Jewish Publication Society's note suggesting an alternate rendering, ''tied it in a bag,'' is as well supported as mine. It would assume the verb *ṣrr* (to bind, tie) with the identical vocalization here as in 2 Kings 5:23 with a noun meaning ''bag, purse'' (which also appears in Isaiah 3:22) but vocalized *ḥarīṭ*. A play on two multivalent roots, one for verb and one for noun, would not be beyond the sophistication of the biblical author who, in a ''gapping'' stratagem (see Index) keeps us in the dark at this point as to what Aaron actually did do. (For the filling of the gap, see below in 32:34.) An analogous word play, preserved for us, like this one, by the Masoretic vocalization, is featured three times in Psalm 2: *nōsdū* ‹ *swd/ysd*, *nāsakti* ‹ verb *nsk* (to pour) and noun *nāsīk* (chieftain) and *tᵉrō'ēm* ‹ verbs *r'h* and *r''*.

4. If, on the other hand, the sense of the Hebrew is that he bound all the earrings in a sack, what is the meaning of the gap between the collection of the gold and the appearance of the manufactured bull?

5. Moberly finds that ''the words of ʏʜᴡʜ in vv. 7–10 fall into two parts'': in verses 7–8 his command to Moses to descend and his description of the people's sin and in verses 9–10 his reaction to the sin and his proposed course of action. Failing to discern a separate episode, beginning with verse 9, he notes that ''a second introductory formula (v. 9a) within the same speech functions to signal the shift of content.'' He thus overlooks that there is a discrepancy between the proposal in verse 10 that Moses stand aside and let him destroy the people and the command to Moses in verse 7 to go down (presumably to take some action), a command unaccompanied by any intent to act on his own part.

6. Note, in keeping with n. 2, that the verb for liberation/deliverance with God as subject in verses 11 and 12 is *hōṣī'*.

7. The paraphrase of *qōl 'ᵃnōt* is unavoidable. ''The sound I hear is indeterminate'' would also be periphrastic—and wooden, at that. The verb *'ānā* (often incorrectly rendered as ''answer, reply'' when no question has been put) means simply ''to make a sound, utter, speak

up, break into song, etc.'' The Hebrew here achieves its effect by contrasting the unmodified *qōl ʿᵃnōt*, which Moses admits to hearing, with the *qōl ʿᵃnōt* "given off in strength or given off in weakness," which Moses would have so identified had he heard such despite Moberly's excursus on the verb (*At the Mountain of God*, 111f.)

8. Moberly cites the frequently noted tension between 32:7–14 and the rest of chapter 32, "particularly in Moses seeming unaware of the people's sin as he descends the mountain despite his having been told by God, and in his seeking forgiveness (vv. 30ff.) as though it had not yet been granted (v. 14)." This last point will be dealt with below in regard to Episode G. In connection with both points, Moberly cites Childs's argument for "the necessity of at least some earlier form of the tradition as integral to the narrative"—yet another example of solving a problem by reference to a nonexistent tradition integrated into the present one by a mindless editor. The narrative strategies made possible by the synoptic/resumptive technique open to us in this story like the unfolding petals of a blossoming rose: it makes it possible for the author to introduce the symbolism of the first set of tablets as against that of the second (see Episode K), the breaking of the tablets in despair over the people's having broken the covenant that the tablets symbolize, and the outbreak of Moses against Aaron—none of which could follow from Moses's descending to the camp with surprise excluded by his foreknowledge of the offense. On the other hand, without Episodes B and C (themselves a single instance of the synoptic/ resumptive) we would not have the theme of these two episodes as the synoptic bottom line, which will be progressively resumed and enlarged upon in Episodes G, H, I, and K.

9. Or, if he made no incision on a gold ingot to which the rings had been reduced but rather bound the rings in a sack, we now realize what he did with that sack of earrings.

10. My translation of this verse is conjectural, as are the various standard renderings. The problems, inhering in the rare root *šmṣ* and the ambiguous connotation of *prʿ*, may reflect a double entendre along the lines suggested in n. 2. The condition of the perpetrators of the offense could have left them vulnerable to attack (as in my rendering) or "out of control . . . so that they were a menace to any who might oppose them" (as in JPS).

11. The logic of Aaron emerging from the golden calf debacle as Israel's first high priest and ancestor of that people's priestly caste requires him to be vindicated in the role he plays in that debacle, as the Levites are vindicated by their championing YHWH's cause against the instigators of the idolatry. As we saw, YHWH in Episode B blames the people, not Aaron, for the making of the calf; and in Episode E we find out why Aaron was blameless. In Episode F Aaron is somehow responsible for the condition of the people—the *somehow* constitutes a narrative gap. In Deuteronomy 9:20 Moses similarly charges the people with the sin of making the calf, while in the preceding verse—without specifying any offense on Aaron's part—he declares that YHWH was enraged with Aaron to the point of wanting to destroy him. What then was Aaron's offense? I would suggest that the offense was (for all his good intentions) that he took it upon himself to champion YHWH against the absurdity of representing him by a humanly manufactured intermediary, going so far as to collect the rings, which he would (as he thought) reduce to a blob in the fire to the consternation of the idolaters. God does not need such defenders: he can argue his own case. For a parallel, compare God's repudiating Job's friends who would have championed the justice of God at the expense of condemning an innocent Job. If this bridging of the gap smacks to the reader of sophistication of a high order, the sophistication is, be it noted, Scripture's not the interpreter's.

12. See Herbert Brichto, "The Case of the *Śōṭā*," HUCA 46(1975): 55–70. On Moses' administration of the potion to the people, Moberly writes, "The superficial similarity of 32:20 to the ordeal in Num. 5:11ff. has frequently been noted. The ritual in Num. 5 does not, however, greatly illuminate the significance of the action in the present context where the concern is to administer punishment rather than determine guilt" (*At the Mountain of God*, p. 199, n. 46). The point here, however, is that the punishment is to be meted out by the

Levites, who would need a method to determine guilt. The narrative strategy here, resorted to elsewhere in Scripture, is the reverse of gapping. Instead of a gap begging to be bridged, we have a bridge (a seemingly pointless detail) over a gap that will only appear later in the story. See, for example, Genesis 20:4, assumption that an entire nation is threatened with extinction; 20:9, attribution of the offense to the entire kingdom; and 20:17–18, revelation that the threat is not to Abimelech alone but to all his subjects, who are included in verse 7's "you and all who pertain to you" (as is pointed up in the following verse, where these subjects are summoned and informed and become duly terror-stricken).

13. The differences between the account here and Moses' description of the events in Deuteronomy 9:8–21 are not to be seen as discrepancies. The writer of these words certainly knew the account in Exodus. For example, Moses' administration of the potion to all of the Israelites is not to be taken literally. The author did not need such experience as we have had in the case of the Reverend Jim Jones in Guiana to realize the logistical problem of force-feeding an entire community. Moses' version in Deuteronomy—that he threw the dust into the brook coming down the mountainside—would seem to be a more literal description of his action: he contaminated the common water supply. Here, too, however, the symbolic overrides the literal; for the fine gold dust would have been swept downstream long before an atom of it could have been guaranteed lodging in the innards of every single Israelite. Note also that the second period of forty days on the mountain is in Deuteronomy the length of time required by Moses to win God over to sparing Israel, whereas in the Exodus account both the sparing of Israel and yhwh's agreement to accompany them is achieved in a few brief passages of dialogue. See also, n. 11.

14. The term *brk* (bless) as an antonym of terms connoting disaster, threat, danger (hence protection or immunity from such) is clear in such contexts as Deuteronomy 29:18 (cf. JPS) and Judges 17:2. Contexts parallel to our text are Genesis 15:1, where Rashi associates God's promise to shield Abram with the possibility of vengeance for the deaths he inflicted in the war described in the previous chapter, and Numbers 25:10–13, where Phineas is promised God's friendship for his act of zealousness that exposed him to vengeance on the part of his victims' kin. In both cases, as here in Exodus 32, the promise is both protection from harm in the near future and reward extending to the protagonist's posterity.

15. From a rhetorical point of view the synoptic/resumptive technique here achieves two effects: it highlights the promptitude of the Levites' response to Moses' call (no intervening words on Moses' part in explication of the niceties of their motivation and action); and by a separate treatment of Moses' charge, it emphasizes the role of the Levites as agents of God and points to the implicit reward (once again, an instance of gapping, to be bridged later, beginning in Numbers 3), namely, the role to be played by the non-Aaronide Levites in the cult.

16. See pp. 184–85.

17. Note "arrived at," not (as JPS) "entered." The verb *bō'* often means "to enter," but that it does not do so here is guaranteed by the cloud pillar's station at the tent's entrance; otherwise, we should have a picture of Moses within the tent, sharing its interior with young Joshua while the Deity, whose symbolic residence the tent is, stands outside.

18. When the elaborate tabernacle (*miškān*), or Tent of Encounter, is made, it will be within the Israelite encampment in a positional sense, with three tribes encamped to the north, three to the south, three to the east, and three to the west. But in a constitutive sense, it will be "outside the encampment" in that it is sacred just as the areas beyond the tribes will be "outside the encampment" in that they are profane. This allows for a play in the books of Leviticus and Numbers on the significance of yhwh's Presence in Israel's midst in one sense and nonpresence in its midst in another sense. An additional function of this digression is to introduce the "face to face" metaphor that characterizes yhwh's way of communicating with Moses (as also in the same metaphor but with a different preposition, with all Israel, in

Deuteronomy 5:4) in anticipation of the altogether different metaphor of "seeing YHWH's face" in Episode J.

19. Moberly, who rightly sees the motif of YHWH's presence in the expressions of "face" and in the tent or tabernacle imagery, also correctly observes that *'et 'ᵃšer tišlaḥ 'immī* in 33:12 "is not specifically personal and could equally well be 'what you will send with me'" (*At the Mountain of God,* 69).

20. See n. 19 above.

21. The consonantal text indicates that *way* is singular while the vocalization treats the noun as plural: *ways.* Either reading is in consonance with our interpretation.

22. For a fuller discussion of this verse, see my article, "On Faith and Revelation in the Bible," HUCA 39(1968): 46, n. 14.

23. The narrative of the theophany at Sinai; the revelation of the divine norms (synoptically in the Decalogue and expanded afterward); and the roles of Moses, Aaron, Joshua, Nadab, and Abihu and the seventy elders in ascending to various elevations on the mount—all these events must also be read in terms of the episodic narrative technique and structured in time in accord with the synoptic/resumptive narrative form. For the present, let us note the vision of the God of Israel beheld on the mount by Israel's representatives in Exodus 24:9–11. Whatever the function of this episode in the revelation and covenant-making narrative, it is not to be assigned to an author other than that of the golden calf narrative or to be read as a contradiction of the theme here that no one can behold God and survive.

24. My rendering of *wᵉḥannōtī 'et 'ᵃšer 'āḥōn wᵉriḥamtī 'et-'ᵃšer 'ᵃraḥem,* for all its departure from the standard translations, requires no justification. I would stress, however, that the nuance I read in this sentence is not to exclude other nuances better expressed by other renderings. This sentence is an expansion of the benevolence (*ṭūb*) that YHWH will parade before Moses, a benevolence that is a general attribute of God's and relevant to many contexts; but in this particular context YHWH's benevolence to Israel is in fine focus.

The multivalence and ambiguity of the imperfect tense in biblical Hebrew is often exploited by Scripture's authors to make a statement in a broadly inclusive sense even while it is addressed to a particular context. Parallel to the sentence that is the subject of this note is the crux in Exodus 3:14, the "name" of God, *'Ehyē 'ᵃšer 'Ehyē.* It conveys all the following senses: "I Am What I Am, I Am What I Was, I Am What I Shall Be, I Was What I Am, I Was What I Was, I Was What I Shall Be, I Shall Be What I Am, I Shall Be What I Was, I Shall Be What I Shall Be." The verse goes on to say that for the Israelites the name 'Ehyē (I Am, I Was, I Shall Be) will suffice. And in the very next verse, in a switch to the third person, the name of this God in all its ontological excess is YHWH (suggestive of Yihyē, who, for all his universality, is, in this context, primarily the God of Abraham, the God of Isaac, and the God of Jacob. The play on the verb *to be* in the *qal* tense does not rule out the possibility that the tetragammaton derives from the *hiph'il* of the verb *to be;* but the shading of the vowel in the imperfect prepositive, the shortening of the tetragammaton to Yā, Yō, Yē, and the play here on the *qal* stem all suggest that little confidence is to be vested in the now general assumption that Yahweh represents a recovery of the name of Israel's God.

25. The parallel texts from Exodus 34 and Exodus 23 will be found on pp.119–21.

26. The New English Bible's rendering is, "All the surrounding peoples shall see the work of the Lord." Such a departure from the text of the original, a text posing a problem for interpretation but not for translation, would be unconscionable even in a version that frankly purported to be a paraphrase—and this without a marginal note to the unwary reader!

27. The metaphor, so baldly put, would seem to be so commonplace as to be congenial even to the most agnostic of humanists. Yet this metaphor reemerges in the extension of the "Old Testament," and to it is traceable the schism among the biblical religionists of the

household of Israel. This reemergent metaphor, taken literally, will be blasphemy to some of the latter and the quintessence of God's revealed truth to others.

28. This verse and its meaning, for all its appearance in a context full of knotty problems, is quite clear. In the preceding verse (13) Jacob's fleeing to Aram and there "keeping, tending [sheep]" in payment of bride-price is the penalty for his attempt to supplant his twin brother Esau, the attempt that began in their mother's womb (verse 4). This penalty is further justified (says the prophet in verse 14) by consideration of Jacob's contentious nature and his preposterous relations with the numen (*'elōhīm*) with whom he struggled for mastery in verse 4b. This numen with whom he struggled, he tells us further (in verse 5) was an angel (*mal'ak*) over whom he prevailed to the extent, indeed, that the angel was reduced to a weeping plea for the mortal's favor. This numen–angel is then identified not with the numen over whom Jacob prevailed at Peniel (Gen. 32:25–33) on his way home from Aram but with the angel standing in for YHWH at Bethel (Gen. 28:10–22, 35:1–8). Yes, continues the prophet (in verse 5), it was at Bethel that he (Jacob) would meet with this numen and there that he (the angel–numen) would hold converse "with us," that is, the collective Israel descended from Jacob—this despite the knowledge we have all been given (verse 6) that "YHWH, God of Hosts—YHWH [only] is his name (*zikrō*)." The idolatrous context of chapter 12 (11:2, 13:2), the kissing of calves in 13:2, the reference to Moses not by name but as "a prophet" in 12:14—all this makes it rather clear that Hosea knew Exodus 32–34 and fully understood it along the lines of my exegesis in this essay. The freedom with which Hosea allows himself to flit from the persona of father Jacob to that of today's collective Israel, back to Jacob, and then back to the people Israel and to blend the narrative themes of Jacob's biography—including his traffic with angelic numina—with the theme of Moses, the prophet who rejected an angel–intermediary and thus became himself the sole stand-in for God on the journey from Egypt to the promised land (see also the discussion of Episode I) should teach us much about the authorial voice that speaks in Scripture even if the writings embodying that voice may date a century or more apart.

Chapter 5

1. There is no question, to my mind, that these last words of God's in this book of prophecy constitute a threat. If Elijah's message will go unheeded, God will deliver a blow with *ḥerem* force. The parents and children in this verse are one and the same, the generation that Malachi is addressing. If, as posterity of former generations, they do not look back to the punishment of their ancestors and if, as parents, they do not look ahead to what their children will suffer if the parents persist in declaring that it is fruitless to serve God (see Mal. 3:14), on receiving the warning that Elijah will bring, then . . . Elijah's oracles are all of doom. There is no context in Malachi's prophecy for reconciliation of parents and children, as there is no context of alienation. The idiom *sīm leb* in biblical Hebrew is to "have [something] in mind" and is usually neutral as far as emotional states are concerned. It is the heart as the seat of the emotions in English idiom (more in terms of love than hate) that lead us to read reconciliation into this passage. (Incidentally, the heart as the seat of hatred or grudge is in Hebrew normally in the context of harboring an emotion within that one does not want to bring out into the open. See p. 44 for this meaning in Leviticus 19:17.)

2. Anyone who has lived through a year of drought in the land of Israel knows that the merism for any precipitation whatsoever is hyperbole. Even in the perennial drought land of the Negev the Nabateans raised crops by ingenious techniques for harvesting the dew. The opening of this narrative with a spoken dialogue (or rather, monologue); the prophet's oath, staking his life on his confidence that God will not grant rain save by his prior announcement; and the

hyperbole of this literary figure—all point (as the author must have intended) to a narrative crafted with an eye to artistic and rhetorical effect rather than to mere chronicle.

3. "What's left," literally just "it," a pronominal object suffixed to the verb with the flour and oil as the antecedent referrent.

4. "Wait for death," literally, "and we shall die." My translation is not less faithful to the Hebrew, whose verbs are rich in the variety of modal meanings they can express.

5. The woman's hyperbolic charge is thus clearly another instance of **oblique expression.** The prophet's presence—yet, for all the innocence of its intention, serving as catalyst to evoke punishment—is pronounced by her as if actually intended to bring about that result. But, one may ask, was the prophet's presence in her household only incidentally related to her son's falling ill? The whole point of this episode, in combination (as we shall see) with the two preceding ones is to elicit the woman's charge, the response of Elijah to that charge, the response of YHWH to Elijah's plea, and the woman's response to the restoration of her son. In the final event, then, her charge, for all its **obliqueness,** turns out not to have been hyperbolic at all. The ability to achieve such intricacy in the seemingly artless **dialogue** of a simple housewife is nothing short of breathtaking.

6. The Hebrew *'ad 'ăšer* "to the point of . . . no breath" is exactly equivalent to our own "I'm all out of breath," a pronouncement that belies both death and literal breathlessness.

7. Herbert Brichto, "On Faith and Revelation in the Bible," HUCA 39(1968): 40–46.

8. Other indications of figurative expression, and of symbolism rather than verisimilitude of plot are (1) the casual way of introducing Jezebel's campaign to exterminate the clergy of the One God of her adopted country (as though the fact itself were a matter of common knowledge) in connection with Obadiah's measures to thwart her campaign; (2) the echoing of the term for extermination, *hakrīt,* in Ahab's words (verse 5) to Obadiah (my translation, rendering the active transitive verb as a passive, is not faithful to the Hebrew, which reads, literally, "lest we be exterminate of cattle"); (3) Jezebel's being charged here with this murderous campaign, as though she needed no leave from her royal husband to persecute his subjects (later Elijah himself charges Israel, not Jezebel with responsibility for this mass slaughter, and Ahab is never charged with even passive complicity); (4) the maintaining of a hundred (round number) succored prophets in two caves by a servant of the crown who, apart from the expectation that he should have had to play a part in the royal campaign, could hardly have carried through the logistics of caring for two communities of hunted fugitives without report of it reaching the ears of the queen; (5) the deliberate contradiction of the success of this rescue effort (as cited earlier) in Elijah's future claim that he is the sole surviving prophetic champion of YHWH's; and (6) the lack of precedent in the ancient world for a religious Kulturkampf featuring any bloodletting (the most extreme precedent in the Western world is the dismissal of thousands of clerics from their benefices—and the judicial murders of a few—in seventeenth-century England). In connection with the last item, it is of ironic interest how often these stories of struggle against Baal worship and the internecine royal conflicts of Judah and Israel were cited in self-justification by the extremist Whigs and Cavaliers, Puritans, Anglicans, Catholics, etc; (7) the power of Ahab to coerce all the neighboring kingdoms to search for a single Israelite fugitive.

9. Students of literary criticism, mindful that the articulation of such categories as "voice" and "point of view" is traceable to writers in the last century such as Henry James and Percy Lubbeck need to remind themselves that these techniques were uncovered or discovered, but not invented, by them. They were known to, and deployed by, such sophisticated storytellers as Homer and, perhaps half a millennium before, the nameless Babylonian genius who authored *The Gilgamesh Epic.*

10. See nn. 12 and 13 on the use of a third-person active verb where the sense calls for a passive (JPS, "that was given to them"). As translated literally here, the *he* can refer only to Elijah, who did indeed grant it to them in that he is pictured as the summoner and host of the

convocation, who grants his guests first choice. So, too, the altar to Baal is one that "he had set up." A similar problem featuring the indirect object of a verb rather than the subject is indicated by the contents of the brackets in our translation of verse 24 above. The address, in context, to Israel rather than the Baal prophets may be a subtle way of lumping together both parties in the anti-YHWH camp.

11. That this is clearly the narrator's intent is confirmed in the stilted Hebrew of verse 16, where the first *liqra't* is inappropriate in context. The verse reads literally, "Obadiah went toward Ahab. He told him. Ahab went toward Elijah." Thus, Obadiah is to Ahab as Ahab is to Elijah.

12. We are given no idea of the depth of the trench, perhaps because even one only a few inches deep in thirsty soil would have soaked up an enormous amount of water before puddles began to form. One measure (Heb. *sĕ'ā*) of seed, equal to about twenty-five pints or pounds would have sufficed for an area of at least a few hundred square feet. The hyperbole of the heat of the blaze is thus almost matched by that of the labors of the bucket brigade required to bring up to the altar (from what drought-defying source?) enough water to overflow the trench. Note again (as twice in verse 26; see n. 10) the active "he filled" rather than (as JPS) "was filled."

13. Here, again, we are confronted by the metaliterary consideration of belief in miracles—more specifically, the question raised on pp. 127–29, of the extent to which faith is aroused, or stimulated, or deepened by the experience of what is perceived as miraculous and whether, in this regard, the experience of people in biblical times differed from that in our own day. When we speak today of the miracles performed in the treatment of disease by antibiotic drugs or transplant surgery, we are aware of our indulging in metaphor; and the prayers we offer are not to Aesculepius or Apollo after the successful treatment but to the God who may or may not respond before the treatment. It would seem that nothing has changed if we judge by the woman of Zarephath. She had faith before the miracle of cornucopia, and that miracle drew no comment from her; but she broke into expression of the depth of her faith when a child carried up to an attic critically ill was brought down a while later, having weathered the crisis. So too, here, we must ask ourselves if tales of the miracles performed daily by fakirs in the Indian subcontinent incline none of us to Hinduism, why should such a tale in antiquity have moved the faithless to faith?

14. In this opinion I am pretty much alone. Modern scholarship seems to agree that there is no reason to doubt the historicity of this tale, at least in the mind of the historian–author. The extermination of Jezebel's spiritual retainers was only seen as evening the score for the elimination of YHWH's prophets—tit for tat, so to speak, on a heroic scale. Unless I seriously misread the religious situation today, such a view seems equal to the ascription to the ancients of a seriousness in matters of formal religion that is equaled today only in concerns for political and economic predominance and racial purity and (to a somewhat lesser extent) in the partisan displays of loyalty that follow triumph and defeat in international soccer competition.

15. If Ashera was the name for a female consort ascribed to YHWH (as in the eighth-century inscription found at Kuntillet Ajrud), the fleeting reference to Ashera prophets may be further indication that what these stories characterize as apostasy from YHWH (or, more accurately, worship of Baal-deities alongside YHWH) was aimed at attribution to YHWH of qualities and imagery proper to pagan deities but loathsome to him.

16. Nor does she ask him of his response to such treatment of her retainers. And if she learned of his celebratory banquet following hard upon the mass slaughter, her pagan lack of sensibility explains her failure to register horror: she, in his place, would have done the same.

17. In the two narratives, Jonah 4 and the present one, we have eloquent instances of the synoptic/resumptive technique. What marks these two apart from other occurrences of this technique is the **verbatim repetition** of **dialogue** in both. It is of interest to note the variations

in the use of this dialogue. In Jonah, the first "Are you really so upset?" is the bottom line of the synoptic episode: YHWH says he could not care less. The second "Are you really so upset" begins YHWH's questioning of Jonah, and the same bottom line is underlined in YHWH's sardonic retort to Jonah's artless confession that he is so upset. In the present Book of Kings story, God's inquiry about Elijah's presence and Elijah's response introduce verbatim the separate answers of God to Elijah's implicit complaint and resignation: (1) that the noise attending YHWH's self-revelation, for all its triumphal note, marks the beginning of history, not its end and that YHWH's patience is unruffled by the impatience of his prophet (be it Elijah here or Jonah in Nineveh); and (2) YHWH accepts the prophet's terminal service, provided (as we shall see) that there are other avatars to represent him in the continuing march of events.

18. See my "Kin, Cult, Land, and Afterlife," HUCA 44(1973): 1–54, esp. 29–32.

19. There is no reason to believe that *gan yārōq* (green garden) is a reference to a vegetable plot as opposed to a grainfield or vineyard. The intent is more likely to be a pleasure park, thus emphasizing Ahab's whimsicality by the withdrawal from cultivation of a fruitful vineyard and converting it into a shady retreat for royalty's ease. As for the title king of Samaria, this anomaly is noted by modern commentators; but I know of none who offers a meaningful explanation.

20. See n. 18.

21. Thus, the words in the mouth of Naboth stress piety as that which dictates his refusal; the narrator, in repeating that piety motif as he gives us the internal dialogue, stresses that piety as itself additional provocation of the king's rage; and the absence of that motif in Ahab's report to Jezebel—a report that in all other respects is accurate—points to its presence (and its function) in Ahab's thought.

22. Note this rendering of the repeated (see verse 8) verb in **paratactic syntax.** If we followed Alter and Sternberg in rendering the waw as "and," we would have her writing the documents twice.

23. See my *Problem of "Curse" in the Hebrew Bible*, SBL Monograph Series, vol. 13 (Philadelphia: SBL, 1963), 127–65. This crux in our Tale 4 is discussed in detail, pp. 159–64. Note also there the discussion on the force of *hayyōš^ebīm* (the magistrates) instead of the rendering still current, "who lived in the same town with" (so, still, JPS), which would indeed constitute a pointless pleonasm.

24. Sternberg (PBN, 408–10) sees in the difference between the formulation of the report of Naboth's death to Jezebel and the formulation of her report of that death to Ahab, a euphemistic purpose on the queen's part: "Having already done the dirty work on her husband's behalf, Jezebel continues to spare his tender conscience by watering down the brutal, 'Naboth has been stoned and he died' into the generalized 'Naboth is not alive, but dead'." Needless to say, my own interpretation is in disagreement with Sternberg's. My point in citing this, however, is not to underline what is essentially an interpretive quibble but rather to show how disagreement with a literary critic one respects underscores one's sharing of methodology with him. Sternberg's discussion of this narrative is in his chapter 11, seventy-five pages of insightful analysis and classifications entitled, "The Structure of Repetition: Strategies of Informational Redundancy."

25. Notice, also, two fine touches in the storytelling. First, verse 19 gives God's words to Elijah, the very words he is to pronounce to Ahab. Verse 20 gives Ahab's answer to these words—but to Elijah. This telescoping serves to merge the identities of God and his prophet and makes it seem as though—in the mind of Ahab—it is indeed the mortal messenger and not his all-seeing Master who has found him out. This same blurring of the line between the principal and agent as speaker occurs between verses 20 and 21, where with no interruption in the dialogue the speaker of the words at the end of verse 20 is Elijah, while the speaker of the words beginning verse 21 is YHWH. Second, God's pronouncement on Ahab is limited to him

and his line, consistent with verse 19. Verse 23, however, telling of the pronouncement of God against Jezebel, is in the voice of the narrator (not, as the Jewish Publication Society, either God or Elijah). It is thus verse 23 which begins the narrator's aside, one in which he speaks of Jezebel's exposure as carrion after her death, continues with the same fate for all of Ahab's kin (including Jezebel as well as Ahab's sons) and then goes on to trace the exceptionally abominable apostasy of Ahab to Jezebel's incitement, an apostasy that is not characterized as worship of Phoenician Baal but of the no-gods of the long extinct Amorites.

26. In 2 Kings 9, where these events are narrated, the events take place in Jezreel as though this were a city like Samaria and not an entire valley. Ahab's son Joram comes out of this palace or city precinct to meet his usurper–assassin in Naboth's field where he dies with an arrow in his back as he seeks to flee from Jehu. Nevertheless, Jehu commands his lieutenant to cast Joram's body into the field of Naboth. In this chapter the territorial name Jezreel appears seven times and Naboth twice, each time identified as the Jezreelite. As if to underline the metaphor, in chapter 10 Jehu sends to Samaria, to the chieftain–elders of Jezreel(!) to send him the remaining seventy sons of Ahab so that he may massacre (and expose) them along with their mentors and tutors in that field or plot.

27. Those readers who would take the third captain's hitting upon the same metaphor as employed by the narrator himself as an improbable narrative strategy and therefore an argument against an intended metaphor should be reminded that the narrator is responsible both for the words he writes himself and those he puts into the mouth of his characters. Such readers must also ask themselves, If the third captain knew of the vaporization of his predecessors, how did he come by this knowledge and how and why was it withheld from the second captain?

28. Actually there is a difference between the words in the first statement of the oracle (addressed by the angel to Elijah) for him to relay and the repetition of it by the courtiers to Ahaziah, as presumably addressed to them by Elijah. In the first address the rhetorical question about the availability of an oracle in Israel is addressed to the messengers, while the prophecy of nonrecovery is addressed in implicit apostrophe to the king who sent them. Thus, the messengers are, in effect, being charged with the responsibility for having accepted the mission without remonstrating with their sender. This charge—addressed to them—as well as the menacing authority of a spokesman for their God who appeared out of nowhere with full knowledge of their mission and a ready answer to the question they had as yet no opportunity to ask sends them packing, back to their master. In the report of these messengers to Ahaziah, the prophecy and the rhetorical question preceding it were both to be conveyed by the messengers returning from their aborted mission. In the third communication, Elijah's words to Ahaziah, the message begins with the ominous reason for the pronouncement to come: "Because you sent messengers to Baal-zebub, god of Ekron." This is then broken off for a sarcastic insertion of the rhetorical question, and then comes the sentence of doom introduced by *lāken*— altogether a virtuoso performance by the author in manipulating free direct discourse and point of view in dialogue, shifting from three explicit speakers and a fourth implicit one: YHWH's angel, Elijah (implicitly), the messengers, Elijah. This also explains why the angel is introduced to begin with; for otherwise we would have only Elijah, messengers, and Elijah as the three voices. Without the angel of God delivering to Elijah the message that he *presumably* retailed verbatim to the messengers, the narrator would, to achieve the same effect, have had to expose the discrepancy between the address of Elijah to the messengers and their report of that address to the king blatantly. The ambiguity created by the gap between communication 1 and 2 is what makes the rich nuancing possible. Another instance of purposeful ambiguity in the interest of nuance is the use of '*ōtō*, the third-person masculine accusative (him) instead of '*ittō*, a preposition plus the dative (with him). But the '*ōtō* may also be used in an anaphoric sense (as for him). Thus, verse 15 may well—perhaps preferably—be translated: "The angel of YHWH spoke to Elijah, 'Go down. As for him, have no fear of him.' He promptly went down [to] him,

[that is,] to the king.'' This subtlety, too, is made possible by the introduction of the angel. (The careful reader will have noted how much this latter item depends on the masoretic vocalization of the consonantal '*tw*.)

29. Thus the Jewish Publication Society: ''When the Lord was about to take Elijah up to heaven in a whirlwind, Elijah and Elisha had set out from Gilgal. Elijah said to Elisha, 'Stay here'.'' The inchoate or inceptive modality of finite verbs in biblical Hebrew occurs frequently, although I am not so sure about this sense in other tenses. In any case, the problem is not with the infinitive *to take up* but with the pluperfect tense *had set out*. It cannot be justified. Nor does this translation solve the problem that led to this solution, for the following command of Elijah to Elisha would leave the latter sitting by the roadside. The New English Bible's similar rendering is open to the second objection as well. And neither translation addresses the singular verb.

30. See J. A. Montgomery, *The Books of Kings* (New York: Scribner's, 1951), 353f. And so another Gilgal is posited and the search for it adds to the gaiety of nations and archaeologists.

31. Perhaps symbolizing, in the emphasis on two sections, that even as he has been granted the double portion, so, too, is he now two people in one: himself and Elijah.

32. It is rather remarkable that virtually all the modern translations into English fail to translate the words '*af hū*' (yes he, even he). The question without these two words might be puzzling enough, but these two syllables bring out the full force of the question's note of daring, triumph, even presumption.

33. That ''Israel's myriads of thousands'' is in apposition to YHWH is concealed from the reader in translation, for all the renderings supply the preposition *to* before ''Israel's myriads.'' Although I came independently to an understanding of the apposition in Numbers 10:36 as a parallel to ''my father, my father, Israel's horse and chariots,'' I have learned since that my friend and colleague Matitiahu Tsevat published this very conclusion (see *''Studies in the Book of Samuel,''* HUCA 36[1965] 49–58) and that both of us were anticipated by Obadiah Sforno in his commentary on 2 Kings 2:12.

Chapter 6

1. Montgomery cites approvingly the following remark of Skinner in support of his own observation that these two chapters stand ''singularly alone in style and novelty of contents'': ''[These chapters] are written from a political rather than a religious standpoint, and exhibiting the character and policy of Ahab in a much more favorable light than is the case in chs. 17–19 or 21.'' The second part of the statement is unquestionably true, but just what ''the political standpoint'' of these stories is lies beyond my grasp. In general, the authors of Scripture included nothing of a political, social, psychological, edificatory, or entertaining nature except as it expressed a significant truth constitutive of, or at least apposite to and concordant with, their religious standpoint.

2. The same phenomenon will be noted in chapter 22, where our royal protagonist is either *''Israel's king''* or *''the king''* twenty-five times and Ahab four times. It must be stressed that the mere usage of a name without a title, such as ''the king, the prophet'' and so on, is not in itself an indication of deprecation. Such people as Moses, Samuel, David, and Elijah are each one of a kind and their names speak for themselves. But the contrast and contexts of usages in regard to Ahab and Ben Hadad place the significance of these usages beyond debate.

3. These occurrences are in 17:2, 8; 18:1; 21:17, 28.

4. Namely, 2 Kings 1:3, 15. In chapter 19, YHWH's angel appears twice to Elijah, and we have *wayyō'mer* YHWH (and YHWH said) a number of times; we do not include these instances

because they are not related to (and hence have no bearing on) the function of such notices as relates to prophetic missions where the dispatch by YHWH is made explicit.

5. See chapter 5 for Elijah's appeal to YHWH to vindicate his claim that "by your command alone have I done these things" (1 Kings 18:36). In the Writing Prophets, of course, this emphasis appears many, many times—indication that one of the crucial questions for prophets and audience in various existential crises was whether indeed the prophet was speaking for YHWH or perhaps only imagined that he was.

6. As in chapter 17 (discussed above in chapter 3, Tale 1). The conformance in structure between the narratives in chapters 17 and 20 is remarkable. In both cases, there are three episodes. The first episode takes place in one locale, the second and third episodes share another locale. The denouement in the third episode emerges organically from the situation in the second episode. At the core of both second and third episodes there is a common phenomenon—miracle in chapter 17, prophecy in chapter 20—with a pointed (if not altogether clear) moral in the third episode and a questionable significant moral in the second episode. The same phenomenon is at the core of the first episode, which, contributing its piece to the total puzzle, is essential to the understanding of the whole.

7. See Jay Holstein, "The Case of the *'îsh hā'elōhîm* Reconsidered", HUCA 48(1977): 69–81.

8. See 2 Sam. 11:27, 12:1–13.

9. The Talmudic formulation of this bit of universal folk wisdom is *shev we'al ta'aše* "Sit tight, do not act."

10. To reject this conclusion without further ado is easy enough, and many will doubtless choose such a course. But a serious rejection would require the putting forward of an alternative explanation of the two pronouncements and arguments as soundly based on Ancient Near Eastern traditions and biblical psychology. I suspect that many who will opt for the first course will do so for various reasons, all of which add up to the same thing: questioning whether the narrator really knew what he was about. To such questioners I would raise (again) the question: Why the preoccupation with searching for any sense at all in the compositions of an inept writer—be that writer a sober historian or an imaginative producer of figments? For those who find themselves with a growing sympathy for the poetical analytic enterprise, I would recall from memory the remark of G. K. Chesterton that Becky Sharp, the heroine of *Vanity Fair*, drank in secret, but Thackeray (the novel's author) did not know it. This deliciously wicked observation was intended facetiously—whether in whole or in part, I am not sure. Superb critic that Chesterton is, he knows that whether or not poor Becky tippled in private is a matter of historic fact: if we can ascertain that she did indeed do so, it is only because her biographer has provided us with ample clues.

11. See my "On Faith and Revelation in the Bible" HUCA 39(1968): 49, esp. n. 22, dealing with the prophetic confrontation in our next story, 1 Kings 22.

12. The poetical consistency and integrity of the prophetic tales in chapters 20 and 22 in the larger framework (not *cycle*) of the Elijah tales is nowhere better attested than in how Elijah's words to Ahab in 21:19 are fulfilled in this verse (22:37). As we pointed out in our discussion of the former passage, the dogs are to lick Ahab's blood in *retribution for*, not *in the place that* they licked Naboth's. Thus, the author was anticipating not the exposure of Ahab's corpse (which never happened) but this metaphoric disrespect for the royal blood, which in the final event is reinforced by the additional metaphor of the whores' bathing in it. It is out of no desire to hector my contemporary colleagues, or my predecessors that I urge them to compare my approach on this passage with that revealed in such commentaries as Montgomery's (ICC). The confidence with which geneticist critics determine what is primary and what is secondary in the text, "correcting" the Hebrew on the basis of the Greek translations (or despite them),

and divining a "formal archival note once preceding v. 39" that has been replaced by the Hebrew (verses 37b–38), bespeak an unearned respect for the achievements of redaction criticism and an unintended but essentially deep and patronizing disrespect for the only Scriptural text we have, which we so pretend to honor with the compliment of our assiduous "research."

13. Neither Gray (*I and II Kings* [Philadelphia: Westminster, 1970], 59) nor Thiele (*Mysterious Numbers of the Hebrew Kings* [Grand Rapids: Eerdmans, 1965], 66) sees any real problem in having the battles of Qarqar and Ramoth–Gilead taking place in 853, the year of Ahab's death. Thiele, however, deems it necessary to render such a reversal in relations between Aram and Israel somehow plausible. D. N. Freedman ("The Chronology of Israel," in *The Bible and the Ancient Near East*, ed. Roland de Baux [New York: Doubleday, 1961], 210) opts for 850 for the year of Ahab's death.

14. See 1 Samuel 28:6 for the clear statement legitimizing these three divinatory practices. For the significance of the following story's featuring of necromancy, see my "Kin, Cult, Land, and Afterlife," HUCA 44(1973): 4–8.

15. The operation of the Urim and Thummim in terms of these three possible responses is a conclusion arrived at by the cumulative and supplementary details of a number of narratives, particularly in the Books of Samuel. Among these are 1 Samuel 10:17–24 and 14:36–42. The Septuagint reading cited in the Jewish Publication Society on 14:41 is not a basis for this conclusion; nor is the assumption of S. R. Driver that this Greek reading represents an original reading lost in the Hebrew by haplography any more convincing than the possibility that the Septuagint is providing (as so often, elsewhere) an explanatory gloss. See S. R. Driver, *Notes on the Hebrew Text of the Books of Samuel* (Oxford: Oxford University Press, 1913).

16. The detail pointing to this conclusion is the anomalous mention of Prince Joash along with Amon, mayor of the city, in verse 26. Why the need for two responsible wardens or, for that matter, the entire detail of Micaiah's being locked up until the outcome of the war? And what indeed was the decision of Prince Joash on Micaiah's fate after his father's death?

This last question relates to the solution of the problem of verse 29b in our text, which I did not include in my translation. In parataxis to Micaiah's last words to the king of Israel are the words, "He said, 'Hear, O peoples, all of them!'" This invocation, identical with that in Micah 1:2, has been duly noted by scholars. Montgomery characterizes these words as "a gloss . . . identifying Micaiah with the canonical Micah" (ad loc.). The author of this narrative knew as well as we do that Micaiah is prophesying about a century before the canonical Micah the Morashtite, whose survival of a disagreeable oracle delivered to Judah's king Hezekiah (two centuries almost after Ahab's death) is told in Jeremiah 26:10–24. Here again, we see the one authorial voice of Scripture, spanning centuries of historical time.

Chapter 7

1. Which of the stones of a city's wall (or gate or palace-administration center) qualifies for the label *foundation stone?* If reference is intended, for example, to the massive slabs supporting major gateways, archaeological practice is to dig around and below these to discover on what they rest, and one would never expect to find a skeleton where one never looks to begin with. Further, would the sacrificial victim have been slain on an altar and removed to the foundation site for burial, or what? We would not labor the point were it not that this category of "human sacrifice" is typical in its revelation of the kind of evidence and use of logic that has been mustered to establish other kinds of human sacrifice in antiquity. And the persistence of the fiction! The Jerusalem Bible, for example, which renders the preposition as "at the price of," nevertheless adds a note, "His two sons were slaughtered as a foundation sacrifice." Con-

sider the contradiction in this explanatory note: the laying of the foundation is the beginning of the building enterprise, the hanging of the doors is the end. How would the younger corpse be made to join its brother's under the foundation? So much for foundation sacrifices. They are a myth in all the negative—and none of the positive—senses of the word. And the persistence of the myth should serve as a warning to every student of cultural history. Every construct inherited from the researchers who have preceded him must be subjected to critical scrutiny; every "fact" of the human past must be reexamined to see whether it does not cloak a fiction of history or derive from a purposeful fiction which has been misread as a historical datum.

2. It is vital that we do not lose sight of the narrative facts in this tale. Elisha is given no efficacious role in the ecological change. He is merely the herald for yhwh's action. Similarly, there is no attribution of efficacy to the salt from a new bowl, it is a symbolic gesture of (as we shall soon see) rich metaphorical meaning. On the subject of metaphor, particularly as related to idiomaticity, let us note that the term *rp'* (*pi'el*), "to heal," is no more proper here than it is in Elijah's "healing" of yhwh's "overthrown altar" on Mount Carmel (see 1 Kings 18:30). Death-dealing waters are detoxified, not healed. But, in both instances, the very impropriety on a literal level makes for the power of the metaphor.

3. An indication of just how uncomfortable this story is for translators is the comment in a note in The Jerusalem Bible on 1 Kings 20:37. On the story of the soldier killed by a lion for his refusal to strike a prophet (see chap. 6 above), the moral of that story is drawn and evaluated: "All who disobey the word of God or a man of God, even for good motives, will be punished. This idea is not perfect and is not that of the great prophets, but it reflects the mentality of the ancient prophetic communities." Thus, the Jerusalem Bible recognizes the problem of inappropriate divine action in this two-verse tale but finds the one before us as altogether too painful for comment. And, I may add, rightfully so.

4. See 2 Sam. 17:8, Hos. 13:8, Prov. 17:12. Note in the Samuel verse the meeting "in a forest."

5. See, in general and ad loc., Herbert Brichto, *Problem of "Curse" in the Hebrew Bible*. SBL Monograph Series, vol. 13 (Philadelphia: SBL, 1963).

6. Lamech boasts that he has slain a mature man (*'īš*) who had wounded him, and a tyro (*yeled*) who had bruised him. The latter would be ludicrously anticlimactic if *yeled* were a child; indeed, part of the boast may lie in the greater vigor ascribed to a young champion. See also Genesis 44:20, where Benjamin, "the child of [Jacob's] old age," is also qualified as *qātān*, "still young, a minor." In Genesis 42:22 Reuben refers to Joseph, old enough to travel alone, as "the *yeled*" to stir his brothers to compassion. Finally, the term is found in the plural in 1 Kings 12 and 2 Chron. 10 for the young and brashly imcompetent courtiers of his own age group, whose advice King Reheboam accepts.

7. "The lot," my rendering of *mēhem* (from among them). As distant as this translation may seem from the literal, its meaning is far more faithful to the Hebrew than is "tore, ripped, mauled *of them* some forty-two children."

8. See Babylonian Talmud, *Sotah*, cols. 46b–47a and Hebrew dictionary *Eben Shoshan*, s.v. *dōb*.

9. Montgomery sees nothing strange in the "move down to the Jordan, where timber was to be had in plenty for a larger conventicle." He goes on to cite the growth there of poplars and tamarisks. This timber growth on the banks of the Jordan is also cited by Gray, ad loc. who notes, "This luxuriant jungle-growth . . . is proverbial in the O.T. as the haunt of wild beasts, e.g. lions . . . and, until recently, wild pig." Only humans are absent from this jungle haunt.

10. The verbs for cutting, lopping, splitting, and hewing (wood) are *krt, gd', bq', and ḥsb*, respectively. Other terms for cutting or dividing that come to mind are *qs', qss, hrs, btq*, and *btr*. The verb *qsb*, translated "cut off [a stick]" (JPS) and "cut off [a piece of wood]" (NEB) is

common in Aramaic dialects as in late Hebrew for butchering, or *cutting up* of meat. The noun *qeṣeb* (possibly, "shape") of the cherubim in 1 Kings 6:25 is the only support in biblical Hebrew for the sense of "whittle" that the context seems to call for. See also (the byform?) *'ṣb*.

11. The text, translation, and a judicious discussion of it is available in S. R. Driver's *Notes on the Hebrew Text . . . of the Books of Samuel* (Oxford: Oxford University Press, (1913), lxxxiv–xciv.

12. Compare Gray's comment on verse 8: "The 'way of the steppe of Edom' cannot be the well-watered western escarpment of Edom, but the desert marches to the east of the land. This is suggested by the length of the detour, a week from the place where the forces of Israel and Judah joined up with those of Edom." See the rest of this comment on the direction of the water coming from Edom in verse 20 and the speculation that an invasion of Moab from the north had been ruled out by Mesha's recovery and refortification of settlements in that area (Gray, I and II Kings, ad loc.)

13. Inasmuch as this detail, for all its probable accuracy, is of little significance for the reader's understanding of the course of Judean history, the authorial (not, as I see it, *editorial*) decision to include it twice is instructive for the piecing together of the stories in the "historical" books in the Bible. Compare, for example, the trivial nature of this historical datum with the omission of such details as Omri's conquests or of a note that the revolt of Moab was as successful as that of Edom, which continues "to this day." The latter notice is in 2 Kings 8:20, the former in 1 Kings 22:47. Also instructive for the student is the kind of jargon (italicized in the following quote) that a rigid historiographical-critical approach to the biblical text entails: "The reference to the 'king' of Edom [in our story's verse 7], is *strictly an inaccuracy,* and seems to us to indicate the *later reworking* of the history of the house of Omri in the *later, rather free,* historical narrative" (Gray, ad loc.). What "history of the house of Omri" do we have at all? And why should anyone want to "rework" it?

14. See Montgomery and Gray on verse 26 for an example of the arrogance of modern Bible critics. Confident that the biblical author was primarily concerned with history, assuming that the text is corrupt (because their interpretation of a biblical expression makes no sense in context), they emend the text and rewrite history so that Mesha attempts to break through the seige lines to make contact with a nonexistent ally, the king of Aram!

15. For example, Deuteronomy 25 begins with a passage dealing with a litigation wherein the verdict calls for a sentence of corporal punishment upon one of the parties. The passage specifies that the flogging must be administered in the presence of the magistrate who pronounces the sentence; it limits the punishment to a maximum of forty strokes, a round number; and it concludes with the reasons for the limitation: To exceed the limit is to risk "your brother's becoming degraded in your sight." The requirements are expressive of the insights that it is easier to mete out pain when one need not witness its infliction and that cruelty may degrade the dispenser more than the victim. The next verse—three words in the Hebrew—is a pithy summation of this insight: "Don't muzzle the ox while it is threshing."

Deuteronomy 22:6–11 provides five additional injunctions whose symbolic thrust is missed if they are taken literally. If heaven sends you a windfall (you come upon a nest with eggs or fledglings and a brooding mother bird), you may take the young, but you must release the mother. The implicit meaning is that she may continue to propagate. If you build a house, you must provide its roof with a parapet. Otherwise, anyone suffering a fall will be charged to your responsibility. Do not seed the aisles of your vineyard with a view to extracting a second crop from it, lest you end up with failures in the harvesting of both. Do not yoke ox and ass together to pull your plow. Wool and linen (animal and vegetable products) are not to be woven together for your garments.

16. See Gen. 4:11, Num. 16:30, Deut. 11:6. In Isaiah 10:14 in the boast that the prophet puts into the mouth of Assyria's king the latter harvests the wealth of various nations as easily as

one might collect eggs from an abandoned nest: there is no parent bird to flap wings (in protest) or *open wide* its beak to squawk! Most interesting for its closeness to our context is Psalm 66:13–15. The psalmist will come to God's shrine to pay his vows, his "lips *proclaimed*[?] and mouth uttered when I was in trouble." I suspect that there may be some tongue in the psalmist's cheek when he promises "holocausts of fatlings . . . rams going up in smoke . . . cattle and billy goats in their prime." If the contrast with Isaiah 1:11's catalogue of scorned offerings is not germane, what about the contrast with Jepthah's niggardly vow? The genealogy (see n. 18) provided Jepthah suggests the speculation that this character is made up of whole cloth, in which case, his name (in Hebrew, He Opens [his mouth?]) is synonymous with his foolish utterance.

17. Note that *substantive* and *stylistic* correspond to metaliterary and literary, respectively.

18. On the other hand, the story, if freely composed several centuries after its temporal placement, may reflect the narrator's freedom to take poetic license with the population and practices of an Israelite community that has long ceased to exist. The hero of our story, Jephthah (his name, literally, He Opens, is a synonym of the verb discussed in n. 16) is a son of Gilead, who is himself son of Machir son of Manasseh. Thus, he is but a generation or two removed from the wilderness generation; yet his parleying with the Ammonites describes the history of the area's conquest as though it were in a far distant past (three hundred years, to be specific). Among other discrepancies relating to the "historic" Manasseh and the Ammonites, see the conflicting genealogies in Numbers 26 and 32 (not to speak of those in 1 Chronicles) and in Judges 11:24, Jephthah's identifying Ammon's god as Chemosh (in seeming confusion with Moab's god) whereas 1 Kings 11:5 and 7 identify Ammon's scorned deity as Milcom and Molech.

19. In the cuneiform inscriptions of Shalmaneser III, $^dAdad-id-ri$. The quote that follows is from ANET, 1955, 280, col. 2. In col. 1 on this page, in face A, lines 99–102, Shalmaneser in his fourteenth year is battling against Hadadezer, while in face B, lines 97–99 he is, in his eighteenth year, warring against Hazael.

20. It is useful, I believe, to raise such metaliterary considerations for deployment in poetical analysis and also to note analogous counterparts as they operate across the borders of such disciplines as physical science, probability theory, and metaphysical philosophy. I have in mind the uncertainty principle of Heisenberg's in physics and its philosophical exploitation in the principle of indeterminacy. See N. H. D. Bohr in *Nature* 121(1928); E. Schrödinger, "Indeterminism and Free Will," *Nature*, 138(1936); W. Seifriz, "Creative Imagination and Indeterminism," *Philosophy of Science* 10(1943). (Pages for these articles are unavailable.) A purely literary variation of our theme is the famous cross-cultural theme of the appointment in Samarra.

21. Such as the Edomites of Psalm 137:7, the Assyria of Isaiah 10:6–7, or the Amalek of Exodus 17 and his descendant Haman the Agagite.

22. In 2 Kings 13 we are told that Jehoahaz, the successor of his father Jehu on Israel's throne incurred YHWH's anger by following in the (cliché-expressed) offense of Jereboam ben Nebat. He therefore "delivered Israel into the power of Hazael, king of Aram and the power of Ben Hadad son of Hazael *kol hayyāmīm*" (verse 3). These last two words are rendered (probably correctly) "repeatedly" by the Jewish Publication Society. A literal rendering "all the days" yields an incomplete sense and is, furthermore, contradicted by the following two verses, which tell us that with YHWH responding to Jehoahaz's entreaty, the Israelites were set free from Aram's tyranny. The italicized words in the following translation are expressions that are in varying degrees, in the Hebrew, poetic, anachronistic, archaizing, otiose—all features inappropriate to straightforward historiography. This would point to this passage's poetical conformance to our story of Elisha and Hazael, providing Elisha's oracular complaint with a

pseudohistorical fulfillment even while it underlines the metaphoric purpose of the story as a whole.

> 4.Jehoahaz *entreated the favor* of YHWH, and YHWH hearkened to him—*he took note of the oppression of Israel, how severely oppressing to them* was the king of Aram. So it was that YHWH provided *a liberator* to Israel, so that they *came up free from under* the power of Aram. The Israelites then resided as in *yestertime in their tents.* (2 Kings 13:4–5)

Not a word of a single battle lost by Israel, of a single town lost to the Arameans; not a single soldier falls in battle, not to speak of that perennial propagandistic ploy of Huns or Boches wantonly dashing infant skulls and ripping unborn babes.

23. Thus, although Jehu and Hazael are both mentioned in the same inscription of Shalmaneser III (see ANET, 1955, p, 280, cols. 1 and 2), the former listed as the son of Omri and paying tribute to the Assyrian king while the latter, defeated in field warfare, withdraws into Damascus which successfully resists a siege, these two kings are not too preoccupied by the threat from Assyria to battle each other for Israel's territories east of the Jordan (2 Kings 10:32–33). Despite Jehu's twenty-two-year reign, granted to him by YHWH for his service against the House of Ahab, Hazael is apparently the victor in the struggle for the eastern marches. In the time of Jehoash, who comes to the throne of Judah in Jehu's seventh year, Hazael of Aram, apparently marching down the coastal road, captures Gath and is bought off from attacking Jerusalem by Jehoash, who empties palace and temple of treasures to do this. Such notices (see 2 Kings 12:1–4, 18–22) of the misfortunes attending a king who does what is pleasing to YHWH should long ago have laid to rest the scholarly staple of a mechanical correspondence in the Book of Kings between the success of YHWH's favorites and the failure of the kings who displease him. Similarly Jehoahaz ben Jehu, is delivered into the power of Hazael of Aram for displeasing YHWH and then delivered from that king's power (2 Kings 13). Again, Jehoash son of Jehoahaz is displeasing to YHWH yet manages (14:7–14) to inflict a crushing defeat on King Amaziah of Judah, who does what is pleasing to YHWH.

24. Note the correct pronoun suffixed to *master,* as opposed to "murderer of *your* master" (JPS and NEB). The original Hebrew strengthens the metaphorical identification of Jehu with Zimri as it makes the regicide into a type rather than a person and at the same time distances the speaker in her abhorrence from the base traitor she addresses in the third person. This care in the use of an impersonal third-person pronoun where a less subtle author would have used the second person was noted in my discussion of Jonah 1:8.

25. In this shout, as in the case of the preceding note, the subtle way in which the narrator signals his intent in dialogue is missed by the mistranslation of *mirmā,* a word which means "guile, trickery, underhandedness," not the substantially different "treason" or "treachery" (JPS, NEB).

26. See 2 Kings 11:18, 21:3–4; 23:4.

27. "The young man, the servant of the prophet" (JPS), "the young prophet" (NEB).

28. On this mark, see chapter 8.

29. The translation "did this [matter] come to pass" is a faithful rendering of the Hebrew. So also would be the rendering "did this word come from YHWH [to Jeremiah]." The absence of the addressee in the Hebrew as well as the departure from such formulations of YHWH's address to this prophet as, for example, in Jeremiah 25:1, 15 (but cf. 21:1, where the addressee is explicit) is, I believe, an ambiguating feature. In view of the conclusion we draw at this essay's end as to its essential kerygma, the point of the ambiguation is not to hint at a possibly unreliable narrator but rather to suggest that for all the wrongness of the indictment, the trial itself and its outcome "came to pass at the direction of" or "by will of" YHWH.

30. See n. 29.

Chapter 8

1. The extent of my indebtedness, in the following discussion to E. A. Speiser can only be gauged after a reading of his *"Palil* and Congeners: A Sampling of Apotropaic Symbols,'' in *Studies in Honor of Benno Landsberger on His Seventy-Fifth Birthday* (AS, no. 16 (1965), 389–93.

My debt to this great teacher, of course, goes far beyond this article or, indeed, the total of his published works. In keeping with the observations with which I conclude this chapter, I should like to pay tribute to this great spirit by quoting from J. J. Finkelstein (epilogue to E. A., *Oriental and Biblical Studies,* ed. by J. J. Finkelstein and Moshe Greenberg, [Philadelphia: University of Pennsylvania Press, 1967], 610): ''Above all, what Speiser strove for in his work in extra-Biblical fields as well as in his Biblical studies was the identification of the human values that were reflected and preserved in the literary remains of these ancient civilizations. Grammatical study, archaeology, even philology on a larger scale, were to him ultimately no more than the necessary means to get at the human, ethical, moral and spiritual values that animated all their activity, and thereby to bring them face-to-face with modern man for all that he could derive from such confrontation.''

2. This sense of the stem *psh* is recognized in the Jewish Publication Society translation notes on Exodus 12:11, 23. The awareness of this sense and the adducing of support for it from Isaiah 31:5 is recorded in the Mekhilta, ad loc.

3. For this important insight, as for many others, into the biblical Weltanschauung I am indebted to Yehezkel Kaufmann's monumental *Toledot Ha'emunah Hayisre'elit.* Recourse to this seminal interpretation is available to those not fluent in Hebrew in the splendid abridgement and translation of Moshe Greenberg, *The Religion of Israel* (Chicago: University of Chicago Press, 1960).

4. See chapter 5, Tale 5.

5. Note with care Exodus 13:8–10. The beginning words of verse 8, ''It will serve you as a sign upon the hand and mark between your eyes,'' refer to the observance of the Festival of Unleavened Bread and the command ''to tell your son at that time period''; the second part of the verse harks back to this telling. Again in 13:16, that which wins you protection (the signs upon hand and forehead) is the sacrifice of firstborn animals and redemption of firstborn males. The second part of this verse is not connected with the immediately preceding words, it is a recapitulation of the command in verse 14, that is, ''Verily (*kî*), it was by dint of great force that YHWH freed us from Egypt,'' despite the translations.

6. A nail is alive when it is free to roll, dead or immobile when it is hammered into a plank. The ''door'' in which the nail is fixed appears in this simile by reason of alliteration.

7. The falsity in the slogan is not in the denial that the Temple is YHWH's palace but in the implication that the presence of his palace in Jerusalem guarantees his protection of the city, come what may. See chapter 7, Tale 8.

Chapter 9

1. See chapter 4. The remark was in reference to the same essay as published in HUCA 54(1983).

2. PI, 16f.

3. ABN, 24.

4. PBN, 29.

5. Ibid., 30.

6. Ibid., 32.

7. Ibid., 31.

8. ABN, 25.

9. Ibid.

10. PBN, 31. The internal quote is from Herbert Butterfield.

11. Edwin M. Good, *Irony in the Old Testament* (Philadelphia: Westminster, 1965).

12. Humor in the satiric vein is, of course, most pervasively present in the book of Jonah (see chap. 3). For sarcasm, see Elijah's characterization of Baal in the duel on Mount Carmel (chap. 5, Tale 2). Sardonic or gallows humor are weak characterizations for the type of humor that informs the tale of Elisha's abetting of regicide (chap. 7, Tale 6). And I suspect that the motive for the near scatology in 1 Samuel 24:1–22 is basically the indulgence of whimsy (see chap. 1, n. 29).

SELECTED BIBLIOGRAPHY

Alonso-Schökel, Luis. *A Manual of Hebrew Poetics*. Rome: Pontifical Biblical Institute, 1988.

Alter, Robert. *The Art of Biblical Narrative*. New York: Basic Books, 1981.

Auerbach, Erich. *Mimesis*. Princeton: Princeton University Press, 1953.

Bar-Ephrat, Shimeon. *Narrative Art in the Bible*. Sheffield: Almond, 1989.

Berlin, Adele. *Poetics and Interpretation of Biblical Narrative*. Sheffield: Almond, 1983.

Brichto, Herbert Chanan. "The Case of the *Śōṭā*." *Hebrew Union College Annual* 46(1975): 55–70.

———. "On Faith and Revelation in the Hebrew Bible." *Hebrew Union College Annual* 39(1968): 35–53.

———. *The Problem of "Curse" in the Hebrew Bible* Society of Biblical Literature Monographs Series, vol. 13. Philadelphia: Society of Biblical Literature, 1963.

———. "The Worship of the Gold Calf." *Hebrew Union College Annual* 54(1983): 1–44.

Childs, Brevard S. *Introduction to the Old Testament As Scripture*. Philadelphia: Fortress, 1979.

Driver, S. R. *Notes on the Hebrew Text of the Books of Samuel*. Oxford: Clarendon, 1913.

Eslinger, Lyle. "Viewpoints and Point of View in 1 Samuel 8–12." *Journal for the Study of the Old Testament* 26(1983): 61–76.

Finkelstein, J. J. Epilogue to *Oriental and Biblical Studies* by E. A. Speiser, ed. J. J. Finkelstein and Moshe Greenberg. Philadelphia: University of Pennsylvania Press, 1967.

Fishbane, Michael. *Text and Texture*. New York: Schocken, 1979.

Freedman, D. N. "The Chronology of Israel." In *The Bible and the Ancient Near East*. ed. Roland de Vaux. New York: Doubleday, 1961.

Fokkelman, J. P. *Narrative Art in Genesis*. Assen: Van Gorcum, 1975.

Good, Edwin M. *Irony in the Old Testament*. Philadelphia: Westminster, 1965.

Gould, Stephen Jay. *Time's Arrow, Time's Cycle*. Boston: Harvard University Press, 1987.

Gray, John. *I and II Kings*. Philadelphia: Westminster, 1970.

Gros Louis, Kenneth R. R., ed. *Literary Interpretations of Biblical Narratives*. Nashville: Abingdon, 1974.

Herodotus. *The Histories*. Trans. Aubrey de Selincourt. Edinburgh: R. & R. Clark, 1954.

The History of Herodotus. Trans. George Rawlinson. New York: D. Appleton, 1862.

Holstein, Jay, "The Case of the *'ish hā' elōhīm* Reconsidered." *Hebrew Union College Annual* 48(1977): 69–81.

Kaufmann, Yehezkel, *The Religion of Israel*. Ed. Moshe Greenberg. Chicago: University of Chicago Press, 1960.

Kugel, James L. *In Potiphar's House*. San Francisco: Harper & Row, 1990.

Landes, George M. "The Kerygma of the Book of Jonah." *Interpretation* 21 no. 1 (January 1967): 3–31.

Moberly, R. W. L. *At the Mountain of God*. Journal for the Study of the Old Testament Supplement Series, vol. 22. Sheffield, UK: JSOT, 1983.

Montgomery, James A. *The Books of Kings*. New York: Charles Scribner's Sons, 1951.

Polzin, Robert. *Moses and the Deuteronomist*. New York: Seabury, 1980.

————. *Samuel and the Deuteronomist*. San Francisco: Harper & Row, 1989.

Price, G., and Eugene A. Vida. *A Translator's Handbook on the Book of Jonah*. Stuttgart: United Bible Societies, 1978.

Pritchard, James B. *Ancient Near Eastern Texts Relating to the Old Testament*. 2d ed. Princeton: Princeton University Press, 1955.

Speiser, E. A. *Genesis*. Garden City, N.Y.: Doubleday, 1964.

————. *Oriental and Biblical Studies*. Ed. J. J. Finkelstein and Moshe Greenberg. Philadelphia: University of Pennsylvania Press, 1967.

————. *"Pālil* and Congeners: A Sampling of Apotropaic Symbols." In *Studies in Honor of Benno Landsberger*. AS, no. 16, 1965.

Sternberg, Meir. "The Bible's Art of Persuasion: Ideology, Rhetoric, and Poetics in Saul's Fall." *Hebrew Union College Annual* 54(1983): 45–82.

————. *The Poetics of Biblical Narrative*. Bloomington: University of Indiana Press, 1985.

Thiele, Edwin R. *The Mysterious Numbers of the Hebrew Kings*. Grand Rapids: Eerdmans, 1965.

Thomas, D. M. *The White Hotel*. New York: Viking, 1981.

Tsevat, Mattitiahu. "Studies in the Book of Samuel." *Hebrew Union College Annual* 36(1965): 49–58.

Dictionaries and Encyclopedias

Millon Hadash. Even Shoshan, Avraham. Jerusalem: Kiryat Sefer, 1958.

The Interpreter's Bible. Nashville: Abingdon, 1952.

The Interpreter's Dictionary of the Bible. Nashville: Abingdon, 1962.

The Random House Dictionary of the English Language. Unabridged ed. New York: Random House, 1966.

Webster's Collegiate Dictionary. 5th ed. Springfield, MA: GSC Merriam, 1936.

INDEX OF
SCRIPTURAL CITATIONS

New Testament

INDEX OF
BIBLICAL HEBREW TERMS

GENERAL
INDEX